William C. Blackman, Jr.

BASIC HAZARDOUS
WASTE MANAGEMENT
Third Edition

William C. Blackman, Jr.

BASIC HAZARDOUS
WASTE MANAGEMENT
Third Edition

CRC Press
Taylor & Francis Group
Boca Raton   London   New York

CRC Press is an imprint of the
Taylor & Francis Group, an **informa** business

Library of Congress Cataloging-in-Publication Data

Blackman, William C.
    Basic hazardous waste management / William C. Blackman, Jr.--3rd ed.
        p. cm.
    Includes bibliographical references and index.
    ISBN 1-56670-533-9 (alk. paper)
    1. Hazardous wastes—United States—Management. I. Title.

TD1040 .B53 2001
363.72′87—dc21   2001020391
                                                                        CIP

Visit the CRC Press Web site at www.crcpress.com

© 2001 by CRC Press LLC

No claim to original U.S. Government works
International Standard Book Number 1-56670-533-9
Library of Congress Card Number 2001020391
Printed in the United States of America   6 7 8 9 0
Printed on acid-free paper

Preface

As the demand for a clean, safe environment grows, so also grows the public demand for protection from the health hazards and environmental horrors of hazardous waste mismanagement. Entrepreneurs of industry and commerce provide daily evidence of the general awakening to the need for reduction or elimination of hazardous waste sources and better management of the wastes that are generated. However, the ever-present drive for new product advantage, competition, budget and capital restraints, and the activities of those who have not yet accepted their environmental responsibilities continue to threaten our environmental well-being. Meanwhile the "not in my backyard" (NIMBY) syndrome has reached the point that almost no site is acceptable as a hazardous waste treatment or disposal facility. This clash of imperatives must be dealt with. We, as a people, cannot permit further episodes of uncontrolled release of hazardous materials/waste to threaten us. We, as a first-world society, cannot tolerate the continuing aftermath of our history of uncontrolled hazardous waste disposal. However, we, as a viable, self-supporting nation, cannot afford to force industry and commerce to their collective knees in the name of environmental purity.

The national conscience, as expressed in the form of research, technological advances, legislative craft, regulatory issue, fiscal support, and public participation, has brought forth great improvement in our hazardous waste management practice. However, most of the easy achievements have been realized. As we embark upon the Third Millennium, the priorities and demands placed upon environmental managers are ever more complex, urgent, and broad in scope. For example, exposure standards for toxic or hazardous chemicals are progressively made more restrictive, but pressures increase for less expensive and intrusive cleanup procedures for sites contaminated with these chemicals. Regulatory agencies seek to eliminate the use of particularly objectionable materials, while the industries traditionally forming the U.S. industrial and labor base, seeking less restrictive operating conditions, flee to neighboring and third-world countries. New generations of hazardous waste managers must acquire a broad-scope understanding of competing interests in scientific, technological, engineering, administrative, political, public health, and environmental issues and the innovations that must be conceived and implemented in order to reconcile these imperatives.

Our traditional approach to the education of future environmental technologists and managers has guided the undergraduate through a basic skill curriculum, then to be followed by a graduate program in engineering or a science. This text is intended to provide an introductory framework which can be the foundation for a program of study in traditional as well as modern hazardous waste management or a component of a related program. It is in an overview format, with many references to more detailed materials, to assist the student or instructor in expansion upon specific topics or to flesh out complex issues. The instructor is encouraged to expand upon issues or topics to meet the perceived needs of students, regions, or industries. Topics for discussion or review are provided at the end of each chapter.

ORGANIZATION AND CONTENT

The first eleven chapters deal with the topics, impacts, technologies, problems, and issues associated with "conventional" hazardous wastes and the management practices and statutory and regulatory controls which have evolved around them. Chapters 12 through 14 introduce the closely related medical/infectious waste, underground storage tank, and radioactive waste management technologies and practices. Chapter 15 introduces the hazardous waste worker health and safety issues and regulatory structures that have become a major focal point and concern for managers and supervisors of hazardous waste facilities and sites.

Objectives are stated as the first element of each chapter. Insofar as is possible or appropriate, the chapters are structured to first outline the issue, subject, or technology, then to describe generic practice, and to then conclude with a summary of the statutory and/or regulatory approach. Historical perspective is provided where appropriate to locale, industry, or other emphasis. The reader who is unfamiliar with the *Federal Register* (FR) and/or the *Code of Federal Regulations* (CFR) should examine these two entries in the Glossary before proceeding with the regulatory material covered in the book.

Acknowledgments

My reviewers have shared generously of their valuable time and expertise to provide insightful and constructive suggestions. I am particularly indebted to Dr. Nicholas R. Hild of the Department of Manufacturing and Industrial Technology, Arizona State University, for his thoughtful and constructive input in reviewing the three editions in their entirety. Reviewers of portions of the first edition were Ms. Pamela R. Jenkins, R.N., of the Environmental Resource Center, Fayetteville, NC — the chapter on medical and infectious waste management; Mr. Arthur C. Gehr, Esq., partner in the firm Snell and Wilmer, Phoenix, AZ — the radioactive waste management chapter; and Ms. Lisa Lund, then Manager, Underground Storage Tank Compliance Section, Arizona Department of Environmental Quality, and later Deputy Director, Office of Underground Storage Tank Programs, U.S. Environmental Protection Agency — the underground storage tank management chapter. Mr. Harold L. Berkowitz, chemical engineer, consultant, and faculty associate of the Department of Manufacturing and Industrial Technology, Arizona State University, provided extensive input and improvements to the new chapter on hazardous waste worker health and safety. The valuable assistance of all of the reviewers is deeply appreciated. Many of their respective contributions are retained in the third edition.

Without the editing and word-processing skills as well as the extraordinary patience of Ms. Cindy Zisner, M.S., and the graphic skills of Mr. Steve Scott, these months of work on the new edition would have been much less pleasant. Ms. Zisner is a private practitioner in Tempe, AZ. Mr. Scott practices in Pasadena, CA. Mr. Jay Carr of the *Dallas Morning News* also made a valued contribution in the graphic presentation of the Yucca Mountain Repository. I sincerely appreciate the time and effort of the many contributors of photographic materials. The illustrations for which no acknowledgment is made are either my own or have been provided to me on earlier occasions. I can only apologize for lack of adequate memory regarding the sources of the earlier contributions.

The Author

William C. Blackman, Jr. is an Environmental Engineer and Professor Emeritus of the Center for Environmental Studies, Arizona State University. Professor Blackman was previously a career engineer and manager assigned to enforcement programs of the U.S. Environmental Protection Agency and predecessor agencies. As Technical Coordinator and Deputy Director of the EPA National Enforcement Investigations Center, he planned and directed early hazardous waste site investigations and participated in the development of the site investigation techniques and site health and safety procedures which have become standard practice.

In 1985 he was appointed Assistant Director, Arizona Department of Environmental Quality, where he managed state and federal RCRA and Superfund programs. He joined the ASU faculty in 1989, teaching undergraduate and graduate courses in hazardous waste management and control of toxic air pollutants. He developed and presented a program of seminars on hazardous waste management, underground storage tank management, emergency planning, and regulation of hazardous materials transportation. He directed ASU participation in the California-Arizona Consortium, presenting OSHA health and safety training for hazardous waste workers and underground storage tank workers. He continues to research and lecture in these programs.

Professor Blackman received his B.S. in Civil Engineering and M.S. in Sanitary Engineering from the University of Missouri at Columbia, his MPA (Environmental Management) from the University of Southern California at Los Angeles, and his DPA (Environmental Management and Public Policy) from the University of Colorado at Denver. He is a Registered Professional Engineer and a U.S. Army Reserve Sanitary Engineer Colonel. He has published a number of papers on water quality and pollution control, and on hazardous waste site investigations and safety procedures.

Table of Contents

FIGURE 1.2 Abandoned hazardous waste site. (Courtesy of Envirosafe Services of Ohio, Inc. (ESOI).)

The ABM-Wade Site. The ABM-Wade site in Chester, Pennsylvania, was typical of dozens of sites throughout the industrialized areas of the nation. During the mid-1970s the operator accepted hazardous wastes; filled the former factory building and pipe tunnels with drums of hazardous waste; filled discarded tank trailers with hazardous waste and parked them on the site; and when aboveground space was filled, underground storage tanks and trenches were filled with wastes.

FIGURE 1.3 Land disposal of hazardous waste. (Courtesy of MAX Environmental Inc., Pittsburgh, VA.)

FIGURE 1.4 Hazardous waste dump site. (From the Arizona Department of Arizona Environmental Quality.)

In February 1978, the site burned. Nearby residents were endangered by clouds of toxic air pollutants, by the proximate natural gas storage tanks, and by contaminated run-off to the Delaware River (Figures 1.5 through 1.9). The site was remediated by a 10-year, $3 million Superfund project (see Chapter 11).

FIGURE 1.5 ABM-Wade site (Chester, Pennsylvania). (From the U.S. Environmental Protection Agency.)

FIGURE 1.10 LaBounty dump site (Salsbury Laboratories, Charles City, Iowa). (From the U.S. Environmental Protection Agency.)

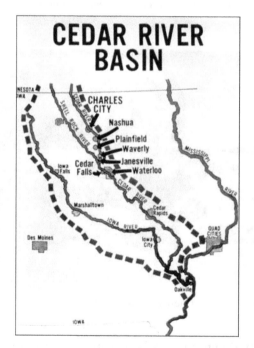

FIGURE 1.11 Cedar River basin downstream of LaBounty (Iowa) dump site. (From the U.S. Environmental Protection Agency.)

leached from the dump to the underlying groundwater and to the Cedar River. The Iowa DEQ issued an executive order requiring Salsbury to remove all hazardous wastes and contaminated materials from the LaBounty site, but Salsbury convinced the courts that the cleanup costs would exceed the company's net worth, and the removal order was stayed (Dahl 1980).

FIGURE 1.12 "Treatment" pond, Union Carbide (Uravan, Colorado). (From the U.S. Environmental Protection Agency.)

Union Carbide — Uravan. In the early 1950s, the Union Carbide Corporation began uranium mining and milling operations on the banks of the San Miguel River at Uravan, Colorado. During the years of operation, the mill produced more than 10 million tons of uranium-vanadium ore, in excess of 10 million tons of tailings, millions of gallons of waste liquid raffinate, raffinate crystal residue, and other milling wastes containing radioactive materials (uranium, radium, and thorium), metals (selenium, aluminum, arsenic, cadmium, zinc, and others), and other inorganic contaminants. Mining, milling, and waste disposal practices have resulted in:

- Wind and surface water dispersal of the tailings materials and the uncontrolled release of radon from the tailings piles
- Seepage of contaminated liquids into soils and groundwater from several areas in the mill complex and waste disposal areas
- Concentrations of large quantities of wastes in locations that pose a risk to public health and the environment

Prior to remediation, soils in the vicinity of the mill contained elevated levels of heavy metals and radionuclides. The San Miguel River, a tributary to the Dolores River, the Colorado River, and Lake Powell, was contaminated with radium 226. For nearly two decades, process wastewaters were discharged to seepage ponds scooped in the alluvium of the San Miguel (Figure 1.12). Contaminated groundwater from the tailings area emerged from the walls of the San Miguel Canyon (Figure 1.13) and continued subsurface movement toward the river.

The site was listed on the National Priorities List (NPL)[2] in 1986, but the massive cleanup was conducted under a consent decree in the U.S. District Court for the

---

[2] *See:* National Priorities List in Glossary.

FIGURE 1.13 San Miguel River Canyon — emergent seepage from tailings ponds, Union Carbide (Uravan, Colorado). (From the U.S. Environmental Protection Agency.)

District of Colorado (State of Colorado v. Union Carbide Corporation and Umetco Minerals Corporation, 1983).

The Kepone Disaster. In 1973, Allied Chemical Company subcontracted the production of the pesticide Kepone to Life Sciences Products, a small company operating out of a converted service station in Hopewell, Virginia, near the James River. Within 2 months following startup, discharges from the plant had killed the aerobes in the Hopewell sewage treatment system. Life Sciences employees were poisoned by Kepone; the site was hopelessly contaminated; fish, shellfish, waterfowl, and a variety of aquatic organisms from the James River and Chesapeake Bay demonstrated classic bioaccumulation of Kepone; Kepone-contaminated sediments accumulated in the James River and Chesapeake Bay; and particulates from air samplers near the plant were found to contain as high as 40% Kepone. Life Sciences was unable to pay for the cleanup, and Allied settled with the court by donating $8 million to establish the Virginia Environmental Endowment. Allied then recouped much of the settlement by taking a tax deduction on the endowment (Wentz 1989, pp. 59–68).

Reilley Tar & Chemical Corporation. The site in St. Louis Park, Minnesota, was the location of the Republic Creosote Company from 1917 until 1972. Reilley Tar & Chemical Corporation operated the facility after 1972. Extensive soil and groundwater contamination has resulted from discharge of contaminated wastewater over land to wetlands adjacent to Minnehaha Creek. Polynuclear aromatic hydrocarbon (PNA) contamination of the Prairie du Chien-Jordan Aquifer has forced closing of seven municipal water supply wells (Beckwith 1990, personal communication). Investigative and cleanup activity has continued for nearly two decades and has included cleanout of two contaminated water supply wells *which had been used for waste disposal*, several pumping wells for removal of contaminants and containment of the contaminant plume, and two granular activated carbon treatment plants. By 1996 an estimated 6.2 billion gallons of groundwater contaminated with PNAs had been pumped and treated. An additional pumping well was constructed in 1997 (EPA ID#MND980609804).

Minamata Bay. As Americans began learning hard lessons from hazardous waste disposal practices and began looking to ocean disposal as an easy solution, the Japanese began learning hard lessons from that very practice. During the period 1932 to 1968, an acetaldehyde production facility discharged mercury-laden wastes to a river which flowed to nearby Minamata Bay. The mercury was converted to the methyl form, in the bottom sediments, and became concentrated in marine organisms, including fish and shellfish. Among nearby villagers, for whom the fish and shellfish were a diet staple, at least 46 died and an estimated 3000 were poisoned. Japanese courts continued to struggle with liability rulings throughout most of the 1990s. In 1997, Governor Joji Fukushima of the state of Kumamoto declared the bay clean and stated that mercury levels were lower than those permitted by government safety standards (Dawson and Mercer 1986, p. 422; *Los Angeles Times,* November 27, 1993; Nebel and Wright 1993, p. 316; *The Associated Press,* July 29, 1997).

Dioxin Discovered. In Italy, near Seveso, Industrie Chemiche Meda Societa Aromia (ICMSA) operated a plant producing 2,4,5-trichlorophenol. A production unit was allowed to overheat and released 2,3,7,8-tetrachlorodibenzo-p-dioxin (2,3,7,8-TCDD) or "dioxin." A small quantity of this highly toxic material was released to the atmosphere and drifted southward toward Milan. This necessitated the evacuation of residents, slaughter of livestock, and condemnation of fruit and vegetable crops. Exposed children developed chloracne and adults were sickened (Wentz 1989, p. 7ff). After 7 years of cleanup activity, the dioxin waste was concentrated into 41 drums. This deadly concentrate passed through the hands of several handlers and was hauled from one end of France to the other, through Paris, and ultimately to a barn in Anquilcourt-le-Sart. Throughout this odyssey the drums were identified only as containing tar (Lesser 1984).

The Stringfellow Acid Pits. From 1956 until 1972, the 17-acre site in a Riverside County, California ravine, known as the Stringfellow Acid Pits, was operated as a state-authorized hazardous waste disposal facility. Evaporation ponds received over 34 million gallons of metal finishing, electroplating, and pesticide production wastes. In 1969, run-off from rainfall caused the ponds to overflow and contaminate a downstream creek. Heavy rains in 1978 threatened another overflow, and 800,000 gal of wastewater were released from the site, in order to prevent further overflow and possible massive release of the ponds contents. Another 500,000 gal were transferred to a federally approved facility. Between 1975 and 1980, approximately 6.3 million gallons of liquid wastes and pesticide contaminated materials were removed from the site. A groundwater plume containing volatile organic compounds and heavy metals threatened private water supply wells in the community of Glen Avon, forcing users to change to public water supplies. A long and costly Superfund cleanup followed, and additional work may be required (EPA ID#080012826).

After 15 years of litigation, an agreement in principle was reached in January 1999 to settle liability at the Stringfellow site. Total cleanup costs are now estimated at $800 million, of which California's share is estimated at $200 million. A California

Environmental Protection Agency official estimated that remediation will take 100 years (*Environment Reporter*, January 15, 1999, p. 1797).

The list could be extended for many pages, with familiar and obscure names such as Times Beach, Missouri, the "Valley of the Drums," USAF Plant Forty-Four (at Tucson International Airport), etc. Indeed the NPL now contains 1217 such sites, another 57 sites have been formally proposed for listing, and 205 sites have been cleaned under Superfund program auspices. Many other sites have been cleaned up under other federal and state regulatory programs and/or without regulatory oversight. Although these brief summaries seem similar to each other, the sites differ greatly in size, on-site activities, complexity, range of remedial options, resources, and time required to accomplish remediation, and extent to which remediation can be achieved. These regulatory actions will be introduced in Chapter 11.

"Take It Out Back and Dump It"

In 1950, before North Korea interrupted my academic endeavors, I took a part-time job at a local dry cleaning establishment. As the owner finished demonstrating how I was to clean the filters, I asked what to do with the filter sludge. His response was "take it out back and dump it." The dump site was easily recognized and frequently used. Later, after some enlightenment regarding such practices, the memory of that scenario returned to haunt me repeatedly. Thirty-five years later, the discovery that a dry cleaner in Phoenix, Arizona, took his used Perc out back and dumped it down a dry well was sad irony indeed. What we did in 1950 was assignable to ignorance. The 1985 episode tells us something about our progress in hazardous waste management and in educating the public regarding hazardous waste management (Author).

Disposal of hazardous waste has been accomplished in every mode of behavior — careful and careless, casual and furtive, clever and mindless, etc. The simple "take it out back and dump it" practice was commonplace. As case histories of contaminated sites continue to be developed, the practice of dumping solvents and other liquid wastes onto the ground, into ditches, (incredibly) into drainage wells, into seepage pits, and into trash dumps is often found to have been the mechanism or practice involved.

"Midnight dumping" became a familiar term, as disposers sought to find ways to "get rid of the stuff." The 55-gal drum rolled into the roadside ditch or onto unwatched property became a commonplace environmental insult. Disposers contrived dump and tank trucks with discharge devices arranged to discharge the cargo of hazardous wastes while in motion. Such "tippers," as they were known in Britain, drove along country roads and dumped the contents of the tanks by tipping them to a high angle. The intent was, of course, to distribute the wastes over a great area and to do so without being detected.

Workers in manufacturing facilities where environmental rules were being enforced discovered drums or other containers of unknown wastes that suddenly appeared on loading docks or elsewhere in the plants. These sudden appearances were referred to as "immaculate conception" events, but created serious hazards and disposal problems. "In-house" incidents of this nature demonstrated the fact that ignorance as well as unsavory and irresponsible attitudes toward hazardous waste management were present among rank-and-file workers, as well as corporate cost-center managers.

Even more reckless schemes were not uncommon. A favorite among criminal elements was the theft of a semitrailer (or tractor and trailer), acceptance of a load of hazardous wastes in drums (for a fee, of course), and abandonment of the stolen truck with load intact. Others carelessly or unknowingly discharged loads of waste into dumps or pits containing incompatible wastes and were overcome or killed on the spot or subjected others to similar fates.

As indicated previously, these examples of earlier activity should not lead to the assumption that such practices have been eliminated. Hazardous waste statutes, codes, and regulations have made it difficult for practitioners of illegal dumping to avoid detection and prosecution. Nevertheless, attempts to avoid the effort and/or cost of proper hazardous waste management continue to be made. A case in point is the recent prosecution and sentencing of two Florida entrepreneurs who attempted to hire an undercover police officer to dump drums of hazardous waste in the Everglades (U.S. v. Rojas, S.D. Florida, No. 98-579, 1999).

"Treatment" and Other Assorted Techniques

Very large volumes of liquid hazardous wastes were provided "treatment" and/or disposal in depressions, impoundments, and excavations which EPA referred to as "pits, ponds and lagoons" or PPLs (Figure 1.14). The expression referred to unlined holding and flow-through facilities that provided no protection against seepage and contamination of groundwater. Most provided no treatment other than settling of

FIGURE 1.14 Surface impoundment. (From the U.S. Environmental Protection Agency.)

solids, radiation from the sun, and some modicum of aeration. Very large numbers of today's remedial sites are former PPLs.

Solid hazardous wastes, wastes that should have been recognized as "hazardous," and wastes of unknown composition were allowed to accumulate in piles and later were

- Awarded legitimacy as treatment units (see Subparts L of 40 CFR 264 and 265)
- Excluded from definition as hazardous wastes because of their "high volume-low toxicity" character (40 CFR 261.43)
- Declared to be materials that were "awaiting recycling"

Many of these waste piles also became today's remediation sites, and some were recycled in a process known as "dump leaching."[3] Great strides were realized in extraction processes during recent years and previously worked ores became lucrative targets for further working. In practice, waste piles containing metals or other values were sprayed with weak acid or other solute-producing liquid, and the leachate was collected for extraction. Without underlining or removal to an impervious surface for leaching, much of the leachate was lost to the groundwater or resurfaced with base-flow of nearby streams. The emerging metals-laden acidic leachate mixed with the natural alkalinity of receiving streams, resulting in precipitation of metallic salts and/or hydroxides. The precipitate, known by practitioners as "Yellow Boy," can destroy a stream by blanketing the streambed as illustrated in Figures 1.15 and 1.16. A related practice, known as "heap leaching,"[4] is used to extract precious metals from low-grade ores. Heap leaching practitioners claim that liners or impervious layers used to capture the valuable leachate also prevent surface or groundwater contamination. Even today, these leaching processes are commonly used in copper mining areas of the southwestern U.S. The strong metals extraction lobby has been highly successful in obtaining exclusions from various of the hazardous waste compliance requirements.[3]

Early and mid-1900s industrial waste generators also enjoyed the option of discharge of hazardous wastes to publicly owned sewerage. Prior to imposition of the National Pollutant Discharge Elimination System (NPDES) permitting, it made little difference whether industrial wastes were discharged directly into surface waters or reached the discharge point via municipal sewerage. As publicly owned, secondary sewage treatment plants came online, it soon became evident that untreated industrial wastes "… can have serious impacts on the ability of systems to operate properly, on options for sludge disposal, and on water quality" (Council on Environmental Quality 1977, p. 37).

Not only were industrial wastes passing through publicly owned treatment works (POTWs) with ineffective or inadequate treatment, the toxic constituents in the industrial wastes were deleterious to the biota in the secondary treatment systems. Moreover, "municipal" sludge from the POTWs receiving industrial waste contained

[3] *See:* Bevill Amendment in Glossary.
[4] *See:* Dump Leaching and Heap Leaching in Glossary.

FIGURE 1.15 Surface stream ruined by inflow from heap leaching operation.

FIGURE 1.16 "Yellow-Boy" blanket on streambed ruined by heap leaching operation.

high concentrations of heavy metals, making digested sludge unsuitable for land disposal or recycling.

The 1972 Federal Water Pollution Control Act Amendments[5] and the 1977 amendments[6] required that EPA establish national standards for "pretreatment of industrial wastes that are not compatible with municipal treatment plants." EPA had, in February 1977, reissued proposed national pretreatment guidelines (42 FR 5986) and in 1978 began adopting pretreatment standards (43 FR 27736). However, major delays in implementation of pretreatment programs, as well as "loopholes" in the RCRA regulations, allowed the continued discharge of some hazardous industrial waste components to POTWs. Pretreatment needs received much-needed emphasis when, in 1981, a 3-mile segment of the Lexington, Kentucky, sewer system exploded. The explosion, caused by excessive discharges of hexane, caused more than $20 million in damages (EPA 1999b).

EPA and some state agencies have recently strengthened enforcement of pre-treatment rules, and EPA has proposed "streamlining" the 40 CFR Part 403 regulations. Industries discharging to municipal systems, as well as the entities responsible for implementing pretreatment programs, are the subject of enforcement actions. A case in point is the recent guilty plea of MRS Plating Inc. of Lockport, New York, and its manager Ron Jagielo, to felony violations of Clean Water Act pretreatment requirements (Title 33, USC Section 1319[c]). The company was alleged to have discharged hazardous waste into the Lockport, New York, sewer system on numerous occasions in 1997. MRS Plating was fined $120,000 and Mr. Jagielo is subject to a 1- to 2-year prison sentence (*Environment and Natural Resources News*, June 1999).

Numbers and Impacts

Uncontrolled releases of hazardous wastes to the environment continued through the 1970s. EPA estimated hazardous waste generation at 20 to 50 million tons annually in 1977, adding the estimate that no more than 10% was disposed of in a manner that was safe to the environment (Council on Environmental Quality, 1977, p. 50). Dams and dikes forming hazardous waste impoundments routinely failed, contaminating streams, rendering water supplies useless, and killing fish. Regulatory gaps and inadequacies continued to provide options for release of hazardous wastes to the air, water, or land. Early remediation of sites having groundwater contaminated with volatile organics frequently involved pumping the contaminated groundwater to the surface and air-stripping the contaminant, without further treatment or capture, to the atmosphere (*see also*: Dawson and Mercer 1986, pp. 3–4).

As was the case with the Love Canal exposures, the Hardeman County contamination of domestic water supplies, and the Minimata Bay fish and shellfish contamination, mismanagement of hazardous waste was clearly exacting a price in human health. The mechanisms were not well understood (and some continue to be so), but there could be no doubt of the connection between exposure and disease

---

[5] 33 USC § 1311(b)(1).
[6] The Clean Water Act, 33 USC § 1314(g)(1).

incidence. Increased incidence of carcinogenic, mutagenic, and teratogenic effects, damage to reproductive systems, respiratory effects, brain and nervous system effects, and many lesser effects were increasingly associated with direct and indirect exposure to hazardous wastes (*see also*: Enger et al. 1989, Chapter 15).

Environmental effects were even more pronounced and frequently better understood. Release by a major chemical manufacturing company of more than 4000 metric tons of DDT[7] manufacturing residue to a marsh near Triana, Alabama, resulted in DDT contamination of Indian Creek and the catfish from the creek. Mallard ducks from a nearby wildlife refuge were found to contain 480 parts per million (ppm) DDT. Entire populations of waterfowl disappeared from the area. Similar environmental tragedies were related to hazardous waste mismanagement throughout the nation and world.

The esthetic effects of hazardous waste mismanagement through the mid-1900s are well understood by most citizens. The rise of the NIMBY (not in my backyard) syndrome exemplifies the fear and revulsion of the public toward hazardous materials[8] in general, and hazardous wastes[8] in particular. The esthetic concerns quickly transition to economic issues as property values are affected, jobs are created or eliminated, and public administrators come under increasing pressures to craft solutions that solve all problems and resolve all issues. Facility siting became a major preoccupation among governments, the regulated industries and facilities, and the public. By the 1980s, public opposition to hazardous waste facilities had become so pervasive and intense that even sites with excellent operating records faced closure, simply because they were "there." Environmental equity issues, usually arising from actual or proposed location of hazardous materials/waste facilities in poor or minority neighborhoods, came to the forefront of the urban agenda in the 1990s. By the late 1990s, the equity issues had become focused in an identified EPA Environmental Justice program.

The economics of hazardous waste management and mismanagement have intruded upon nearly every facet of life in America. From the relatively minor costs of the local dry cleaner's shift from dumping to recycling to the projected $230 billion for cleanup of the nation's nuclear weapons facilities, management of currently generated and earlier mismanaged hazardous waste has become a major element of the economy.

By 1979, the EPA estimate of hazardous *waste* generation was 51 million tons per year (Council on Environmental Quality, 1979, p. 181). Estimates by the chemical industry and various federal agencies in subsequent years (Table 1.1) indicate that generation of hazardous wastes continued in the 200 to 300 million ton range. In more recent years, the EPA has established two inventories that include data on hazardous waste quantities managed in the U.S. The National Biennial RCRA

---

[7] DDT (dichlorodiphenyltrichloroethane) is a highly persistent insecticide that was banned for use in the U.S. in 1972.

[8] As is developed in later chapters, the distinction between hazardous *materials* and hazardous *waste* is derived largely from (1) lists of each that are published by various agencies and (2) whether a material is used or unused (virgin). However, much of the lay public generally perceives no distinction between the two.

TABLE 1.1
Estimated Quantities of Hazardous Waste Generated in the U.S.

Source	Million Tons	Comment
1981 EPA National Survey	247 (135–402)[a]	Sample survey
1981 Office of Technology Assessment (OTA)	255–275[a]	State data
1983 Congressional Budget Office (CBO)	266 (223–308)[a]	Industry data
1984 Chemical Manufacturers Association (CMA)	247	Chemical industry survey
1985 EPA National Survey	272	Census of treatment, storage, and disposal facilities
1986 CMA	220.5	Chemical industry survey
1989 EPA	198	EPA 530-S-94-039
1991 EPA	306[b]	EPA 530-S-94-039
1993 EPA	258[b]	EPA 530-S-97-022
1995 EPA	214[b]	EPA 530-S-97-022
1997 EPA	40.7[c]	EPA 530-S-99-036

[a] Range.
[b] EPA attributes this increase to newly regulated wastes.
[c] Not a directly comparable number. EPA excluded wastewaters in the 1997 inventory.

Hazardous Waste Report[9] and the Toxics Release Inventory[10] provide useful data for evaluation of waste minimization, comparisons of industrial categories, etc. As noted below, both have significant shortcomings with respect to year-to-year comparisons of hazardous wastes managed. Nevertheless, these data provide a gross estimate indicating that some reduction in quantities managed may have occurred in the 1996–1997 biennium.

     Table 1.1 statistics do not include other important data. For example, by the mid-1980s the amounts of hazardous waste "created" by site cleanup activity had become significant, and these quantities continue to be significant. In the 1991 through 1997 Biennial RCRA Reports, the quantities managed "on-site" are not included, nor are they included in the TRI. Moreover, until 1998, the TRI reported only releases by manufacturing industries listed in the 20 Standard Industrial Classification (SIC) codes[11] 20 through 39. Thus, an unknown quantity of the waste created by site cleanup and other activity is not "captured" by the reporting system. Between 1980 and 1995 the numbers of generators which were brought under regulation increased greatly, but the number of reporting generators fell sharply in 1997 due to the exclusion of

[9] U.S. EPA, The Biennial RCRA Hazardous Waste Report (Based on 1991 Data), EPA 530-S-94-039; U.S. EPA, The National Biennial RCRA Hazardous Waste Report (Based on 1995 Data), Executive Summary, EPA 530-S-97-022; U.S. EPA, The National Biennial RCRA Hazardous Waste Report (Based on 1997 Data), Executive Summary, EPA 530-S-99-036.

[10] U.S. EPA, Toxics Release Inventory Public Data Release, EPA 745-R-96-002; U.S. EPA Toxics Release Inventory 1987–1994, EPA 749-C-96-003; U.S. EPA, Toxics Release Inventory 1987–1995, EPA 749-C-97-003; U.S. EPA, 1987–1996 Toxics Release Inventory Reporting and the 1997 Public Data Release, EPA 749-C-99-003.

[11] Established and Published by the Office of Management and Budget (OMB).

wastewaters that are regulated under the CWA.[12] Then in 1998, EPA added industries represented by six SIC codes to the reporting lists.[13] An interesting aside is that when/if petroleum products and or other mining wastes are brought under the hazardous waste regulations, the quantities of "hazardous waste" generated may again rise sharply.

Early Efforts — What Worked/Didn't Work

Any summary of "What Worked/Didn't Work" is incomplete without mention of the key role played by Murray Stein and his Enforcement Conference procedure, conducted under the authorities of the Federal Water Pollution Control Act (FWPCA) (33 U.S.C. 466 et. seq.), the precursor to the Clean Water Act. Stein, a lawyer, was Chief Enforcement Officer of EPA's predecessors the Federal Water Quality Administration and, earlier, the Division of Water Supply and Pollution Control, U.S. Public Health Service.

The FWPCA contained provisions for collective action to abate interstate pollution or contamination of fish or shellfish. The extreme limitations (compared to present legislative and regulatory powers) of those authorities notwithstanding, Stein convened Enforcement Conferences on all manner of environmental issues from industrial wastes in Raritan Bay to radioactivity in the Colorado River Basin and from mining waste in the Great Lakes to petrochemical wastes in Galveston Bay.

Stein gathered federal, state, and local officials, media, scientists, and industrialists together, orchestrated the Conferences as if they were judicial proceedings, extrapolated from the most minimal of technical "evidence," and extracted wide-ranging concessions, compromises, and commitments from polluters. The early cleanup achievements of Stein and his Conferences and the impacts of these pioneering enforcement procedures deserve a prominent place in whatever form is eventually given to the environmental history of America.

Legislation/Litigation

As the hazardous waste mismanagement outrages were thrust upon an unknowing public, governments — local, state, and federal — began attempts to coerce and force perpetrators, variously, to clean up contaminated sites, stop dumping, stop generating, treat properly, recycle, reclaim, or destroy the waste. Individuals and governments sought relief in the courts by tort actions. Nuisance is the most common of tort claims in the field of environmental law.

---

[12] *See:* Bevill Amendment and Benson Amendment in Glossary.
[13] The newly listed categories and SIC Codes are Metal Mining (10), Coal Mining (12), Electric Utilities (49), Chemical Wholesalers (5169), Petroleum Bulk Terminals (5171), RCRA/Solvent Recovery (4953/7389).

Nuisance is defined as ... the class of wrongs that arise from the unreasonable, unwarrantable, or unlawful use by a person of his own property either real or personal or from his own lawful personal conduct working an obstruction of or injury to the right of another or of the public and producing material annoyance, inconvenience, discomfort, or hurt (Sullivan 1985, p. 10).

It should be noted that prior to 1972 with the promulgation of the Clean Water Act Amendments, most environmental remedies were conducted under nuisance legislation, and nuisance law was only minimally helpful in addressing the broad range of environmental insults. Two pre-RCRA cases are illustrative.

- In a 1973 case, Harrison v. Indiana Auto Shredders, the Seventh Circuit Court of Appeals refused to permanently enjoin operation of an automobile shredding and recycling plant based on a nuisance action. The court held that under the evidence presented and in the absence of an imminent hazard to health and welfare — none of which was established — the defendant could not be prevented from continuing to engage in its operation (Sullivan 1985, p. 15).
- The Earthline Corporation attempted to operate an industrial waste recovery, treatment, storage, and disposal site on a 130-acre site in Illinois. Ninety acres are located within the village of Wilsonville and the remaining acres are adjacent to the village. The operation accepted hazardous wastes and toxic substances. In 1979, the village sued Earthline to stop the operation and also to require the removal of those hazardous wastes and toxic substances that had been deposited on the site. The court ruled that the site was a public/private nuisance, issued an injunction against Earthline's further operation of the site, and required them to remove all wastes and contaminated soil (Sullivan 1985, p. 16).

In the Stringfellow Acid Pits settlement in principle, noted earlier in this chapter, one of several key elements was the settlement of a 1984 toxic tort action filed by 3800 Glen Avon residents (Newman v. Stringfellow, Cal. Super. Ct., Riverside Cty., No. 165994). After a jury awarded $160,000 to 6 of 17 test plaintiffs on September 15, 1993, most of the defendants paid residents a combined $109 million to settle the tort case (*Environment Reporter*, January 15, 1999, p. 1798).

In 1970 the federal government resurrected the long-dormant 1899 Rivers and Harbors Act[14] and established a permit system for control of water pollution. The efficacy of the act and the permit system for control of hazardous wastes was limited in that it applied only to "navigable waters." (Debate over the "navigable waters" designation became the focus of many of the actions.) The Act was successful in beginning the long-overdue elimination of indiscriminate dumping of all wastes into

---

[14] Also known as the "Refuse Act" (33 U.S.C. 407).

the streams and rivers of the nation, but the enforcement process was cumbersome, and the Federal Water Pollution Control Act (FWPCA) and the successor Clean Water Act processes soon obscured the earlier mechanism.

The Atomic Energy Act of 1954 vested authority for management of most *radioactive* wastes in the Atomic Energy Commission (AEC) and the Nuclear Regulatory Commission (NRC). The AEC was responsible for managing nuclear wastes generated by the 17 nuclear weapons production facilities that it operated. Nuclear weapons development and production activities were assumed by the U.S. Department of Energy (DOE) in 1977. The NRC has responsibility for nuclear waste management oversight at the nation's nuclear-generating plants. For many years, the AEC gave weapons production the highest priority, generating huge quantities of radioactive process wastes. These wastes were stored or disposed of without treatment, or with inadequate treatment, at the production sites, some of which are now so badly contaminated that they may require 50 to 60 years to clean up, and "... there are some sites we are not going to be able to clean up" (Satchell 1989, p. 20). Radioactive waste management is the subject of Chapter 13.

The Solid Waste Disposal Act of 1965 was the first federal government attempt to improve solid waste disposal practice. It provided funding for development of state solid waste management programs and began the regulation of municipal waste management. The increasing concerns for human health and environmental protection led to amendment of the Act with the 1970 Resource Recovery Act, but neither the 1965 SWDA nor the 1970 RRA had significant impact on the emerging problems of hazardous waste management. These two statutes were the predecessors to the Resource Conservation and Recovery Act (RCRA) of 1976, discussed later in this overview.

The Clean Air Act (CAA) of 1970 was, as implied by the name, directed at controlling sources of air pollution and improving ambient air quality. The 1970 Act did little to enhance hazardous waste management. In fact, emission control equipment installed by industry to meet CAA-required National Ambient Air Quality Standards (NAAQS), as well as CAA-required State Implementation Plans, has generated large volumes of hazardous and toxic material that must be managed as hazardous waste. By 1990, National Emission Standards for Hazardous Air Pollutants (NESHAPS), required by Section 112 of the Act, had been established for only seven chemicals. The 1990 CAA Amendments required EPA to issue standards for 189 additional NESHAPS by November 2000. In December 1999, EPA had promulgated 42 of the standards.

The National Environmental Policy Act of 1970 (NEPA) required agencies and activities of the federal government to prepare an Environmental Impact Statement (EIS) for any project that might affect the environment. Unquestionably, NEPA stimulated some hazardous waste management activity and improvement by many agencies. Numerous lawsuits have been filed against the agencies; however, the suits did not directly compel cleanup nor compliance with environmental law. The nature of the suits has been largely to the effect that (1) no EIS was prepared, (2) the EIS was inadequate, or (3) the environmental assessment performed was inadequate (Wentz 1989, pp. 39–40). The Departments of Defense and Energy have, until recently, largely ignored NEPA requirements. To their credit, both departments now have extensive site remediation programs under way *(see also:* Vanderver 1985, pp. 374–375; Satchell 1989, pp. 20–22; and Chapters 11 and 13, this text).

The Federal Insecticide, Fungicide, and Rodenticide Act of 1972 (FIFRA) as amended regulates storage and disposal of pesticides and requires informative and accurate labeling of pesticide products. In May 1974, EPA promulgated regulations for the storage and disposal of pesticides (39 FR 15236, May 1, 1974) (40 CFR 165) and proposed others that were never implemented (39 FR 36874, October 15, 1974). These regulations detailed the appropriate conditions for incineration, soil injection, and other means of disposal of pesticide wastes. They devote considerable attention to the disposal of pesticide containers, which have caused a significant proportion of accidental poisonings (M. Miller 1985, p. 430).

The Marine Protection Research and Sanctuaries Act of 1972 (MPRSA) or "Ocean Dumping Act" regulates the dumping of materials at sea and was intended to prevent or limit the dumping of materials that would have adverse effects. However, ambiguities in the original language of the Act left unclear the intent of Congress as to ending all ocean dumping or allowing it only where it is the best disposal alternative and will not unduly affect the marine ecosystem. This ambiguity was heightened by Congress and EPA permitting ocean dumping of sludge from POTWs along the East Coast [33 USC Section 1412(a)]. Municipal sludge, particularly that generated by older cities in the East, may be laced with heavy metals from industrial sources on the collection systems. The 1988 amendment to MPRSA, the "Ocean Dumping Ban Act" (33 USC Chapter 27) effectively ended ocean dumping of municipal sludges and industrial wastes after December 31, 1991 (EPA, 1991, p. 1, 40).

The Federal Water Pollution Control Act of 1972[15] was significantly modified in 1977 to deal with toxic water pollutants and was renamed the Clean Water Act (CWA). The Act has five main elements:

- The National Pollutant Discharge Elimination System (NPDES) permit program which requires all municipal and industrial waste sources that discharge wastes into "waters of the United States" to have a permit
- A system of minimum national effluent standards for each industry
- Receiving water quality standards
- Specific provisions for oil spills, discharges of toxic chemicals, and non-process discharges such as contaminated plant site run-off
- A construction grant program to assist in funding POTW and related construction (The grant program was replaced by the State Revolving Loan Fund (SRLF) Program, established per Title VI of the CWA.)

Similar to the Clean Air Act, implementation of the Clean Water Act achieved the intended removal of waste materials from municipal and industrial wastewater streams, but created new waste management challenges. Sludges and residues from

---

[15] The 1972 FWPCA was also the first federal environmental statute to use "technology-forcing" standards. Where existing treatment technology did not meet industrial treatment needs, Congress reasoned that it could "force" the desired technological development and achieve the "zero discharge" goal by 1983. This was highly significant lawmaking and a harbinger of laws of the future.

pretreatment and treatment systems were not regulated initially and were frequently mismanaged.

Industrial facilities that discharge into collection systems serving POTWs are required to meet pretreatment standards. Thus, the CWA controls and limits discharges of hazardous wastes directly to surface waters and indirectly through POTWs. The CAA controls (at least some) emissions of hazardous wastes to the atmosphere. MPRSA limits dumping of hazardous wastes at sea. Through these various statutes, Congress had, by the mid-1970s, theoretically and stat-utorily shifted the entire burden of ultimate disposal to the land.

The SDWA of 1974 embodied the potential to provide significant protection to underground sources of drinking water through two major provisions:

- The EPA was authorized to designate individual areas as having an aquifer which is the sole source of water supply to the area and which would create a significant hazard to public health if contaminated. Once an area is so designated, no federal assistance may be provided for any project in the area that the EPA determines may contaminate the aquifer. Any "person"[16] may petition the EPA for a Sole Source Aquifer designation (U.S. EPA 1987, p. 1). By December 1999, EPA had applied this provision to 74 areas.
- The Act regulates underground injection to protect usable aquifers from contamination. Underground injection is usually thought of as being accomplished with a deep, high-pressure pumped injection well. The 40 CFR 144 standard defines underground injection as the subsurface emplacement of fluid through a well or dug-hole, whose depth is greater than its width. This definition encompasses five classes of injection wells used for a variety of purposes including the disposal of hazardous wastes. The EPA designated all states as requiring underground injection control (UIC) programs, and the agency must promulgate the program where a state fails to do so. The EPA missed the 1979 deadline for these promul-gations, and the program moved forward slowly in the early 1980s (J. Miller 1985, pp. 199, 204–205). A new 40 CFR 148 brought injection of hazardous wastes in Class I (deep) wells under the land disposal restrictions, with implementation dates ranging from August 1988 through June 1995, and a new standard for "high risk" Class V shallow injection wells was proposed on July 29, 1998 (63 FR 40586).

The TSCA of 1976 provides the EPA with authority to require testing of chemical substances, both new and old, entering the environment and to regulate them where

---

[16] Individual, corporation, company, association, partnership, state municipality, or federal agency.

necessary. It may also be used to regulate the development of biotechnology and genetic engineering (M. Miller 1985, pp. 141–142). The TSCA empowers the EPA administrator to place restrictions on the production, distribution, use, and disposal of toxic substances. The TSCA was the first legislation in which Congress singled out a specific substance, by name, for regulation. Section 2605(e) directed the EPA to phase out the manufacture and use of polychlorinated biphenyls (PCBs) according to a statutorily regulated timetable. Subchapter II provides the authority and structure for regulation and control of asbestos.

The RCRA of 1976, as indicated earlier, was the outgrowth of the SWDA of 1965 and the RRA of 1970. RCRA Subtitle D continued and expanded the solid waste management programs of the 1965 act, but the Act and implementing programs remained relatively weak on hazardous waste management. Subtitle C, the framework for the federal hazardous waste management program, was strengthened somewhat by amendments in 1980, but more significantly by the Hazardous and Solid Waste Amendments of 1984 (HSWA). The HSWA required the EPA to develop and implement the land disposal restrictions, created the new small quantity generator (SQG) and "conditionally exempt" small quantity generator categories, and in Subtitle I authorized the underground (petroleum) storage tank regulations. Subtitle J, the Medical Waste Tracking Act (MWTA), was enacted in 1988, but was allowed to expire at the end of the 5-year statutory life.

In RCRA, Congress attempted to provide "cradle-to-grave" management of hazardous waste by imposing regulatory requirements upon generators, transporters of hazardous wastes, and owners and operators of treatment, storage, and disposal facilities. Although this text deals with hazardous waste management, in general, the focus will be upon RCRA Subtitle C, which is the statutory authority for federal regulation of the control and management of hazardous waste in the U.S. and its territories (*see also:* Hall and Bryson 1985, Chapter 2; U.S. EPA 1990, Section I).

The RCRA did not address the equally serious problem of abandoned and inactive hazardous waste sites. Legislation establishing remedies and allocating responsibilities for correcting problems at these sites is contained in CERCLA of 1980, commonly known as Superfund. CERCLA provides federal funding for response and site remediation where responsible parties cannot be identified or are unwilling or unable to accomplish the necessary cleanup. The EPA then may sue identified responsible parties for recovery of funds expended in the remediation. CERCLA was extensively modified by the Superfund Amendments and Reauthorization Act of 1986 (SARA). In addition to renewal of Superfund authorities and funding, SARA significantly broadened the reach of CERCLA. Title I of SARA required the Secretary of Labor to issue workplace health and safety standards for hazardous waste workers. Title III of SARA, known as the Emergency Planning and Community Right-to-Know Act (EPCRA), imposed an emergency planning regime upon states and communities and required community right-to-know and toxic release reporting. The intent was to make information regarding chemical use and storage information available to communities and to require states and communities to prepare and implement planning for chemical disasters such as the disaster that devastated Bhopal, India.

The Hazardous Materials Transportation Act of 1975 (HMTA) authorizes the regulation of marking, labeling, and packaging of hazardous materials for transportation and thereby includes the transportation aspects of hazardous waste management. In RCRA, Congress recognized the potential for overlap of regulatory issue by EPA and the Department of Transportation (DOT) and specifically instructed EPA to coordinate all transportation-related hazardous waste management regulations with DOT. The HMTA was substantially amended by the Hazardous Materials Transportation Uniform Safety Act of 1990 (HMTUSA). The 1990 amendments were implemented in the form of a general overhaul of the packaging standards, to bring American practice into accord with international standards.

Political

Early efforts at hazardous waste management, by state and local governments, ranged from effective to nonexistent. The early federal legislation was intended to provide funding, technical assistance, and moral support to state and local governments and to prod them into more effective actions and postures. The federal grant-in-aid was the most common mechanism. Typically, the grant provided start-up funds to staff, train, and equip state and local agencies enabling them to launch specific programs. As time passed, the federal statutes, regulations, and policies increasingly tied funding to levels of performance. Federal programs were "delegated" to states and funding depended upon the favorable outcomes of periodic reviews of program accomplishments. RCRA delegation and state program oversight by EPA continues to this day. Most states have opted for delegation, and many have adopted RCRA regulations "by rule." The state agency director thereby takes on the role of the EPA administrator in implementation of RCRA within the state.

States' rights and police powers issues helped to shape the form of early legislation and policy toward environmental management and cleanup of hazardous waste sites. It was widely held that environmental problems were best managed at the state and local level, and that view was reflected by Congress in the federal legislation and by the federal agencies in the implementation thereof.

Although there were obvious exceptions, the state and local agencies were frequently inadequately funded and hampered by limited staff and equipment. EPA and its predecessor agencies, although similarly limited, were able to focus staff and laboratory capability on specific problems. Congress sought to assist state and local governments in the development of decentralized programs by funding technical assistance capability in the federal agencies. Technical assistance programs were frequently helpful in source identification, problem definition, and impact assessment, but somewhat less helpful in securing prevention and remediation of hazardous waste problems.

The early legislation, programs, and policies were too frequently ineffective in securing the intended protection of public health and the environment. Activists and the public demanded direct and rigorous action to bring hazardous wastes under control.

A careful review of the progression of federal legislation, from the 1965 Solid Waste Act through the 1984 Hazardous and Solid Waste Amendments, shows increasingly direct and detailed involvement by Congress in hazardous waste management in the U.S. As will be shown later in this text, this same impatience with progress was not ameliorated by the direct involvement of federal agencies. In HSWA, Congress took the then unprecedented step of writing regulatory language, standards, and calendar deadlines into the statute.

By 1994, the extent, range, and detail of congressional and federal agency involvement in hazardous waste management, and environmental management in general, had burgeoned to the point that public and political sentiment had turned against these regulatory programs. Federally imposed "unfunded mandates" were a major issue of the 1994 election campaigns, and the new Congress promised to "review" many of these programs. Environmental equity issues arose in poorer urban neighborhoods wherever hazardous waste management activity was proposed. Individual members of Congress were frequently involved in efforts to prevent construction or enlargement of such facilities. Nevertheless, with the exception of the 1990 CAA Amendments, Congress did little to strengthen, weaken, or in any way change environmental statutes during the 1990s. Superfund reauthorization was allowed to languish through most of the 1990s, and the "Brownfields"[17] concept was left to the Administration to develop.

Administrative

As the "Environmental Decade" (1970s) progressed, federal, state, and local agencies and officials learned that large numbers of constituents have lively interests in environmental matters. Public hearings, and similar forums, became procedurally ingrained in most legislation, regulation, and policy. It is now taken as routine that no significant environmental decision is made at any level of government, without full public participation, usually through a hearing process. In countless instances, these hearings have sharply affected the course of resolution of major issues.

In recent years, several federal agencies have been shown to be among the worst offenders of hazardous waste management statutes, regulations, and policies. In 1978, President Carter ordered federal agencies to comply with the nation's environmental laws, but his executive decree had little effect. In 1980, Congress passed CERCLA, but exempted federal government facilities. Not until 1986 were federal agencies brought under Superfund rules. During this same period, the Departments of Energy and Defense hid their hazardous waste practices behind the national security curtain, and the true picture of these practices is now emerging. As noted earlier, these practices have so severely contaminated some sites that officials of the Departments of Energy and Defense have been quoted to the effect that there may

---

[17] A "Brownfield" is a site, or portion thereof, that has actual or perceived contamination and an active potential for redevelopment or reuse (see: Glossary).

be no way to clean them up (Satchell 1989). In these matters, public meetings, hearings, and forums were ineffective, and administrative approaches to hazardous waste management actually served to conceal the extent of the problems.

"Administrative" actions and policies brought about serious delays in the implementation of the newly enacted RCRA and CERCLA, under the direction of EPA Administrator Anne Gorsuch (Burford), during the early 1980s. Gorsuch and her hazardous waste program manager Rita Lavelle held strongly negative views toward environmental regulation in general and toward hazardous waste regulation in particular. These ideologies were further strengthened by political activism that put party advantage far above environmental urgency. In sworn testimony before a congressional committee, Lavelle offered a "frankly political motive for shutting off Superfund help to western mining states. She was afraid that the mining states would resent the federal intrusion and that the 1984 election campaigns of western Republican senators might suffer." Gorsuch testified that "she had held up a $6 million grant to clean up the huge Stringfellow acid pits in Riverside, California because Jerry Brown, then governor of California and Democratic candidate for the Senate, might get the credit" (Lash et al. 1984, pp. 82–83). The serious student of the politics of the environment should read the referenced book in its entirety (*see also*: Davis 1993).

Technical

Approaches to definition of hazardous wastes, their identification, impacts, and remedies have progressed along similar lines to the developments cited previously. The initial environmental and public health concerns with hazardous waste sites had a "fires and explosions" focus. The early history is replete with cases and episodes wherein sites burned and/or exploded, releasing huge amounts of heat energy, toxic vapors, and particulates. Many were confirmed as arson cases and many more were suspect. Other sites emitted toxic vapors without fire or explosion.

Such events were often spectacular, frequently hazardous to human health and/or safety, and always frightening to the public. However, the exposures they created tended to be short-lived (i.e., a few hours or days), with a relatively small number of acute health effects, fewer fatalities, and even fewer cases of chronic health effects.

In 1975, a senior engineer, assigned to an EPA field office, was detailed to the Office of the Assistant Administrator for Water and Hazardous Materials and given the task of evaluating the Agency's groundwater management efforts. When he had completed his report, he made the rounds briefing the program managers on his findings. As he attempted to explain to a Deputy Assistant Administrator (DAA) the threat to groundwater quality posed by toxic metals and organics leaching from land disposal of hazardous wastes, the DAA interrupted him and sneered, "Ah b___ s___, the rocks and the sand strain that stuff out."

A drinking water program official declared, "... we don't have a groundwater problem."

Not until the January 1977 rendition of *The Report to Congress: Waste Disposal Practices and Their Effects on Groundwater*, prepared by a major consulting firm under contract to the EPA Offices of Water Supply and Solid Waste Management Programs, did the magnitude of the groundwater impacts begin to be understood by policy makers. The report confirmed what a few professionals in the Agency had been saying — the real human health and environmental impacts of hazardous waste mismanagement were (are) to the groundwater resources of the nation.

The groundwater resource was then known to supply drinking water to over half the populace and to be the source of water for 30% of the domestic water systems (EPA 1977, p. 1). Aquifers were being contaminated with a wide range of soluble and leachable inorganic and organic pollutants, many of which are toxic.

Whereas the release of toxic air pollutants and heat energy from fires and explosions could be measured in days, hours, minutes, or even seconds, the impacts of groundwater contamination may persist for decades or centuries. As experience was gained, it was seen that similar differences of scale prevailed in cleanup costs. Congress and the regulatory agencies began to shift their focus to prevention of groundwater pollution and remediation of contaminated aquifers. This shift of focus became most evident in the 1984 HSWA provisions, which will be a frequent topic throughout this text.

Large East Coast cities, for many years, barged their industrial waste-laden sewage sludge to ocean dumping areas. During the 1970s the ever-tightening restrictions of the CAA and CWA regulatory programs, together with the disappearing land disposal sites, caused ocean disposal of hazardous wastes to become a popular alternative. Opposition to these practices mounted, and in 1977 Congress amended MPRSA to require EPA to "end the dumping of sewage sludge and industrial wastes into ocean waters." EPA had never developed criteria for safe ocean disposal, and the new amendment contained language defining sewage sludge and industrial wastes as materials which unreasonably degrade or endanger human health or the environment (J. Miller 1985, pp. 465–467). The lack of a more precise definition of "industrial waste" hampered the effectiveness of the amended MPRSA and several court decisions allowed some dumping to continue. Public pressure and the Ocean Dumping Ban Act of 1988 effectively ended the practice (EPA 1991, p. 40).

Thus, the legislation and implementation of the 1970s had the effect of diverting most hazardous wastes onto the land or beneath the land surface. Surface impoundments and land "farming" were offered up as "treatment," and landfills and deep well injection systems were regarded as acceptable disposal techniques. The realities were that the impoundments and land "treatment" facilities provided little treatment and made their hazardous constituents available to the groundwater resource. The landfills similarly made leachable and liquid hazardous constituents available to the groundwater, and the environmental safety of deep well disposal was in doubt among many professionals. These considerations were paramount when Congress enacted HSWA in 1984 (*see also*: Piasecki and Davis 1987, Chapter 3).

By the early 1990s, a wide range of treatment and destruction technologies had evolved, and prospects for significant reductions in quantities of hazardous waste managed in land "treatment" and disposal were good. However, increasingly vigorous activism and public opposition to siting of new treatment and destruction facilities caused widespread withdrawal of RCRA permit applications, cancellation of construction plans and contracts, and a marked reluctance toward corporate involvements in major new facilities. Old line waste management companies adopted strategies of buying out smaller companies, expanding and upgrading existing facilities, and generally lowering profiles. These developments were both cause and effect of the rise of the environmental equity/justice issues and of the brownfields concept, both of which are discussed in Chapters 10 and 11.

International Aspects

Western Europeans have been perceived to be several steps ahead of the U.S. in the evolution of hazardous waste management. The long-standing perception has been that land has not been available in Europe for land disposal to have its day as the alternative of choice. Incineration became the early choice and remains so to this day, but land disposal is practiced in varying degrees. Britain is a case in point, with 85% of hazardous waste disposed of in landfills (Skinner 1987, p. 7). Denmark, (Muller 1987, p. 118), France (Leroy 1987, p. 144), and the former West Germany (Sierig 1987, pp. 128–130) utilize land disposal, but clearly emphasize treatment and destruction processes.

The West German chemical industry began ocean incineration of waste chlorinated hydrocarbons to avoid costly scrubbers for land-based incinerators. High-temperature incineration effectively destroys the wastes, but there is insufficient space on ocean-going incinerators for stack gas treatment equipment, and the units emit hydrochloric acid (HCl) in the exhaust gases. The buffering capacity of the limitless sea water was counted upon to neutralize the HCl emitted by ocean-going incinerators. Europeans became disenchanted with this technology and, in two international agreements, forced the phase-out of ocean incineration (Piasecki and Davis 1987, p. 68). As is discussed in Chapter 7, a U.S. company made major investments of money and resources in an ocean incineration venture. The project was eventually abandoned (*see also*: Piasecki and Davis 1987, Chapters 3 and 4; U.S. EPA 1991, Chapter 5).

The shortage of land for land disposal notwithstanding, Europeans made some costly errors in land treatment/disposal, as did their U.S. counterparts. In Britain, an early cornerstone of the toxic waste plan was co-disposal — the deliberate mixing of hazardous wastes with conventional municipal wastes in permeable landfills. The practice was based upon the belief that the leaching of toxic chemicals would change the wastes into nontoxic substances over time by dilution and biological degradation (Piasecki and Davis 1987, p. 193). Skinner (1987, p. 7) stated ... "there is little evidence of problem landfill sites in the UK ... This is attributed to comprehensive land use controls, favorable geology and control of groundwater usage." Nevertheless, the U.K. government subsequently found it necessary to adopt more aggressive definitions of the key terms "contaminated land" and "harm" and to give higher priority to groundwater protection in the Environmental Protection Act of 1990 and the Environment Act of 1995 (Petts et al. 1997, pp. 4–5; *see also:* Wilson 1987, pp. 256–257).

In Holland, much of which is below sea level, fill is often required before construction can take place on land. For more than 40 years, solid wastes containing hazardous chemicals were utilized as fill material. Authorities estimate that up to 8 million metric tons of hazardous chemical waste may be buried in that small country (Enger et al. 1989, p. 379). The government estimates that there are now nearly 5000 leaking waste sites. The small village of Lekkerkerk is the Dutch version of Love Canal, where local government allowed dumping of chemical wastes as fill material (Piasecki and Davis 1987, pp. 190–191; USEPA, 1992). Some 250 houses had to be abandoned temporarily and approximately 156,000 tonnes (nearly 175,000 tons) of contaminated fill was removed and barged to Rotterdam for high temperature incineration (Petts et al., 1997, p. 17; *see also*: Page 1997, Chapter 10)

Denmark, notably, places very heavy emphasis on waste minimization, recycling, and incineration in combined power and heating plants. Landfilling of wastes (municipal and hazardous) has been reduced to 20%. Since 1997, it has been forbidden to landfill waste that could have been incinerated for energy recovery (Veltze 1999, p. 78). Pump-and-treat methods for cleanup of contaminated aquifers has been widely used, apparently with success at some sites in Denmark (Madsen 1998, p. 257).

Western European hazardous waste management practice has been generally optimal compared to practices in the former Soviet Bloc/Warsaw Pact nations. As the Soviet Union disintegrated, Russian military units withdrawing from their Central and Eastern European training sites, garrisons, airfields, and seaports left thousands of contaminated sites. These include artillery ranges, training grounds, above- and below-ground storage areas and tanks, fueling areas, and maintenance and repair areas. The sites are contaminated with hydrocarbons, heavy metals, acids, PCBs, PAHs, solvents, ordnance, biological agents, and numerous other wastes. The former occupants have accepted no financial responsibility for remediation of these sites and the host countries must now find ways and means of responding to the potential hazards (Voss 1995, p. 7). Similar conditions exist in Latvia, Estonia, and Lithuania (Hadonina 1998, p. 63ff; Tammemae 1998, p. 305ff). The Baltic countries have also found it necessary to clean up their polluted harbors, bays, and estuaries, as well as the open Baltic Sea, as a result of Russian naval activity. The ports, bays, fuel stations, communications centers, ammunition depots, and rocket bases were contaminated with a wide variety of fuels, munitions, and chemical residues. The seawater areas are also polluted with the wrecks of sunk and scuttled vessels (Tammemae 1998, p. 306–308; *see also*: Page 1997, Chapter 12; Krosshavn and Fonnum 1998, pp. 343–345).

The North Atlantic Treaty Organization (NATO) has long recognized that the mismanagement of radioactive waste is an international problem, but had little information on the extent of the problem in the former Soviet Union states. In an effort to gain better understanding of this Cold War legacy, in 1997 NATO made funds available to researchers of the Environmental Technology Program at the East Campus, Arizona State University (ASU), to conduct joint studies at missile sites near St. Petersburg and Moscow, Russia. Their preliminary field investigations indicated widespread disposal of radioactive wastes at the former military installations; however, Russian government officials are reluctant to allow thorough investigation and analysis of the problems. This unofficial reluctance has brought about a halt in the investigations. The ASU professors (Drs. Nicholas Hild, Larry Olson, and Danny Peterson) believe that the potential problem areas may rival the U.S. Department of Energy weapons facilities in both magnitude of wastes and geographic areas requiring cleanup.

American military forces have created unique contamination problems along the former Distant Early Warning (DEW) Line, which was constructed across the North American Arctic in the 1950s. The contaminants, primarily fuel, lubricants, and PCBs, pose especially difficult remediation problems due to the remoteness of sites, the harshness of the environment, and the short fieldwork seasons. The Canadian government established the National Contaminated Sites Remediation Program (NCSRP) to oversee a methodical decommissioning procedure for industrial and military sites. In 1998, 3 of the 42 DEW Line sites had been restored to a state that allows natural attenuation to take place (Reimer and Zeeb 1998, pp. 41–54).

Canadian provincial hazardous waste management programs parallel, to some extent, those in the U.S. Primacy for regulatory programs resides with the provinces; however, the federal government has some interprovincial authority (Dawson and Mercer 1986, p. 37; Assante-Duah and Nagy 1998, p. 11). Canadian national capacity for off-site treatment and volume reduction of hazardous wastes is growing, with modern, integrated treatment facilities in operation or planned. The Bovar facility at Swan Hills, Alberta, which recently began accepting hazardous waste imports, is a case in point. Therefore, extensive and prescriptive regulations governing treatment and disposal of hazardous wastes are not in effect. The Canadian Environmental Protection Act (CEPA) lists nine hazardous classes of materials which are either prohibited or release is controlled. Recent amendments add a Leachate Extraction Procedure similar to that of RCRA (Krieger and Austin 1995, p. 90). In April 1999, the Canadian Environment Minister announced new regulatory action by the federal government to achieve reductions in dichloromethane (methylene chloride) releases by 63% by 2002 and 85% by 2007 and in hexavalent chromium releases by 75% by 2005 (*Environment Reporter,* April 16, 1999, p. 2486).

In the outlying areas of Mexico, discharge of toxic materials goes on virtually uncontested (Dawson and Mercer 1986, p. 37). This absence of controls invited the import of hazardous wastes from the U.S., but RCRA requires exporters to provide detailed notification to EPA and to obtain the consent of the receiving country (40 CFR 262,263). Burgeoning industrial development along the U.S.-Mexican border has brought hazardous waste management sharply into focus. The maquiladora[18] industries supposedly bring their hazardous wastes from their Mexican facilities into the U.S., to ensure safe treatment and/or disposal and to meet "duty-free" requirements of U.S. and Mexican customs regulations. RCRA (40 CFR 262) also regulates import of hazardous waste into the U.S. As discussed later in this text, accountability of maquila-generated hazardous wastes has not proven satisfactory, and the environmental impacts of these activities in some of the industrialized border areas are severe.

In April 1999, a Mexican attorney and private consultant, lecturing in a series organized by the government's National Ecological Institute stated that "... Mexican trade officials are giving scant attention to environmental protection in their trade development policies ... ." He cited as an example the actions of the government to reconsider the environmental impact statement it had originally rejected, for a joint venture between a Mexican company, Exportadora de Sal, and Japan's Mitsubishi. The company, which pays 20% of its earnings to the Mexican government in taxes, wants to expand operations of one of the largest salt-producing enterprises in the world, in a natural [sic] protected area in which thousands of gray whales arrive each year. The attorney gave other examples and noted passage in December 1998 of an amendment to the Mexican constitution that requires government consideration of "sustainable development" in all trade affairs (*Environment Reporter,* April 16, 1999, p. 2487).

Hazardous waste management in Central and South America is primitive at best. Management is hampered by less than vigorous enforcement. Page quotes a (World Bank, 1992) report stating: "The worst toxic pollution in the developing world comes from heavy metals from smelters and manufacturing plants, especially in Eastern Europe, and from chemical and fertilizer plants in Latin America, Asia, and Eastern Europe ... Even though the volume of toxic waste being produced in the developing countries is increasing rapidly, it is still below the level found in the industrial economies" (Page 1997, p. 106; *see also*: Assante-Duah and Nagy 1998, Chapter 4).

The Japanese environmental focus has been on protection of worker health, but new more comprehensive legislation has been enacted. More recently, Japanese laws and regulations, as well as enforcement mechanisms are said to be "... comparable to those found in other industrialized countries" (Assante-Duah and Nagy 1998, p. 17). Other Pacific Rim countries' hazardous waste management programs have generally lagged industrial development (Krieger and Austin 1995, pp. 92–99).

The United Nations Environment Programme (UNEP) has limited global environmental oversight functions. Acting through various suborganizations and in concert with international organizations, such as the World Health Organization (WHO), UNEP organizes various conventions and manages environmental databases, some of which are specific to hazardous waste management. Pertinent conventions include the London Convention on Ocean Dumping, the Oslo Convention on Incineration-At-Sea, and the Basel Convention on the Control of Transboundary Movements of

---

[18] Maquila (twin) industrial facilities are established on both sides of the U.S.-Mexican border, usually by U.S. companies, to take advantage of significantly less stringent regulatory burdens and labor costs on the Mexican side. Work that is labor-intensive or that may involve use of toxic chemicals is performed in the Mexican facility. Subassemblies, etc. are shipped to the U.S. side for final assembly, inspections, distributions, etc. (*see:* "maquiladora" entry in the Glossary).

Hazardous Wastes and their Disposal. A database specific to international hazardous waste management is the International Register of Potentially Toxic Chemicals (IRPTC) (U.S. EPA 1991, pp. 9ff; UNEP 1992, pp. 28–29, 42).

TOPICS FOR REVIEW OR DISCUSSION

1. The hazardous waste mismanagement episodes summarized in the "Early Hazardous Waste Management" section of this chapter suggest several regulatory measures that might be needed to protect human health and the environment. Discuss a few such possibilities.
2. Discuss two undesirable effects of the discharge of toxic industrial wastes to municipal sewerage.
3. Review the progression of environmental laws and regulations that culminated in the need to regulate hazardous waste management in the U.S.
4. Identify at least two provisions of the 1984 Hazardous and Solid Waste Amendments that had major significance or impact.
5. What event(s) of the mid-1980s has/have caused, and will continue for some time to cause, annual increases in the quantities of hazardous wastes generated?
6. SARA Title I was an unusual legislative step. Discuss.
7. Public hearings on environmental issues are now taken for granted. Discuss any that may have been held in local jurisdictions and any impacts they may have had.
8. If major and minor world powers suddenly began development of new weapons systems that threatened world peace, or even human survival, would the U.S. be justified in again postponing environmental controls on weapons production in order to quickly build an offsetting defensive arsenal of these weapons?

REFERENCES

Assante-Duah, D. Kofi and Imre V. Nagy. 1998. *International Trade in Hazardous Waste.* E & FN Spon-Routledge, New York.

The Associated Press, July 29, 1997.

Beckwith, Douglas C. 1990. Minnesota Pollution Control Agency. Personal communication.

Council on Environmental Quality. 1977. Environmental Quality: The Eighth Annual Report of the Council on Environmental Quality. U.S. Government Printing Office, Washington, D.C., Stock No. 041-011-00035-1.

Council on Environmental Quality. 1979. Environmental Quality: The Tenth Annual Report of the Council on Environmental Quality. U.S. Government Printing Office, Washington, D.C., Stock No. 041-011-00047-5.

Dahl, Thomas O. 1980. "Salsbury Laboratories — LaBounty Site, Phased Approach to a Hazardous Waste Disposal Problem." U.S. Environmental Protection Agency, National Enforcement Investigations Center, Denver, CO.

Davis, Charles E. 1993. *The Politics of Hazardous Waste,* Prentice-Hall, Englewood Cliffs, NJ.

Dawson, Gaynor W. and Basil W. Mercer. 1986. *Hazardous Waste Management.* John Wiley & Sons, NY.

Enger, Eldon D., J. Richard Kormelink, Bradley F. Smith, and Rodney J. Smith. 1989. *Environmental Science: The Study of Interrelationships.* Wm. C. Brown, Dubuque, IA.

Environment and Natural Resources News, June, 1999, p. 5. Law Offices of Kevin J. Brown, Syracuse, NY and Harrisburg, PA.

Environment Reporter, January 15, 1999, pp. 1797–1798. Bureau of National Affairs, Washington, D.C.

Environment Reporter, April 16, 1999, p. 2487. Bureau of National Affairs, Washington, D.C.

Hadonina, Dzidra. 1998. "Environmental Situation and Remediation Plans of Military Sites in Latvia," in *Environmental Contamination and Remediation Practices at Former and Present Military Bases,* F. Fonnum, B. Paukstys, B.A. Zeeb, and K.J. Reimer, Eds., Kluwer, Dordrecht.

Hall, Ridgeway M. and Nancy S. Bryson. 1985. "Resource Conservation and Recovery Act," in *Environmental Law Handbook,* Eighth Edition, Government Institutes, Inc., Rockville, MD.

Krieger, Gary R. and Ian Austin. 1995. "International Legal and Legislative Framework," in *Accident Prevention Manual for Business and Industry — Environmental Management,* Gary R. Krieger, Ed., National Safety Council, Itasca, IL.

Krosshavn, M. and F. Fonnum. 1998. "Nuclear Accidents and Radionuclide Transport," in *Environmental Contamination and Remediation Practices at Former and Present Military Bases,* Kluwer, Dordrecht.

Lash, Jonathan, Katherine Gillman, and David Sheridan. 1984. *A Season of Spoils.* Pantheon Books, NY.

Leroy, Jean-Bernard. 1987. "Hazardous Waste Management in France," in *International Perspectives on Hazardous Waste Management,* William S. Forester and John H. Skinner, Eds., Academic Press, NY.

Lesser, George H. 1984. "Trends and Problems in International Legislation on Transportation of Hazardous Materials," in *Atmospheric Dispersion of Hazardous/Toxic Materials from Transport Accidents.* Elsevier, NY, p. 2.

The Los Angeles Times, November 27, 1993

Madsen, Bjarne. 1998. "Groundwater Contamination Cleanup — Reality or Illusion," in *Environmental Contamination and Remediation Practices at Former and Present Military Bases,* F. Fonnum, B. Paukstys, B.A. Zeeb, and K.J. Reimer, Eds., Kluwer, Dordrecht.

Miller, Jeffrey G. 1985. "Marine Protection, Research and Sanctuaries Act," in *Environmental Law Handbook, Eighth Edition,* Government Institutes, Inc., Rockville, MD.

Miller, Marshall Lee. 1985. "Federal Regulation of Pesticides," in *Environmental Law Handbook,* Eighth Edition, Government Institutes, Inc., Rockville, MD.

Muller, Klaus. 1987. "Hazardous Waste Management in Denmark," in *International Perspectives on Hazardous Waste Management,* William S. Forester and John H. Skinner, Eds., Academic Press, NY.

Nebel, Bernard J. and Richard T. Wright. 1993. *Environmental Science.* Prentice-Hall, Englewood Cliffs, NJ.

Page, G. William. 1997. *Contaminated Sites and Environmental Cleanup.* Academic Press, NY.

Petts, Judith, Tom Cairney, and Mike Smith. 1997. *Risk-Based Contaminated Land Investigation and Assessment.* John Wiley & Sons, NY.

Piasecki, Bruce W. and Gary A. Davis. 1987. *America's Future in Toxic Waste Management — Lessons from Europe.* Quorum Books, NY.

Reimer, K.J. and B.A. Zeeb. 1998. "Environmental Risk Assessment and Management in Evaluation of Contaminated Military Sites," in *Environmental Contamination and Remediation Practices at Former and Present Military Bases,* F. Fonnum, B. Paukstys, B.A. Zeeb, and K.J. Reimer, Eds., Kluwer, Dordrecht.

Satchell, Michael. 1989. "Uncle Sam's Toxic Folly." *U.S. News and World Report*, March 27, pp. 20–22.

Sierig, Gerhard. 1987. "Hazardous Waste Management in The Federal Republic of Germany," in *International Perspectives on Hazardous Waste Management,* William S. Forester and John H. Skinner, Eds. Academic Press, NY.

Skinner, John H. 1987. *International Perspectives on Hazardous Waste Management,* William S. Forester and John H. Skinner, Eds., Academic Press, NY.

Sullivan, Thomas F. P. 1985. "Environmental Law Fundamentals and the Common Law," in *Environmental Law Handbook*, Eighth Edition, Government Institutes, Inc., Rockville, MD.

Tammemae, Olavi. 1998, "Remediation of Polluted Environment at Naval Ports of the Baltic Sea," in *Environmental Contamination and Remediation Practices at Former and Present Military Bases,* F. Fonnum, B. Paukstys, B.A. Zeeb, and K.J. Reimer, Eds., Kluwer, Dordrecht.

United Nations Environmental Programme. 1992. *Two Decades of Achievement and Challenge.* Information and Public Affairs Branch, Nairobi, Kenya.

U.S. Environmental Protection Agency. 1977. The Report to Congress: Waste Disposal Practices and Their Effects on Groundwater. Office of Water Supply and Office of Solid Waste Management Programs, Washington, D.C.

U.S. Environmental Protection Agency. 1987. Sole Source Aquifer Background Study: Cross-Program Analysis. Office of Ground-Water Protection, Washington, D.C., EPA 440.6-87-015.

U.S. Environmental Protection Agency. 1990. RCRA Orientation Manual, 1990 Edition. Office of Solid Waste, Washington, D.C.

U.S. Environmental Protection Agency. 1991. Report to Congress on Ocean Dumping 1987-1990. Office of Water, Washington, D.C., EPA 503/9-91/009.

U.S. Environmental Protection Agency. 1994. The Biennial RCRA Hazardous Waste Report (Based on 1991 Data). Office of Solid Waste and Emergency Response, Washington, D.C., EPA 530-S-94-039.

U.S. Environmental Protection Agency. 1996a. Toxics Release Inventory Public Data Release. Office of Pollution Prevention and Toxics, Washington, D.C., EPA 745-R-96-002.

U.S. Environmental Protection Agency. 1996b. Toxics Release Inventory 1987-1994. Office of Pollution Prevention and Toxics, Washington, D.C., EPA 749-C-96-003.

U.S. Environmental Protection Agency. 1997. The Biennial RCRA Hazardous Waste Report (Based on 1995 Data) Executive Summary. Office of Solid Waste and Emergency Response, Washington, D.C., EPA 530-S-97-022.

U.S. Environmental Protection Agency. 1997. Toxics Release Inventory 1987–1995. Office of Pollution Prevention and Toxics, Washington, D.C., EPA 749-C-97-003.

U.S. Environmental Protection Agency. 1999. 1987–1996 Toxics Release Inventory Reporting and the 1997 Public Data Release. Office of Pollution Prevention and Toxics, Washington, D.C., EPA 749-C-99-003.

U.S. Environmental Protection Agency. 1999a. The Biennial RCRA Hazardous Waste Report (Based on 1997 Data) Executive Summary. Office of Solid Waste and Emergency Response, Washington, D.C., EPA 530-S-99-036.

U.S. Environmental Protection Agency. 1999b. Introduction to the National Pretreatment Program. Office of Wastewater Management, Washington, D.C., EPA-833-B-98-002.

Vanderver, Timothy A., Jr. 1985. "National Environmental Policy Act," in *Environmental Law Handbook, Eighth Edition,* Government Institutes, Inc., Rockville, MD.

Velze, Susanne Arup. 1999. "Waste Management in Denmark." *Waste Management & Research,* Internation Solid Waste Association, Copenhagen K, Denmark.

Voss, Charles F. 1995. "Strategy for Identifying and Evaluating Site Remediation Approaches for Former Soviet Military Bases in Central and Eastern Europe," in *Cleanup of Former Soviet Military Installations. Identification and Selection of Environmental Technologies for Use in Central and Eastern Europe,* R. C. Herndon, P.I. Richter, J.E. Moerlins, J.M. Kuperberg, and I.L. Biczo, Eds., Springer-Verlag, Berlin.

Wentz, Charles A. 1989. *Hazardous Waste Management.* McGraw-Hill, NY.

Wilson, David C. 1987. "Hazardous Waste Management in The United Kingdom," in *International Perspectives on Hazardous Waste Management,* William S. Forester and John H. Skinner, Eds., Academic Press, NY.

World Bank. 1992. *World Development Report: Development and the Environment.* Oxford University Press, NY.

Worobec, Mary Devine. 1986. Toxic Substances Control Primer. The Bureau of National Affairs, Washington, D.C.

2 Definition of Hazardous Waste

OBJECTIVES

At completion of this chapter, the student should:

- Understand the generally accepted definitions of "hazardous waste" and why the *definition* is of singular importance.
- Understand the Resource Conservation and Recovery Act (RCRA) definition of "hazardous waste" and the importance, application, and limitations thereof.
- Understand the relationship of RCRA "solid waste" and RCRA "hazardous waste."
- Have an overview familiarity with the perspective of various professionals in the management and control of hazardous wastes.
- Understand the differences in perception of hazardous *waste* and hazardous *materials* management by regulators, environmentalists, the public, and the media.
- Be familiar with other definitive approaches — state and foreign — and their strengths and weaknesses.

INTRODUCTION

If every person who creates, handles, or manages hazardous waste was sufficiently knowledgeable, motivated, capable, and unfailingly trustworthy regarding roles and responsibilities, regulation of hazardous waste management would not be necessary. Unfortunately, we live in an imperfect world, and it has become obvious that the practice of hazardous waste management must be regulated. Clearly, if a regulatory agency is to regulate something, there should be an unambiguous means of identifying and describing that something which is to be regulated.

One source tells us that:

… The definition of hazardous waste varies from one country to another. One of the most widely used definitions, however, is contained in the U.S. Resource Conservation and Recovery Act of 1976 (RCRA). RCRA considers wastes toxic and/or hazardous if they "cause or significantly contribute to an increase in mortality or an increase in serious irreversible, or incapacitating reversible illness; or pose a substantial present or potential hazard to human health or the environment when improperly treated, stored, transported, disposed of, or otherwise managed." Having read this definition you can begin to appreciate the complexity in regulating the problem" (Enger et al. 1989, p. 372).

Imagine having to determine whether or not the contents of a truckload of drums meet this criteria, while the driver waits, and other trucks are lined up behind it.

Countless such scenarios hang upon the legal *definition* of hazardous waste, and the importance of a workable definition cannot be overemphasized. In this chapter we will explore this matter of definition and identification of "hazardous waste." In the study and management of hazardous waste, the terms "hazardous" and "toxic" are frequently used interchangeably. There is a technical difference, and it is important, as well, to recognize that distinction.

"Toxic" commonly refers to poisonous substances which cause death or serious injury to humans and animals by interfering with normal body physiology. The term is properly used to describe a pure substance, whether or not it has become a waste (i.e., "toxic substance" or "toxic chemical"). A toxic effect is imposed intrinsically.

"Hazardous," a broader term, refers to all wastes that are dangerous for any reason, including those that are toxic (i.e., flammable, explosive, or reactive). A hazardous waste may impose the effect intrinsically or extrinsically.

THE CHEMIST

The analytical chemist, perhaps to a greater extent than others, must deal with the definition of hazardous waste from a number of standpoints. He/she may be called upon to define hazardous waste in terms that will enable analytical determinations and/or screening procedures to be carried out expeditiously, at reasonable cost, and to be sufficiently comprehensive that definitional loopholes are not created. He/she may be called upon to develop analytical or screening procedures or to select the most appropriate option from several procedures. The chemist may find it necessary to configure a laboratory to most efficiently handle the analytical requirements of a particular source. He/she may be involved in manufacturing or treatment process control where wastes may vary from hazardous to nonhazardous as a result of control factors.

The chemist is particularly concerned with the safety of analytical procedures. Where screening techniques are employed, for decision making in the field or on-site, the chemist must devise procedures that enable the decision to be made without exposing the analyst and/or others to hazards. He/she is expected to define "hazardous waste" in chemical terms that are sufficiently simple so that needed tests can be performed safely, in the field, by semiskilled workers, yet be sufficiently precise to withstand the rigors of the courtroom. This dichotomy is made more pronounced by the fact that many of the analytical methods prescribed by SW 846, (U.S. EPA 1986) are highly complex, requiring sophisticated instrumentation and procedures that usually are incomprehensible to courts, the media, and the lay public.[1]

Analytical chemists are frequently called as expert witnesses or to testify regarding chemical determinations. The regulatory definition, grounded in the statute, is the criterion against which the hazardous or nonhazardous status of a sample is judged. Ambiguity or unnecessarily complex definition can cause the testimony to be beyond the capability of the nonlawyer and can make credible enforcement actions difficult or impossible. Needless to say, the findings in such cases can have enormous significance.

THE LIFE SCIENTIST/HEALTH PROFESSIONAL

The roles of the life scientist and the health professional, in hazardous waste management, are closely related and deal with the biological impacts of exposure of living cells to hazardous wastes. The life scientist is primarily concerned with the exposure impacts upon nonhuman cells, as indicator organisms. The health professional is concerned with the incidence of disease or genetic effect, the hazardous waste constituents that cause the disease or genetic effect, and the pathway(s) or means by which the waste constituent impacts the human target.

The life scientist may be called upon to develop or improve bioassay procedures that will be used to establish or modify exposure criteria or to evaluate a consignment or category of waste against established criteria. He/she may be called upon to evaluate rates and/or impacts of bioaccumulation of toxic constituents of hazardous wastes, to evaluate a given waste treatment process in terms of biopopulations, or to prescribe a bioremediation process that may be expected to meet a cleanup criterion.

The health professional may be assigned the task of translating the life scientists' data, regarding nonhuman exposure, to human exposure criteria. Other responsibilities may include establishing a threshold level based upon morbidity statistics and measured exposure level or providing expert testimony regarding cause and effect in exposure cases.

The life scientist and the health professional are expected to define "hazardous waste" or evaluate a waste material in terms of an established life science or health standard. As before, circumstances rarely permit real-time, detailed, or complex scientific evaluations of a waste shipment or a collected batch of waste. The challenge, also as before, is to define "hazardous waste" in terms that will meet environmental and human health protection goals, without significant failure, yet keep the procedure simple and timely.

THE ENVIRONMENTALIST

The broad context of the environmentalist's concern with hazardous waste releases is any alteration of the environment caused or induced by such releases. Specifics

---

[1] SW 846 — a massive document, published by EPA and available from the Government Printing Office, detailing the analytical procedures that are "approved" for use in identifying hazardous wastes. The document is also available on CD-ROM from NTIS. *See:* Glossary.

of his/her concern lie in acute and chronic toxicity to organisms, bioconcentration, biomagnification, genetic change potential, etiology, pathways, change in climate and/or habitat, extinction, persistence, and esthetics such as visual impact. More broadly still, the environmentalist seeks to protect the environment from hazardous waste impacts by education, activism, statutory and/or regulatory development, and advocacy.

For the environmentalist, derivation of a workable *definition* of hazardous waste is critical and frustrating. The DDT issue was resolved by the clear association of the material with bioaccumulation, thinning of egg shells, and threatened extinction of important species. DDT was a specific chemical for which substitutes were available and which could be banned and eventually purged from the environment. Few such possibilities exist among the innumerable wastes, constituents, combinations, and concentrations which may be released or may occur subsequent to release.

Criteria which may be suggested or proposed by the environmentalist are certain to be the subject of challenge by special interests demanding proof of direct cause-and-effect. The actual process of determining the environmental impact of a substance may be obscured in a variety of sub-processes and may require years to run its course. Sadly, the committees, hearing boards, bureaucracies, legislatures, and courts which must find words to construct the definition, continue to fall back on the nebulous "harmful-to-the-environment" generalities. Those who must make the definition work, if the environment is to be protected, are frequently hard pressed to do so (*see also*: Nebel and Wright 1993, Chapter 14).

THE LEGISLATOR/LAWYER/ADMINISTRATOR/DIPLOMAT

Perhaps without significant distinction from what was previously stated, legislators, lawyers, administrators, and diplomats are concerned with the "workability" of the definition. Statutes must provide the basis for regulations. Regulations must be understandable and enforceable. Administrators of regulatory agencies must have the statutory and regulatory authority and the financial resources provided to protect the public from exposure to harmful concentrations or quantities of hazardous waste. Workable approaches to definition clearly do not include development of proof, in every situation that may arise, that the substance in question has "... cause(d) or significantly contribute(d) to an increase in mortality or an increase in serious irreversible or incapacitating reversible illness; or pose(d) a substantial present or potential hazard to human health or the environment ..." [RCRA Section 1004(5)].

Diplomatic efforts to achieve international and/or regional hazardous waste management agreements and treaties are continuously preoccupied with sorting out each participating government's notion of what wastes are being discussed. The United Nations Environmental Programme (UNEP) makes exactly the point ... "Off-site recycling is widely utilized to achieve waste minimization, but ill-defined and ill-specified exports of wastes destined for recovery open the door to illegal traffic" (UNEP 1994, p. 2). As noted below, various nations may work with highly sophisticated definitions, while others may simply resort to the rationale that any chemical that is discarded is a hazardous waste.

The RCRA regulations (40 CFR 261 and 262) specify that a solid waste is a hazardous waste if it is not excluded from regulation, and meets any of the following conditions:

- Exhibits any of the *characteristics* of a hazardous waste
- Has been named as a hazardous waste and *listed* as such in the regulations
- Is a *mixture* containing a listed hazardous waste and a nonhazardous solid waste
- Is a waste *derived-from* the treatment, storage, or disposal of a listed hazardous waste

---

FIGURE 2.1 Identification of RCRA hazardous wastes.

Responsible officials, generators of hazardous waste, and/or owners of hazardous waste facilities expect their regulatory requirements to be understandable and workable and their efforts at compliance to be measurable without ambiguity. Lay citizens expect regulatory agencies to protect them from exposure to harmful substances by *preventing* the release thereof. We expect contaminated sites to be cleaned up without prolonged exposure of test organisms to prove the contaminant(s) to be hazardous. It is not difficult to envision the absurd scenarios that could arise from the RCRA definition, if left standing without workable implementing language in the regulations.

IMPLEMENTING THE RCRA DEFINITION OF "HAZARDOUS WASTE"

Congress defined "hazardous waste," but left it to EPA to develop the regulatory framework that would *identify* those solid wastes that must be managed under Subtitle C of RCRA (U.S. EPA 1990, p. III-4). Some European countries began identifying hazardous wastes by drawing up lists of known wastes that present no significant short-term handling or long-term environmental hazards and defined hazardous waste by exclusion, i.e., as any wastes not listed. In the U.K. the exclusive list was employed until 1972 (World Health Organization 1983, p. 12). The *exclusive* list has obvious shortcomings in application in regulatory programs.

The inclusive list is more commonly used, either with or without accompanying criteria. This approach was employed, in the 1980s, in Belgium, Denmark, France, West Germany, The Netherlands, Sweden, and the U.K. (World Health Organization 1983, p. 12). Eventually, more than 20 member countries of the Organization for Economic Cooperation and Development (OECD) produced lists of potentially hazardous wastes, no two of which were identical.[2] In 1988 OECD produced a "Core List" of hazardous wastes that require control when proposed for disposal following transfrontier movements (Assante-Duah and Nagy 1998, pp. 89–90).

Other nations and UNEP apparently consider any toxic chemical a hazardous waste when "thrown away" (UNEP 1992, pp. 28–29). The EPA adopted the listing approach, but also defined "characteristics" and conditions under which wastes become or remain "hazardous." The four methods prescribed by RCRA for identification of hazardous wastes are highlighted in Figure 2.1.

---

[2] *See:* OECD in Glossary.

Hazardous. Any solid waste that exhibits one or more of these characteristics* is classified as hazardous under RCRA:

• Ignitability
• Corrosivity
• Reactivity
• Toxicity*

* As determined by the Toxicity Characteristic Leaching Procedure (TCLP), which is described in EPA Publication SW 846, Test Methods for Evaluating Solid Waste, Physical/Chemical Methods.

FIGURE 2.2   RCRA hazardous waste characteristics.

The first step in identifying a RCRA hazardous waste is the determination that a waste meets the RCRA definition of a *solid waste.* Section 1004(27) of the statute defines solid waste as: any garbage, refuse, sludge from a waste treatment plant or air pollution control facility and other discarded material, including solid, liquid, semisolid, or contained gaseous material resulting from industrial, commercial, mining, and agricultural operations, and from community activities, but does not include solid or dissolved material in domestic sewage, or solid or dissolved materials in irrigation return flows or industrial discharges which are regulated by the clean water Act or Nuclear Regulatory Commission. The EPA interpretation of this language in the regulatory language of 40 CFR 261.2 speaks of *discarded material* which is *abandoned, recycled,* or considered *inherently waste-like.* Each of these terms has specific meanings which are detailed in 40 CFR 261 and which should be studied by the newcomer to the practice.

Hazardous Waste Characteristics

By mid-2000, the EPA had established four characteristics for hazardous waste identification (Figure 2.2). The EPA applied two criteria in selecting these characteristics:

• The characteristic must be defined in terms of physical, chemical, or other properties that cause the waste to meet the definition of hazardous waste in the Act.
• The properties defining the characteristics must be measurable by standardized and available testing protocols.

The second criterion was adopted because the primary responsibility rests with generators for determining whether a solid waste exhibits any of the characteristics. EPA regulation writers believed that unless generators were provided with widely available and uncomplicated methods for determining whether their wastes exhibited the characteristics, the identification system would prove unworkable (U.S. EPA 1990, pp. III-4, III-5; *see also*: discussion of carcinogenicity, mutagenicity, bioaccumulation potential and phytotoxicity, U.S. EPA 1990, p. III-5).

Ignitability. A solid waste that exhibits any of the following properties is considered a hazardous waste due to its ignitability:

- A liquid, except aqueous solutions containing less than 24% alcohol that has a flashpoint less than 60°C (140°F)
- A nonliquid capable, under normal conditions, of spontaneous and sustained conbustion
- An ignitable compressed gas per DOT regulations
- An oxidizer per DOT regulation

(40 CFR 261.21) (D001)

FIGURE 2.3 Ignitability characteristics.

The EPA has studied several other characteristics, including an "organic toxicity" characteristic, but the four described in 40 CFR 261 continue in use. The agency has assigned hazardous waste identification numbers prefixed by the letter D to the four. The four characteristics and their respective rationales are summarized as follows.

Ignitability. The EPA's reason for including ignitability as a characteristic (Figure 2.3) was to identify wastes that could cause fires during transport, storage, or disposal. Many used solvents are ignitable wastes.

Corrosivity. The EPA chose pH as an indicator of corrosivity (Figure 2.4) because wastes with high or low pH can react dangerously with other wastes or cause toxic contaminants to migrate from certain wastes. It chose steel corrosion because wastes capable of corroding steel can escape from their containers and liberate other wastes. Examples of corrosive wastes include acidic wastes and used pickle liquor (employed to clean steel during its manufacture) (U.S. EPA 1990, pp. III-5, III-6).

Reactivity. Reactivity was chosen as a characteristic (Figure 2.5) to identify unstable wastes that can pose a problem, e.g., an explosion, at any stage of the waste management cycle. Examples of reactive wastes include water from TNT manufacturing operations, contaminated industrial gases, and deteriorated explosives.

Corrosivity. A solid waste that exhibits any of the following properties is considered a hazardous waste due to its corrosivity:

- An aqueous material with pH less than or equal to 2, or greater than or equal to 12.5
- A liquid that corrodes steel at a rate greater than 0.25 inch per year at a temperature of 55°C (130°F)

(40 CFR 261.22) (D002)

FIGURE 2.4 Corrosivity characteristics.

Reactivity. A solid waste that exhibits any of the following properties is considered a hazardous waste due to its reactivity:

- Normally unstable and reacts violently without detonating
- Reacts violently with water
- Forms an explosive mixture with water
- Generates toxic gases, vapors, or fumes when mixed with water
- Contains cyanide or sulfide and generates toxic gases, vapors, or fumes at a pH of between 2 and 12.5
- Capable of detonation if heated under confinement or subjected to strong initiating source
- Capable of detonation at standard temperature and pressure
- Listed by DOT as Class A or B explosive

(40 CFR 261.23) (D003)

FIGURE 2.5 Reactivity characteristics.

Toxicity. The term toxicity refers to both a characteristic of a waste and a test. The Toxicity Characteristics Leaching Procedure (TCLP)[3] is designed to produce an extract simulating the leachate that may be produced in a land disposal situation. The extract is then analyzed to determine if it includes any of the toxic contaminants listed in Table 2.1. If the concentrations of any of the Table 2.1 constituents exceed the levels listed in the table, the waste is classified as hazardous. Toxicity of a waste may also be declared by the generator based upon knowledge of the waste and/or the generating process (EPA 1996).

Listed Hazardous Wastes

The inclusive listing adopted by EPA includes separate lists of nonspecific source wastes, specific source wastes, and commercial chemical products. These lists are described briefly, as follows:

- *Nonspecific source wastes,* also called "F" wastes because their EPA waste identification codes begin with the letter F, are generic wastes, commonly produced by manufacturing and industrial processes. Examples from this list include spent halogenated solvents used in degreasing and wastewater treatment sludge from electroplating processes as well as dioxin wastes, most of which are "acutely hazardous" wastes due to the danger they present to human health and the environment (40 CFR 261.31).
- *Specific source wastes* ("K" code) are from specially identified industries such as wood preserving, petroleum refining, and organic chemical manufacturing. These wastes typically include sludges, still bottoms, waste-

---

[3] The Toxicity Characteristic Leaching Procedure replaced the formerly specified "Extraction Procedure Toxicity" (EP Tox).

TABLE 2.1
Maximum Concentration of Contaminants for the Toxicity Characteristics

EPA HW Number	Contaminant	Regulatory Level (mg/L)
D004	Arsenic	5.0
D005	Barium	100.0
D018	Benzene	0.5
D006	Cadmium	1.0
D019	Carbon tetrachloride	0.5
D020	Chlordane	0.03
D021	Chlorobenzene	100.0
D022	Chloroform	6.0
D007	Chromium	5.0
D023	o-Cresol	200.0[a]
D024	m-Cresol	200.0[a]
D025	p-Cresol	200.0[a]
D026	Cresol	200.0[a]
D016	2,4-D	10.0
D027	1,4-Dichlorobenzene	7.5
D028	1,2-Dichloroethane	0.5
D029	1,1-Dichloroethylene	0.7
D030	2,4-Dinitrotoluene	0.13[b]
D012	Endrin	0.02
D031	Heptachlor (and its epoxide)	0.008
D032	Hexachlorobenzene	0.13[b]
D033	Hexachlorobutadiene	0.5
D034	Hexachloroethane	3.0
D008	Lead	5.0
D013	Lindane	0.4
D009	Mercury	0.2
D014	Methoxychlor	10.0
D035	Methyl ethyl ketone	200.0
D036	Nitrobenzene	2.0
D037	Pentachlorophenol	100.0
D038	Pyridine	5.0b
D010	Selenium	1.0
D011	Silver	5.0
D039	Tetrachloroethylene	0.7
D015	Toxaphene	0.5
D040	Trichloroethylene	0.5
D041	2,4,5-Trichlorophenol	400.0
D042	2,4,6-Trichlorophenol	2.0
D017	2,4,5-TP (Silvex)	1.0
D043	Vinyl chloride	0.2

[a] If o-, m-, and p-cresol concentrations cannot be differentiated, the total cresol (D026) concentration is used. The regulatory level of total cresol is 200 mg/L.

[b] Quantification limit is greater than the calculated regulatory level. The quantification level therefore becomes the regulatory level.

Source: 40 CFR 261.24.

waters, spent catalysts, and residues, e.g., wastewater treatment sludge from pigment production (40 CFR 261.32).
- *Commercial chemical products* ("P" and "U" codes) include specific commercial chemical products or manufacturing chemical intermediates. This list includes chemicals such as chloroform and creosote, acids such as sulfuric and hydrochloric, and pesticides such as DDT and Kepone (40 CFR 261.33).

The EPA makes an important additional distinction, among the listed wastes — one which may easily be overlooked by the newcomer to the hazardous waste management practice. Certain wastes have been identified by the EPA as being so dangerous that small amounts are regulated in a manner similar to larger amounts of other hazardous wastes and are designated as *acutely* hazardous. They are the F020-F023 and F026-F028 wastes listed in 40 CFR 261.31 and the "P" wastes listed in 40 CFR 261.33. The *acute* designation has major significance in the determination of the categories to which hazardous waste generators are assigned, the definition of "empty" containers, and limits placed upon accumulation and storage.

The EPA developed the lists by examining different types of wastes and chemical products to determine whether they met any of the following criteria:

- Exhibit one or more of the four characteristics of a hazardous waste
- Meet the statutory definition of hazardous waste
- Are acutely toxic or acutely hazardous
- Are otherwise toxic

The "Mixture" and "Derived-From" Rules

EPA has also ruled that most mixtures of solid wastes and listed hazardous wastes are considered hazardous wastes and must be managed accordingly. This applies regardless of what percentage of the waste mixture is composed of listed hazardous wastes. Without such a regulation, generators could evade RCRA requirements simply by mixing or diluting the listed wastes with nonhazardous solid waste. Wastes derived from hazardous wastes, such as residues from the treatment, storage, and disposal of a listed hazardous waste are considered a hazardous waste as well. *Caution: The "mixture" and "derived-from" rules contain a variety of conditions, exceptions, and exclusions. The student or reader should carefully examine the text of 40 CFR 261.3 before reaching conclusions regarding the applicability of these rules.*

Hazardous Waste Identification Rule Development

A series of related and somewhat parallel events and actions has caused the original listing/characteristics/mixture rule/derived-from rule approach to hazardous waste identification to be caught up in a prolonged state of uncertainty. The caution suggested in the paragraph above should extend also to the need for practitioners to stay informed on the progress of the proposed Hazardous Waste Identification Rule (HWIR). A summary of the situation follows, but may have changed by the time of publication of this edition.

A December 6, 1991 decision of the U.S. Court of Appeals for the District of Columbia vacated the "mixture" and "derived-from" rules due to procedural deficiencies in the 1980 promulgation of these rules (*Shell Oil Company v. EPA,* 950 F .2d 741 CA DC, 1991). EPA subsequently reinstated the rules on an interim basis and solicited comment thereon (57 FR 49278). A very large volume of comment and technical material was received by the agency, and the review and analysis of these materials caused the agency to exceed the "sunset" provisions applicable to the interim rulemaking.

Further litigation ensued, with the EPA again being challenged on procedural grounds (*Mobil Oil Corp. v. EPA,* CA DC, 1994). After Mobil's challenge was filed, Congress intervened with legislation stating that the interim "mixture" and "derived-from" rules were not to be terminated or withdrawn until revisions are promulgated and become effective. Congress imposed a deadline of October 4, 1994. That deadline was also missed and the EPA was again sued in separate actions by the Chemical Manufacturers Association and other industry groups and by the Environmental Technology Council. Both actions sought court-ordered immediate action by the EPA to reinstate the rules (57 FR 49278; *see also:* McCoy, May/June, 1992, pp. 2.1ff).

During the same time period, the EPA began an extensive review of the rules for identification of hazardous wastes. A Hazardous Waste Identification Rule (HWIR) was proposed on May 20, 1992 (57 FR 21450). The rule embodied two general concepts:

- A concentration-based exclusion criteria (CBEC) would exempt wastes from RCRA identification as a hazardous waste if concentrations were less than technology-based exemption levels. The criteria were to be based upon concentrations achievable by proven technologies.
- An expanded characteristics option (ECHO) would have provided "entry" to the regulatory system, as before, but would now provide "exit" from the system as well. The four existing characteristics would have remained in place, but the number of constituents listed in Table 2.1 would be greatly expanded or a similar table would be added.

Based upon extensive criticism of the rule, and upon the realization that a new rule must deal with the remanded "mixture" and "derived-from" rules, the EPA withdrew the proposed hazardous waste identification rule on October 30, 1992 (57 FR 49280) and began a series of outreach conferences and "round-table" meetings in an attempt to reach consensus on a workable approach. In April 1997, the HWIR development effort became the subject of a consent decree (*Environmental Technology Council v. Browner,* CA No. 94-2119, 94-2346), which required EPA to propose revisions to the mixture and derived-from rules by October 31, 1999. Accordingly, EPA formally proposed a new rule which actually embodies a new HWIR-Media in addition to the new HWIR-Waste. The thrust of the proposal is

- HWIR-Waste retains and amends the mixture and derived-from rules to ensure that hazardous wastes that are mixed with other wastes or remain following a treatment process do not escape regulation as long as they are reasonably likely to continue to pose threats to human health and the environment. The proposal discusses two regulatory options for concentration-based exemptions, a "generic" exemption and a "landfill-only" exemption.

• HWIR-Media proposes modified Land Disposal Restriction (LDR) treatment requirements and permitting procedures which would replace technology-based treatment standards with risk-based standards.[4]

The consent decree requires the EPA to promulgate a final rule by April 30, 2001 (U.S. EPA 1999).

The "Contained-In Policy"[5]

The "contained-in policy" was first announced in a November 1986 EPA memorandum, "RCRA Regulatory Status of Contaminated Groundwater," and has been updated many times in *Federal Register* preambles, EPA memos, and correspondence. The policy states that media containing a listed hazardous waste are also a hazardous waste once excavated or otherwise brought under management. The EPA generally considers contaminated environmental media to contain hazardous waste when they (1) exhibit a characteristic of a hazardous waste or (2) are contaminated with concentrations of hazardous constituents from listed hazardous wastes that are above health-based levels (U.S. EPA 1998). As the mixture and derived-from rules events unfolded, a new regulatory imperative required attention. The EPA had regulated contaminated media removed from remediation sites by imposing the contained-in policy. The agency had recognized that the policy brought significant quantities of slightly contaminated material under regulation. With increasing numbers of site remediation projects producing growing quantities of waste, the need to correct the problem took on new urgency. This brought about the coupling of the HWIR-Waste and HWIR-Media proposals.

TOPICS FOR REVIEW OR DISCUSSION

1. As noted in this chapter, some nations have rationalized identification of hazardous wastes by simply declaring any discarded chemical as "hazardous." Is this workable in the U.S.? If so, how? If not, why not?

2. In describing wastes, the scientific and technical communities assign a clear difference to the meanings and applications of the terms "toxic" and "hazardous." Provide a short definition of each, making this distinction clear.

3. What is the rationale for the distinction, made by RCRA, between "hazardous waste" and "acutely hazardous waste?"

---

[4] The term "media" identifies contaminated soil, groundwater, or sediment that contains hazardous waste. *See:* Glossary.

[5] It is important to note that the "Contained-In Policy" has not been codified as a regulation.

4. Why is the "mixture rule" of such great importance to practitioners and regulators?
5. Similarly, why is the "derived-from rule" important?
6. Why is a scheme such as the "characteristics" necessary? Why not rely entirely on lists?

REFERENCES

Assante-Duah, D. Kofi and Imre V. Nagy. 1998. *International Trade in Hazardous Waste,* E & FN Spon-Routledge, NY.

Enger, Eldon D., J. Richard Kormelink, Bradley F. Smith, and Rodney J. Smith. 1989. *Environmental Science: The Study of Interrelationships.* Wm. C. Brown, Dubuque, IA.

McCoy and Associates. 1992. *The Hazardous Waste Consultant.* May/June, Lakewood, CO.

Nebel, Bernard J. and Richard T. Wright. 1993. *Environmental Science, Fourth Edition,* Prentice-Hall, Englewood Cliffs, NJ.

UNEP. 1992. *UNEP Two Decades of Achievement and Challenge.* Information and Public Affairs Branch, United Nations Environment Programme, Nairobi, Kenya.

UNEP. 1994. *Environmentally Sound Management of Hazardous Wastes Including the Prevention of Illegal International Traffic in Hazardous Wastes.* United Nations Environment Programme, Nairobi, Kenya.

U.S. Environmental Protection Agency. 1986. *Test Methods for Evaluating Solid Waste: Physical/Chemical Methods, Third Edition.* Superintendent of Documents, Government Printing Office, Washington, D.C., EPA/SW-846.

U.S. Environmental Protection Agency. 1990. *RCRA Orientation Manual,* 1990 Edition. Office of Solid Waste, Washington, D.C., 20406.

U.S. Environmental Protection Agency. 1996. *Understanding the Hazardous Waste Rules. A Handbook for Small Businesses.* Office of Solid Waste and Emergency Response, Washington, D.C., EPA 530-K-95-001.

U.S. Environmental Protection Agency. 1998. *Management of Remediation Waste under RCRA.* Solid Waste and Emergency Response, Washington, D.C., EPA 530-F-98-026.

U.S. Environmental Protection Agency. 1999. *Hazardous Waste Identification Rule: Proposed Rule.* Solid Waste and Emergency Response, Washington, D.C., EPA 530-F-99-046.

World Health Organization. 1983. Management of Hazardous Waste. Copenhagen, Denmark.

3 Pathways, Fates, and Disposition of Hazardous Waste Releases

OBJECTIVES

At completion of this chapter, the student should:

- Understand basic theories of movement, mobility, dispersion, and natural breakdown mechanisms.
- Gain overview familiarity with the generally accepted and established pathways and measurements of releases to the environment.
- Be able to relate some important pathways and movement mechanisms to impacts on human health, the environment, land and marine life, and global changes.

INTRODUCTION

Although two decades of cradle-to-grave management of hazardous wastes are behind us, it remains difficult to quantify the relative contributions of particular source categories. The U.S. Environmental Protection Agency (EPA) Biennial Report[1] estimates of total quantities generated have continued in the 200 to 300 million tons per year range until 1997 (see Table 1.1). The 1997 data excluded hazardous wastes contained in wastewater discharges, causing an apparent large decrease in the total generated. Moreover the total generation was from generators only.[2] There are other significant possibilities for error. For example, wastes exhumed from a remediation site and transported to an approved treatment, storage, and disposal facility may or may not be counted as newly generated.

The Emergency Planning and Community Right-to-Know Act (EPCRA) of 1986 required *manufacturing* facilities to report the disposition of more than 300 "toxic"

---

[1] The National Biennial RCRA Hazardous Waste Report (Based on 1997 Data), EPA 530-S-99-036.
[2] Generators that generate more than 1000 kg of hazardous waste or more than 1 kg of acutely hazardous waste per month (*see:* Chapter 5 or Glossary).

chemicals, including the quantities released to the environment or sent off-site to waste treatment or disposal facilities. In 1989, the EPA began publishing summaries of data from these reports in the publication Toxic Release Inventory: A National Perspective (TRI) (U.S. EPA 1989).[3] By 1995, the TRI list of released toxic chemicals had more than doubled to 648 (U.S. EPA 1999b). This increase makes comparisons of 1988 to 1997 TRI data nearly impossible. The EPA has, however, compared a list of "core" chemicals, which infers that between 1988 and 1997, industrial on- and off-site releases have decreased by approximately 43%. The 1997 TRI data do not include waste chemical releases from manufacturing facilities having fewer than 10 full-time employees, companies that used less than 10,000 lb of chemicals during the year, nonmanufacturing facilities, nor the thousands of small businesses such as dry cleaners, paint shops, and service stations.

While much hazardous waste is generated by other than manufacturing facilities, the TRI provides some insight to the kinds and relative magnitude of sources of hazardous waste releases to the environment. Our reference throughout this chapter to the TRI is not in the context of overall quantification of hazardous waste generation, but is to provide insight regarding entry into the pathways of human exposure and environmental impact.

RELEASES OF CHEMICALS TO THE ENVIRONMENT

Releases to the Atmosphere

Atmospheric releases may be thought of as being either controlled or uncontrolled. Open burning of wastes is no longer condoned in most jurisdictions, but legal and illegal burning occurs. RCRA regulates both the generation, marketing, and burning of "hazardous waste fuel" and the destruction of hazardous wastes by permitted combustion facilities. The overall thrust and objective of these regulations is to ensure that combustion of hazardous waste is accomplished under conditions which ensure their destruction and that hazardous waste residues are captured and managed effectively. The practice of mixing hazardous waste with other fuels, followed by burning in boilers and industrial furnaces (BIFs), became a highly contentious issue during the early 1990s and continues to be so. The practice, technologies, regulations, issues, and policies pertaining to hazardous waste combustion are discussed in Chapters 7, 9, and 10.

Automobile wrecking yards routinely spill fluids from fuel tanks, transmissions, engine blocks, radiators, and brake systems. They similarly release chlorofluorocarbons (CFCs) from automotive refrigeration systems. Land "farming" operations, involving bulk disposal, make volatiles available for evaporation, and the heavier fractions are left to percolate into the soil. Hazardous waste impoundments, by design or default, release their volatiles to the atmosphere. As late as 1980, technical papers

---

[3] Prior to 1998, the TRI reported on *manufacturing* industries of the Standard Industrial Codes (SIC) 20-39 only. In 1998, the EPA added six industrial categories to the required reporting list: Metal Mining (10), Coal Mining (12), Electric Utilities (49), Chemical Wholesalers (5169), Petroleum Bulk Terminals (5171), RCRA/Solvent Recovery (4953/7389).

describing optimized evaporation facilities for pesticide wastes were being reprinted in EPA publications (Egg and Reddell 1980; Hall 1980).

Some remediation projects continue to air-strip contaminated groundwater without capture or destruction of the stripped volatiles. Land "treatment" facilities continue the practice of thinly spreading hazardous wastes containing volatiles on the land surface where the intent is to enhance vaporization. Sewage treatment plants regularly release or flare digester gases containing volatiles from sludge. Activated sludge and aeration basins, trickling filters, and holding ponds in publicly owned treatment works (POTWs) very effectively strip volatiles from the sewage and, unless captured, release them to the atmosphere. Hazardous waste tank storage facilities similarly release the lighter volatiles as vapor pressure in the tank varies. The 1990 Clean Air Act Amendments (CAAA) (PL 101-549, November 15, 1990) placed new emphasis on such releases, and as the new National Emission Standards for Hazardous Air Pollutants (NESHAPS) are promulgated, some will require new or more sophisticated controls on point and area sources such as those mentioned here.

The TRI reports the 1997 release by manufacturing facilities of SICs 20-39 of more than 1.3 billion pounds of toxic chemicals to the atmosphere. Atmospheric releases thus amount to approximately 50% of the total 2.6 billion pounds of TRI-reported chemical releases and transfers into the environment in 1997. The chemical products industry was the source of nearly 342 million pounds or nearly 26% of the total 1.3 billion pounds of TRI atmospheric releases. It was followed by the primary metals industry which released 132 million pounds (10% of the total). The chemical releases to the atmosphere reported in greatest quantities were methanol, ammonia, and toluene, totaling 194, 156, and 113 million pounds, respectively. These were followed by xylene, n-hexane, chlorine, and hydrochloric acid, each totaling 60 to 74 million pounds (U.S. EPA 1999b, Chapters 3 and 4).

The 1997 TRI data indicates that of the reported 1.3 billion pounds of toxic chemical emissions to the atmosphere, 1.0 billion pounds (77%) were point source emissions, while 317 million pounds (23%) were fugitive emissions (U.S. EPA 1999b, Chapter 4). The latter statistic is highly suspect. The reader should consider how many fugitive sources are actually measured or even reported. The point to be understood is that very large quantities of hazardous wastes continue to be emitted to the atmosphere via controlled and uncontrolled sources.

The TRI-reported atmospheric releases of "core chemicals" in 1997 were reduced by 1.2 billion pounds (55%) from the 1988 TRI Report totals (U.S. EPA 1999b, Chapter 3). This apparent improvement probably reflects some combination of improved control technology, regulatory activity, waste minimization efforts, and more complete reporting. Waste minimization/pollution prevention programs (covered in a later chapter) continue to be the best hope for reducing the transfer of hazardous waste constituents to the atmosphere.

As discussed in Chapter 1, until 1990 only seven NESHAPS had been finalized by the EPA, leaving hundred of hazardous chemical constituents uncontrolled by the CAA. State and local agencies had regulated some of those, but the limits of their jurisdictions left many hazardous emissions uncontrolled. Title III of the Clean Air Act Amendments of 1990 (CAAA) listed 189 additional hazardous air pollutants

(HAPs) and established a schedule according to which the EPA must promulgate emission standards for the listed HAPs.[4]

While the HAP control promulgations will eventually control some or most of the emissions of toxic organics from hazardous waste management facilities, the EPA has authority under RCRA Section 3004(n) to control organics emissions from specific sources.[5] The implementing regulations are found in Subparts AA, BB, and CC of 40 CFR 264 and 265. Section AA provides controls on process vents, and BB attempts to control leaks from valves, flanges, etc. Subpart CC applies to hazardous waste in tanks, surface impoundments, or containers. Applications of the controls imposed on the point sources and area emissions that are subject to the CAA and RCRA authorized regulations will be discussed in Chapters 7 through 14. There are, however, many exclusions which limit the effectiveness of both the HAP and specific source controls. These exclusions enable many hazardous waste management sources, operations, and practices to emit unknown but possibly significant quantities of toxic and hazardous pollutants to the atmosphere.

Releases to Surface Waters

The TRI-reporting facilities released 218 million pounds of toxic chemicals to surface water, such as rivers, lakes, ponds, and streams and transferred 266 million pounds of toxic chemicals to publicly owned (sewage) treatment works (POTWs). The two release categories were approximately 8.4 and 10% of the total releases and transfers of toxic chemicals to the environment in 1997. As with the TRI atmospheric releases, comparisons with 1988 through 1997 TRI data are difficult because of changes in reporting requirements. Releases of TRI "core chemicals" to surface waters infer a decrease of nearly 63%, while core chemical transfers to POTWs decreased by nearly 75%, from 1988 to 1997 (U.S. EPA 1999b, Chapter 3). By chemical category, the highest volume releases to surface waters in 1997 were of manufacturing wastes containing ammonia nitrate compounds, phosphoric acid, methanol, and ammonia. The greatest volumes of metals in manufacturing wastes transferred to POTWs in 1997 were of barium, zinc, copper, and manganese and their compounds.[6]

The greatest quantities of releases of chemicals to surface waters, by industrial category, and their percentages of the total releases to surface water in 1997 were from chemical manufacturing (49), primary metals (21), paper (9), and food processing (7) categories (U.S. EPA 1999b, Chapter 4). Again, the TRI statistics summarized in these paragraphs are from the *manufacturing* categories 20 through 39 of the SIC.

Pretreatment regulations, standards, and codes are intended to compel the removal or destruction of the hazardous constituents in industrial waste streams prior to discharge to publicly owned sewerage. Significant amounts of the liquid compo-

---

[4] Although the new NESHAP list was mandated in Title III of the 1990 Clean Air Act Amendments, these provisions amended Title I of the CAA. Therefore, the correct citation for the NESHAP program is Title I of the CAA.

[5] Generators and permitted treatment, storage, and disposal facilities (*see also*: EPA 530-R-97-064).

[6] The 1997 TRI report excluded discharges of chemicals to POTWs other than metals.

nent in sewage are discharged to surface waters. Chemicals in the sewage may pass through or interfere with treatment processes, thereby escaping removal in the sewage treatment plant. Chemicals that are removed by sewage treatment processes may be transferred to the environment in the form of air emissions or as sewage sludge. State and local regulatory agencies have come under increasing pressure by the EPA to improve compliance with pretreatment requirements and to strengthen pretreatment ordinances and regulations. Industrial waste discharges to POTWs must receive appropriate pretreatment if surface water and urban air quality objectives are to be achieved.

Ocean dumping has been curtailed by the Marine Protection Research and Sanctuaries Act (MPRSA), but significant quantities of hazardous wastes have continued to find their way into the marine environment. The implementation of the Act was made uncertain by a U.S. district court ruling growing from a City of New York lawsuit, alleging that the EPA had incorrectly implemented the law (Dawson and Mercer 1986, p. 417). Ocean disposal of sludge and permitted dumping of hazardous waste continued in the face of strong opposition by environmentalists and the public. The Ocean Dumping Ban Act of 1988 made ocean dumping of industrial waste and municipal sludge unlawful after December 31, 1991 (U.S. EPA 1991, p. 1, Chapters 1, 4, and 5). Both practices were halted somewhat earlier, but some permits for "emergency" dumping of industrial wastes have been issued since that date.

The toxic chemicals that pass through primary sewage treatment plants along coastlines are ultimately deposited in the oceans. Past abuses, such as the dumping of 38,000 drums of chlorinated hydrocarbon wastes in the North Sea during 1963 to 1969 (Piasecki and Davis 1987, p. 68), the uncontrolled dumping in U.S. coastal waters prior to MPRSA, continued dumping by other nations, together with ongoing discharges to coastal waters (foreign and domestic), require us to consider the oceans as waste sinks and exposure pathways.

Releases to Land

Earlier, the point was made that the evolution of environmental legislation in the U.S. has had the effect of driving hazardous waste treatment and disposal to the land. The continuing evolution, including RCRA, the Hazardous and Solid Waste Amendments (HSWA), and the site remediation rules and policies of the Comprehensive Environmental Response, Compensation, and Liability Act (CERCLA), has focused upon reducing the human health hazards and environmental impacts of the land treatment and disposal practices. Nevertheless, large quantities of hazardous waste continue to be treated and/or deposited upon and beneath the land surface. The EPA has recently given tentative blessing, in the form of a policy directive, to a remediation technique which allows contaminated soil or groundwater to remain in place while "natural attenuation" reduces mobility or toxicity of the contaminant(s).[7]

The EPA 1977 report, mentioned earlier, concluded that 48.3% of hazardous wastes went into unlined surface impoundments and 30.3% was deposited in non-secure landfills. A 1981 survey indicated radical changes in the predominant *methods*

---

[7] *See:* "Monitored Natural Attenuation" in Chapter 11 or Glossary.

of land disposal, but more than 90% went to land disposal (Dawson and Mercer 1986, pp. 125–126). By 1991, the EPA reported that 76% of the national total was managed in aqueous treatment units, and that land disposal accounted for only 9% (U.S. EPA 1994a, p. ES-4). The EPA reported that 1995 land disposal practices accounted for 12.3% of the hazardous waste management total (U.S. EPA 1997, p. ES-5). The 1997 RCRA Biennial report data are not helpful in this chronology because of the exclusion of wastewaters from the statistics (U.S. EPA 1999a). The 1997 TRI data infer that approximately 17% of SIC 20-39 manufacturing hazardous wastes were land disposed (U.S. EPA 1999b). Nevertheless it is apparent that a shift away from land disposal has occurred. That, of course, was the intent of the Congress in the HSWA land disposal restrictions. The 1993 suspension of permitting of combustion facilities, embodied in EPA's proposed "Combustion and Waste Minimization Strategy," had only minimal potential to affect the trend, since only 1.1% of the 1991 national total was being managed in thermal units at that time (U.S. EPA 1994a, p. ES-4).

Regardless of trends, significant quantities of hazardous wastes continue to be released to the air, the surface water (directly and indirectly), the land, and the subsurface. It is now important to gain an understanding of the fates of the waste constituents following these dispositions.

MOVEMENT, FATES, AND DISPOSITION

It is an oversimplification, but conceptually useful to say that waste constituents achieve their impacts according to the concentrations present at the point and time of exposure, the extent to which the concentrations increase or decrease during the exposure, and the time over which the exposure continues (more on this in Chapter 4). With this concept in mind, rates of dispersion, accumulation and decay rates, and residence times take on great importance. The concept can be extended to say that human health and environmental impacts of a hazardous waste release may be greatly dependent upon the medium to which it is released and to the forces that act upon the waste following release.

Behavior of Waste Constituents Released to the Atmosphere

The earth's atmosphere extends several hundred miles above its surface, but about 95% of the total air mass is concentrated in a layer some 12 mi deep. The lower part of that 12-mi layer containing most of the air mass is called the troposphere. The troposphere is about 5 mi thick at the poles and about 10 mi thick at the equator. Due to the force of gravity and the compressibility of gases, the troposphere contains about 75% of the total mass of the atmosphere. It is the behavior of the troposphere and the forces acting upon it that govern behavior of the pollutants released to the atmosphere. The sun's energy warms the tropospheric air by radiation, conduction, and convection. All play a role in the behavior of air pollutants, but convection heating and the rotation of the earth are basically responsible for the pollution dispersing winds that blow across continents (Hare 1989, Chapter 7; see also: Nebel and Wright 1993, p. 359ff).

Air is a mixture of gases of which nitrogen (78%) and oxygen (21%) are the major components. Air, in the natural state, also contains particulates and water vapor. These constituents, together with heat energy and photochemical activity, can and do bring about an endless variety of chemical reactions involving the hazardous wastes that are released to the atmosphere. Although the roles of sulfur releases in the formation of sulfur oxides and acid rain, the conversion of emitted nitric oxide (NO) to nitrogen dioxide (NO_2) and nitric acid (HNO_3) in the atmosphere, and the formation of carbon monoxide (CO) and carbon dioxide (CO_2) in combustion processes are widely publicized, other chemical reactions in the atmosphere are of concern to the hazardous waste manager. In later chapters, the releases of heavy metals and aromatic hydrocarbons will occupy our attention. These releases are subject to further physical and chemical change in the atmosphere.

Releases to the unconstrained atmosphere are more likely to be dispersed quickly than are releases to the land or water. Atmospheric releases usually result in concentrations at the point and time of exposure that are greatly less than at the point of release. Concentrations of atmospheric pollutants usually diminish with time following their release. These characteristics tend to impose chronic rather than acute effects. R. A. Horne states the problem clearly: "... in dispersing a pollutant, two things happen: (a) control is lost over the pollutant and the capability of surveillance over the pollutant diminishes, and (b) the pollutant is really 'thrown away,' and any possibility of future utilization of the pollutant is lost" (Horne 1978, p. 121). The latter might also be couched in terms of treatment or destruction, but the point is well-made.

Releases trapped under an inversion may accumulate to dangerous levels in relatively short times, may persist through the period of inversion, and may produce acute human health impacts. Indoor and workplace releases can produce both acute and chronic health impacts. Some of the important health effects of atmospheric releases will be reviewed in Chapter 4.

The nearly uninhibited movement, activity, and reactivity of hazardous chemicals in the atmosphere has been clearly shown, is well established, and movement from one media to another is evident. Polychlorinated biphenyls (PCBs) and the banned pesticide DDT have migrated from contaminated soils into the air and eventually have accumulated in the fish and wildlife of the Great Lakes (Poje et al. 1989, p. 5). Among the most-reported TRI chemicals released, acetone, toluene, and xylene react with other compounds to form ozone, a lower atmospheric pollutant. Methanol, which has a vapor density of 1.11, tends to remain at or near the surface as it either disperses or collects in surface depressions. In either case it remains a fire hazard until dispersed. Released 1,1,1-trichloroethane is highly persistent and can migrate to the upper atmosphere where it becomes a contributor to the depletion of the protective ozone layer in the stratosphere. Ammonia is an extremely dangerous irritant and asphyxiant when encountered in high concentrations, but disperses rapidly into the atmosphere.

Interstate, transborder, and even intercontinental movement of acid rain components, with fallout upon the land and water surfaces is well documented (Enger et al. 1989, pp. 405–406; Nebel and Wright 1993, pp. 360–368). Movement of ozone-depleting waste CFCs, from the earth's surface to the outer fringes of the atmosphere, is established (Poje et al. 1989, p. 5; Nebel and Wright 1993, pp. 377–381). Heavy metals including vanadium, manganese, and lead are transported from industrialized areas of Europe and Russia across and around the north pole. Radioactive debris from the 1986 Chernobyl nuclear power plant explosion was carried and scattered across Yugoslavia, France, Italy, Germany, and Scandanavia (Cunningham and Saigo 1997, pp. 394, 478).

Movement of Hazardous Waste Constituents in Surface Waters

Movement of released chemical constituents in surface streams is somewhat more constrained than in the atmosphere, due to the confining effect of the stream channels. Volatile organic compounds (VOCs) are released at the water-air boundary. Higher molecular weight organics and soluble inorganics are available for transfer to the groundwater from losing streams. Downstream diversions may transfer pollutants to the land surface (there, subject to further transfer to the atmosphere or the groundwater) or to domestic or industrial water supplies. Water treatment plants may precipitate inorganics and deposit them upon the land or return them to the source stream in concentrated form or may strip the organics in aeration processes and vent them to the atmosphere.

Two major classes of chemical waste constituents in surface waters are major environmental detriments. They are the heavy metals and their compounds and nonbiodegradable synthetic organics. Even in minute concentrations, these chemicals may be concentrated to pathological or lethal levels as they ascend the food chain. The Minimata Bay (methyl mercury) and Life Sciences Products/Allied Chemical Company (Kepone) cases are classic examples of these processes. Movement of hazardous waste constituents in surface streams is significant because of the ultimate flow to the fragile environments of the coastal waters (Figure 3.1).

Releases to impounded surface waters may have even greater concentrating effects and may similarly transfer hazardous constituents to the atmosphere or to groundwater. Surface impoundments which are artificially aerated also transfer VOCs to the atmosphere. A survey by the EPA in the late 1970s counted 132,709 sites having waste impoundments, of which 75% contained industrial wastes (U.S. EPA 1978, p. 32). This number is undoubtedly reduced greatly by now, but the newly found focus upon aqueous treatment may involve similar losses to the atmosphere. Many of these ponds were designed to dispose of wastewater by evaporation or seepage or both (U.S. EPA 1978, pp. 10–11). The EPA, elsewhere, using standard leakage coefficients, estimated that more than 100 billion gallons of industrial wastewaters had entered the groundwater system from these impoundments (U.S. EPA 1977, p. 108).

Pathways of Hazardous Waste Constituents Reaching Groundwater

Groundwater constitutes a very large percentage of the freshwater supply of the U.S. More than 50% of the populace is dependent upon groundwater for domestic pur-

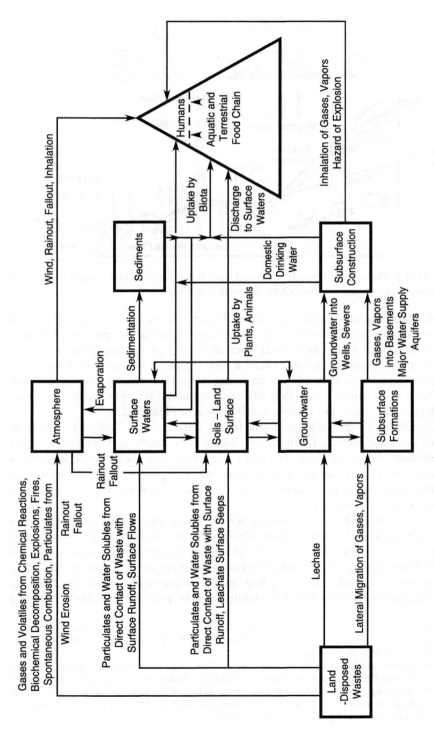

FIGURE 3.1 Flow of land-disposed waste contaminants through the environment. (From the U.S. Environmental Protection Agency.)

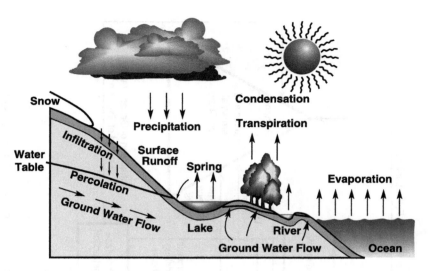

FIGURE 3.2 The hydrologic cycle. (Adapted from David Keith Todd, *Groundwater, Volume 1, Hydrology*. By permission from John Wiley & Sons, NY.)

poses. Water-bearing formations of the earth's crust act as conduits for transmission and storage of water. Water enters these formations from the ground surface or from bodies of surface water, after which it travels slowly for varying distances until it returns to the surface by action of natural flow, plants, or man or until it percolates to deeper formations. Groundwater emerging into surface stream channels as "base flow" aids in sustaining stream flow when surface run-off is low or nonexistent (Todd 1960, pp. 5–7). These phenomena are illustrated in Figure 3.2.

Practically all groundwater originates as surface water. Principal sources of natural recharge include precipitation, stream flow, lakes, and reservoirs. Other contributions, known as artificial recharge, occur from man's activities such as excess irrigation, but more to the point, from liquid waste disposal in pits, impoundments, landfills, and other land applications and from leaching of solid or semisolid hazardous wastes (Figure 3.3).

Constituents of hazardous wastes may be transported underground by one or more of several mechanisms. Vapors of volatile contaminants may disperse through voids in the soil above the water table where they may dissolve in water contained in or infiltrating through the voids. Chlorinated solvents and petroleum products that do not mix easily with water[8] may flow on the surface of the groundwater or sink, depending upon their densities.[9] Contaminants may sorb onto colloids or naturally

---

[8] Non-aqueous-phase liquids (NAPLs): if more dense than water, they are known as DNAPLs; if less dense than water, they are LNAPLs.

[9] A case in point, which is currently the subject of much investigative effort, is the behavior of methyl tertiary butyl ether (MTBE) a gasoline additive. The MTBE is miscible and easily mixes with groundwater while the remaining gasoline components float on the surface of the groundwater. This characteristic causes difficulty in monitoring, controlling and remediating leaked product from petroleum storage and dispensing facilities (Stocking et al., 1999; *see:* Chapter 14).

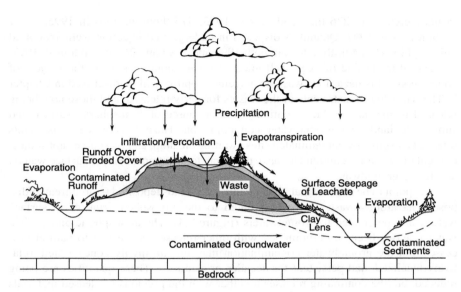

FIGURE 3.3 Hydrologic pathways for contamination by waste disposal sites. (From the U.S. Environmental Protection Agency.)

occurring organics and be transported to the extent that the hosts are mobile. Metals may precipitate and become similarly mobile. These mechanisms may change or vary as saturation of the soil particles and voids occurs or as pH of the subsurface environment changes (NAS 1997, pp. 24–27).

Implementation of HSWA land disposal restrictions (with help from other state and federal regulations) has brought about significant reductions in the quantities of wastewaters being treated, stored, or disposed of in unsealed surface impoundments. In 1989, the EPA estimated that approximately 31 million gallons (129,000 tons) of liquid hazardous wastes were being committed to surface impoundments. These impoundments were required to meet the no-migration standards of RCRA Section 3005(j) or be taken out of service (FR November 22, 1989, p. 48472). As implementation of the land disposal restrictions progressed, the EPA estimated that alternative capacities had to be found for:

- *Treatment* of 29 million gallons per year of liquid waste then being "treated" in waste piles
- *Storage* of 76 million gallons per year of liquid waste then stored in waste piles
- *Disposal* of 240 million gallons per year then disposed of in landfills, 6 million gallons per year then disposed of in land treatment units, and 5164 million gallons (21 million tons) per year then disposed of by underground injection (FR, November 22, 1989, p. 48472).

The RCRA biennial reports indicate that in the ensuing years, the quantities of hazardous wastes committed to permitted (presumably sealed) surface impoundments increased to 276 thousand tons in 1993, 575 thousand tons in 1995, and 1 million tons in 1997. Quantities disposed in underground injection wells increased only slightly, i.e, 24 million tons in both 1993 and 1995 and 26 million tons in 1997.

The ultimate fate of deep well-injected wastes continues to be the subject of controversy. This hazardous waste management option will be discussed in Chapter 7. The hazardous constituents of the wastes formerly released, and those now being released to the land surface, continue to be a source of atmospheric, surface and subsurface land, and groundwater contamination. Hazardous waste constituents released by spills, by contaminated sites awaiting cleanup, and by sources not subject to regulatory control continue to make their way to the atmosphere, to surface waters, and to the groundwater.

The point to be made is that great quantities of land-deposited (stored/treated/disposed) hazardous wastes have evaporated to the atmosphere, run-off to surface waters, and percolated to groundwaters (Figure 3.4). The atmospheric and surface water releases become commingled with other releases or are lost to natural processes, but the groundwater contamination may remain highly concentrated, relatively localized, and persistent for decades or centuries. These quantities are being reduced, but the continuing releases together with the previously released materials have contaminated and are contaminating aquifers in many areas, and many groundwater supplies have been degraded or ruined.

Chemical Transformations

Recent findings of chemical transformations of groundwater pollutants are disturbing if not alarming. Historically, two of the most ubiquitous hazardous waste releases have been of trichloroethylene (TCE) and tetrachloroethylene (PCE) to the land surface, to landfills, and to surface impoundments. Both have been widely used as cleaning solvents and degreasers and, until serious controls were applied, were disposed of by the most readily available method. Both are considered hazardous wastes (40 CFR 261.33); both are liver, kidney, reproductive system, and central nervous system hazards; they are ranked number 2 and 4, respectively, on the ATSDR[10] list of most frequently found Confirmed Exposure Pathways (CEPs)[10] at hazardous waste sites (U.S. Department of Health and Human Services 1999); PCE is a confirmed carcinogen; and TCE is a suspect carcinogen (Lewis 1993, pp. 998, 1264). In the early 1980s, EPA investigators noticed that where TCE, PCE, and other chlorinated compounds were detected in soils, in the presence of bacteria from sewage or septic tank leachate, vinyl chloride (VC) could usually be detected (Vincent 1984; Science Applications International Corporation 1985). It was soon established that in anaerobic conditions, dechlorination of TCE and PCE can be expected to progress to VC (Pavlostathis and Zhuang 1997). VC is an established, synthetic carcinogen (Manahan 1994, p. 667), having a drinking water risk level approximately two orders of magnitude greater than that of trichloroethylene (Science Applications International Corporation 1985, p. 1), and is ranked number 4 on the CERCLA List

[10] Agency for Toxic Substances and Disease Registry, Public Health Service, U.S. Department of Health and Human Services; 1999 CERCLA List of Priority Hazardous Substances and 1999 Substances Most Frequently Found in Completed Exposure Pathways (CEPs) at Hazardous Waste Sites.

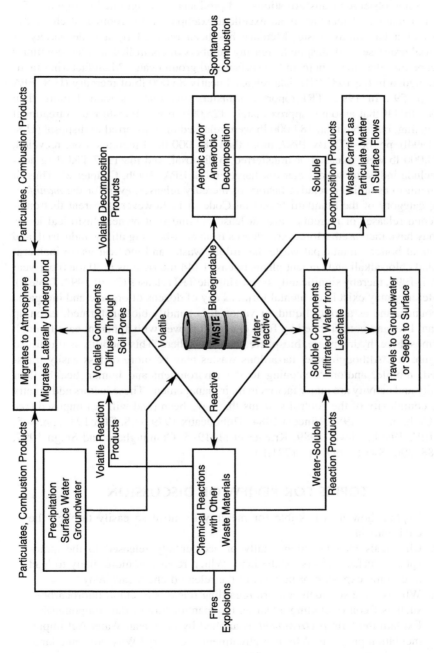

FIGURE 3.4 Initial transport processes at waste disposal sites. (From the U.S. Environmental Protection Agency.)

of Priority Hazardous Substances and number 21 on the most frequently found CEPs[10] at hazardous waste sites (U.S. Department of Health and Human Services 1999). Some observed transformations and products are diagramed in Figure 3.5.

The behavior of mercury is an excellent example of environmental chemodynamics of a hazardous waste. Mercury has been released by a wide variety of industrial processes and, despite increasing numbers of controlled sources, continues to escape into the atmosphere and to surface and groundwater. Manufacturing facilities included in the 1987 TRI data released nearly 86,000 lb of mercury (U.S. EPA 1989, p. 59). In 1992, TRI reported transfers and releases totaled more than 310,000 lb. Of that amount, approximately 125,000 lb was transferred to treatment or recycling, but more than 184,000 lb were released or transferred to disposal (U.S. EPA 1994b, pp. 66–67). By 1997, more than 485,000 lb of mercury were recycled, but 73,000 lb were released or transferred to disposal, and the 1997 TRI data does not include industrial wastewater discharges (U.S. EPA 1999b, Chapter 2). Thus, it is difficult to establish a relative history of mercury releases, even for the *manufacturing* category of the Standard Industrial Code. It is, however, apparent that environmental releases of mercury have, at least, continued at or near historical levels and may have increased. Moreover, releases from coal-burning utility, industrial, and residential boilers; municipal waste, hazardous waste, and medical waste incinerators; and chlor-alkali and cement plants released 144 metric tons (130 tons) per year in 1994–1995, thereby significantly surpassing the TRI releases (Swift 1999, Table 1).

Mercury may exist in elemental form, as any of dozens of organic and inorganic compounds, and as a solid, liquid, or vapor. It is handily biotransformed, taken up by plant life, and concentrated by food chains. It moves with apparent ease through the atmosphere, hydrosphere, biosphere, and lithosphere. This mobility is illustrated in Figure 3.6. Although other hazardous wastes may be more or less easily transformed, mobile, and/or threatening to the environment and human health, their mobility and activity are major factors in their management. These factors add greatly to the complexity of the control systems that have been (and will be) imposed (*see also*: Cothern et al. 1986; Vincent 1984; Thibodeaux 1979; U.S. EPA 1982, pp. 4–5; U.S. EPA 1999b, Meyer 1989; Krieger et al. 1995; Cunningham and Saigo 1997, pp. 388–390; Swift 1999, pp. 1721ff.)

TOPICS FOR REVIEW OR DISCUSSION

1. Explain how it is possible for mercury to move so easily through the environment.
2. Chemicals may be intentionally or accidentally released to the atmosphere, surface waters, or the land. Which release is more likely to lead to chronic exposure of humans to the released chemical? Why?
3. Why is there so much concern regarding releases of chlorofluorocarbons such as freon (including a ban on the manufacture of the compound)?
4. Explain the term *pretreatment* as applied by the Clean Water Act implementation program. Why is pretreatment necessary? Why are some large coastal cities major sources of toxic chemical discharges to the oceans?

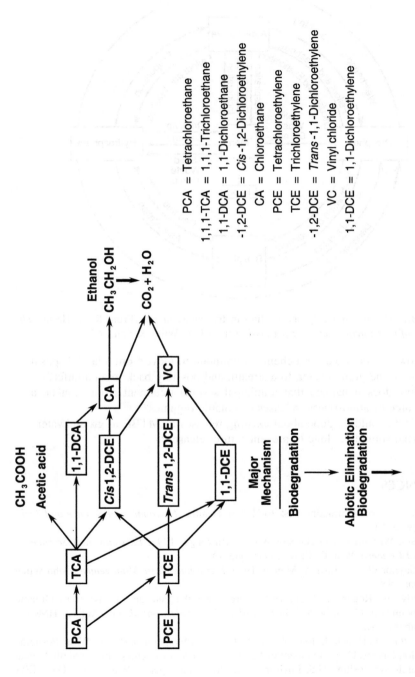

FIGURE 3.5 Transformation of aliphatic hydrocarbons. (Adapted from Andy Davis and Roger L. Olsen, Predicting the fate and transport of organic compounds in groundwater, *Hazardous Materials Control*, July/August 1990.)

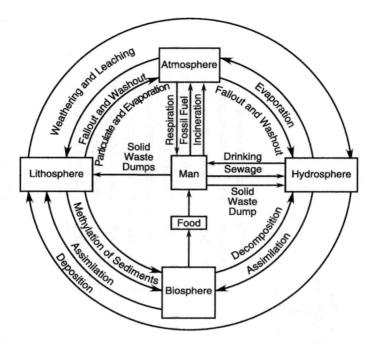

FIGURE 3.6 Mercury mobility and cycling in the environment. (From R. A. Horne, *The Chemistry of Our Environment.* By permission from John Wiley & Sons, NY.)

5. How is it possible for a chemical pollutant to move from a land disposal site to the groundwater, to a stream, and possibly back to an aquifer?
6. How does it happen that municipal sewage treatment plants can be a source of atmospheric releases of volatile organics?
7. What should be done about existing releases of MTBE to groundwater? What should be done to prevent future releases?

REFERENCES

Cothern et al. 1986. "Estimating Risk to Human Health," *Environmental Science and Technology,* 20(2).

Cunningham, William P. and Barbara Woodworth Saigo. 1997. *Environmental Science: A Global Concern.* Wm. C. Brown, Dubuque, IA.

Dawson, Gaynor W. and Basil W. Mercer. 1986. *Hazardous Waste Management.* John Wiley & Sons, NY.

Davis, Andy and Roger L. Olsen. 1990. "Predicting the Fate and Transport of Organic Compounds in Groundwater," in *Hazardous Materials Control,* July/August, HMCRI, Greenbelt, MD.

Egg, Richard P. and Donald L. Reddell. 1980. "Design of Evaporative Pits for Waste Pesticide Solution Disposal," in Treatment of Hazardous Waste, Proceedings of the Sixth Annual Research Symposium, U.S. Environmental Protection Agency, Washington, D.C., EPA 600-9-80-011.

Enger, Eldon D., J. Richard Kormelink, Bradley F. Smith, and Rodney J. Smith. 1989. *Environmental Science: The Study of Interrelationships.* Wm. C. Brown, Dubuque, IA.

Hall, Charles V. 1980. "Holding and Evaporation of Pesticide Wastes," in Treatment of Hazardous Waste, Proceedings of the Sixth Annual Research Symposium, U.S. Environmental Protection Agency, Washington, D.C., EPA 600-9-80-011.

Hare, F. Kenneth. 1989. Chapter 7, "Climatology and Meteorology," in *Environmental Science and Engineering*, J. Glynn Henry and Gary W. Heinke, Prentice-Hall, Englewood Cliffs, NJ.

Horne, R. A. 1978. *The Chemistry of Our Environment.* John Wiley & Sons, NY.

Institute of Chemical Waste Management. 1989. *Managing Hazardous Waste: Fulfilling the Public Trust.* National Solid Wastes Management Association, Washington, D.C.

Lewis, Richard J. 1993. *Hazardous Chemicals Desk Reference, Third Edition.* Van Nostrand Reinhold, NY.

Krieger, Gary R., Mark J. Logsdon, Christopher P. Weis, and Joanna Moreno. 1995. Chapter 5, "Basic Principles of Environmental Science," in *Accident Prevention Manual for Business & Industry — Environmental Management,* Gary R. Krieger, National Safety Council, Itasca, IL.

Manahan, Stanley E. 1994. *Environmental Chemistry, Sixth Edition.* CRC Press, Boca Raton, FL.

Meyer, Eugene. 1989. *Chemistry of Hazardous Materials.* Prentice-Hall, Englewood Cliffs, NJ.

National Academy of Sciences. 1997. *Innovations in Ground Water and Soil Cleanup.* National Academy Press, Washington, D.C.

Nebel, Bernard J. and Richard T. Wright. 1993. *Environmental Science,* Fourth Edition. Prentice-Hall, Englewood Cliffs, NJ.

Pavlostathis, Spyros G. and Ping Zhuang. 1997. "Effect of Sorption on the Microbial Reductive Dechlorination of Soil-Bound Chloroalkenes," in *Emerging Technologies in Hazardous Waste Management 7*, D. William Tedder and Frederick G. Pohland, Eds., Plenum Press, NY.

Piasecki, Bruce W., and Gary A. Davis. 1987. *America's Future in Toxic Waste Management — Lessons from Europe.* Quorum Books, New York.

Poje, Jerry, Norman L. Dean, and Randall J. Burke. 1989. *Danger Downwind.* National Wildlife Federation, Washington, D.C.

Stocking, Andrew, Stephen Koenigsberg, and Michael Kavanaugh. 1999. A Remediation and Treatment of MTBE, *Environmental Protection,* April.

Summary of Available Information Related to the Occurrence of Vinyl Chloride and Ground Water as a Transformation Product of Other Volatile Organic Chemicals. 1985. Science Applications International Corporation, McLean, VA, NTIS PB86-117868.

Anon., *Suspect Chemicals Sourcebook, Sixth Edition.* 1987. Roytech Publications, Bethesda, MD.

Swift, Byron. 1999. "A Better, Cheaper Way to Regulate Mercury," in *Environment Reporter,* 29(34).

Thibodeaux, Louis J. 1979. *Chemodynamics.* John Wiley & Sons, NY, pp. 1–5.

Todd, David K. 1960. *Ground Water Hydrology.* John Wiley & Sons, NY.

U.S. Department of Health and Human Services, Public Health Service, Centers for Disease Control and Prevention, National Institute for Occupational Safety and Health. 1999a. NIOSH Pocket Guide to Chemical Hazards, Washington, D.C.

U.S. Department of Health and Human Services, Public Health Service, Agency for Toxic Substances and Disease Registry. 1999b. "Substances Most Frequently Found in Completed Exposure Pathways (CEPs) at Hazardous Waste Sites" (1999 ATSDR CEP Site Count Report), Washington, D.C.

U.S. Department of Health and Human Services, Public Health Service, Agency for Toxic Substances and Disease Registry. 1999c. "1999 CERCLA List of Priority Hazardous Substances," Washington, D.C.

U.S. Environmental Protection Agency. 1977. The Report to Congress: Waste Disposal Practices and Their Effects on Groundwater, Office of Water Supply and Office of Solid Waste Management Programs, Washington, D.C.

U.S. Environmental Protection Agency. 1978. Surface Impoundments and Their Effects on Ground-Water Quality in the United States — A Preliminary Survey, Office of Drinking Water, Washington, D.C., EPA 570/9-78-004.

U.S. Environmental Protection Agency. 1982. Handbook for Remedial Action at Waste Disposal Sites, Washington, D.C., EPA 625/6-82-006.

U.S. Environmental Protection Agency. 1989. The Toxics-Release Inventory: A National Perspective, Office of Pesticides and Toxic Substances, Washington, D.C., EPA 560/4-89-005.

U.S. Environmental Protection Agency. 1991. Report to Congress on Ocean Dumping 1987-1990, Office of Water, Washington, D.C., EPA 503/9-91/009.

U.S. Environmental Protection Agency. 1991. Toxics in the Community National and Local Perspectives, Office of Pesticides and Toxic Substances, Washington, D.C., EPA 560/4-91-014.

U.S. Environmental Protection Agency. 1994a. The Biennial RCRA Hazardous Waste Report (Based on 1991 Data), Solid Waste and Emergency Response, Washington, D.C., EPA 530-S-94-039.

U.S. Environmental Protection Agency. 1994b. 1992 Toxics Release Inventory Public Data Release, Office of Pollution Prevention and Toxics, Washington, D.C., EPA 745-R-94-001.

U.S. Environmental Protection Agency. 1997. The National Biennial RCRA Hazardous Waste Report (Based on 1995 Data) Executive Summary, Washington, D.C., EPA 530-S-97-022.

U.S. Environmental Protection Agency. 1999a. The National Biennial RCRA Hazardous Waste Report (Based on 1997 Data) Executive Summary, Washington, D.C., EPA 530-S-99-036.

U.S. Environmental Protection Agency. 1999b. Toxics Release Inventory Reporting and the 1997 Public Data Release, Washington, D.C., EPA 749-C-99-003.

Vincent, J. R. 1984. South Florida Drinking Water Investigation, U.S. Environmental Protection Agency, National Enforcement Investigations Center, Denver, CO, EPA 330/1-84-001.

4 Toxicology and the Standard-Setting Processes

OBJECTIVES

At completion of this chapter, the student should:

- Understand the basic mechanisms of human exposure.
- Be able to relate the exposure mechanisms to the pathways overviewed in Chapter 3 and to the common release mechanisms.
- Be able to locate appropriate data on the toxicology of the chemical constituents of hazardous wastes.
- Know the components of the general risk assessment process and understand their relationship to each other.
- Understand how toxicological and human health considerations have been addressed in RCRA and how RCRA measures, regulates, and attempts to minimize toxic and health impacts of hazardous wastes.

INTRODUCTION

Living organisms are composed of cells, and all cells must accommodate and facilitate a variety of chemical reactions to maintain themselves and perform their functions. Introduction of a foreign chemical into a cell may interfere with one or more of these cellular reactions, leading to impaired cell function or viability. All chemicals are toxic, but the concentration, route of entry, and time of exposure are factors that determine the degree of toxic *effect*.

Toxicology is the study of how specific chemicals cause injury to living cells and whole organisms. Such studies are performed to determine how easily the chemical enters the organism, how it behaves inside the organism, how rapidly it is removed from the organism, what cells are affected by the chemical, and what cell functions are impaired. A risk assessment process is used to derive a reliable estimate of the amount of chemical exposure which is considered acceptable for humans or other organisms. Risk-based exposure limits are then rationalized in the form of risk-based standards. The alternative form of exposure limits is the technology-based standard, in which the goal is to minimize exposure by the imposition of control technologies.

In recent years, important advances have been achieved in toxicology and in the research methods that are employed by toxicologists. Nevertheless, for many chemicals, current toxicological knowledge is insufficient to provide the basis for quantitative toxicity assessments. Similarly, analytical techniques for risk assessment have been evolving toward attainment of greater sophistication and precision, but nonrepresentiveness, inconsistency, uncertainty, and/or absence of input data[1] continue to limit the utility of these techniques (*see:* Johnson and DeRosa, 1997, Tables 3 and 4 and discussion). It is these very limitations that cause the standards-setting process to be exceedingly lengthy and/or seemingly endless.

PUBLIC HEALTH IMPACTS

Toxicity Hazard

In the hazardous waste context, toxicity is the ability of a chemical constituent or combination of constituents in a waste to produce injury upon contact with a susceptible site in or on the body of a living organism. *Toxicity hazard* is the risk that injury will be caused by the manner in which a waste is handled.

Acute Toxicity: Adverse effects on, or mortality of, organisms following within hours, days, or no more than 2 weeks after a single exposure or multiple brief acute exposures, within a short time, to a chemical agent.

Chronic Toxicity: Adverse effects manifested after a lengthy period of uptake of small quantities of the toxicant. The dose is so small that no acute effects are manifested and the time period is frequently a significant part of the normal lifetime of the organism.

(Adapted from Hodgson and Levi 1987, pp. 357, 360.)

Chemical constituents of wastes may be acutely or chronically hazardous to plants or animals via a number of routes of administration. Phytotoxic wastes can damage plants when present in the soil, atmosphere, or irrigation water. Phytotoxicity is the result of a reduction of chlorophyll production capability, overall growth retardation, or some specific chemical interference mechanism.

Chemical components that are acutely toxic to mammals may be injurious when inhaled, ingested, and/or contacted with the skin. Symptoms resulting from acute exposures usually occur during or shortly after exposure to a sufficiently high concentration of a contaminant. The concentration required to produce such effects varies widely from chemical to chemical. Data pertinent to a single route of administration may not be applicable to alternative routes. For example, asbestos dust is

---

[1] "Data Gaps" is a problem addressed by the Comprehensive Environmental Response, Compensation, and Liability Act (CERCLA), and a focus of the Agency for Toxic Substances and Disease Registry (ATSDR) and the EPA, which are jointly tasked by CERCLA with elimination of the data gaps. The topic is discussed later in this chapter.

toxic at very low levels when present in air, but asbestos particles in water are believed to pose no ingestive threat at low concentrations.

"Acute exposure" traditionally refers to exposure to "high" concentrations of a contaminant and/or short periods of time. "Chronic exposure" generally refers to exposure to "low" concentrations of a contaminant over a longer period. Chemical contaminants may be chronically toxic to mammals if they contain materials that (1) are bioaccumulated or concentrated in the food chain or (2) cause irreversible damage that builds gradually to a final, unacceptable level. Heavy metals and halogenated aromatic compounds are classic examples of chronic toxicants (HHS 1985, p. 2-1; Dawson and Mercer 1986, p. 62; Kamrin 1989, p. 134; Manahan 1994, Chapters 22 and 23).

The U.S. Environmental Protection Agency (EPA) has classified some 35,000 chemicals as either definitely or potentially harmful to human health. A number of them, including some heavy metals (cadmium, arsenic) and certain organic compounds (carbon tetrachloride, toluene), are carcinogenic. Others, like mercury, are mutagenic and may tend to induce brain and bone damage (mercury, copper, lead), kidney disease (cadmium), neurological damage, and many other problems. Multiple exposures can be additive or synergistic, but in many cases, the risk resulting from simultaneous exposure to more than one of these substances is not known.

A wide variety of reference materials are available which provide basic toxicity data on specific chemicals. The Registry of Toxic Effects of Chemical Substances (RTECS) has been widely used and quoted (HHS 1975). In recent years, the "Health Assessment Guidance Manual," published by the Agency for Toxic Substances and Disease Registry (ATSDR) has become widely accepted among toxicologists and related practitioners (HHS 1990). Moreover, ATSDR is preparing individual toxicological profiles for 275 hazardous substances found at Superfund sites. A 1997 publication states that the agency is concentrating on filling 194 data gaps for 50 top-ranked CERCLA hazardous substances (Johnson and DeRosa 1997). These profiles may be obtained from NTIS[2] as they become available. *A Textbook of Modern Toxicology,* by Hodgson and Levi (1987), is an excellent introductory text and provides a wealth of references on individual topics. The *Handbook of Toxic and Hazardous Chemicals and Carcinogens (Third Edition)* by Marshall Sittig is an authoritative source. The NIOSH "Pocket Guide to Chemical Hazards" is a handy, quick-reference guide to chemical hazards (HHS 1997). The American Conference of Governmental Industrial Hygienists (ACGIH 1997) publishes a handbook of Threshold Limit Values (TLVs) and Biological Exposure Indices (BEIs) for a variety of chemical substances and physical agents. The EPA operates a database — "Integrated Risk Information System" (IRIS)[3] — containing up-to-date health risk and EPA regulatory information pertaining to numerous chemicals. Other new databases, with current toxicology data and search capabilities, are becoming available.

For a chemical to exert a toxic effect on an organism, it must first gain access to the cells and tissues of that organism. In humans, the major routes by which toxic chemicals enter the body are through ingestion, inhalation, and dermal absorption.

---

[2] The National Technical Information Service, 5285 Port Royal Road, Springfield, VA.
[3] *See:* IRIS entry in the Glossary.

The absorptive surfaces of the tissues involved in these three routes of exposure (gastrointestinal tract, lungs, and skin) differ from each other with respect to rates at which chemicals move across them.

Ingestion. Ingestion brings chemicals into contact with the tissues of the gastrointestinal (GI) tract. The normal function of the tract is the absorption of foods and fluids that are ingested, but the GI tract is also effective in absorbing toxic chemicals that are contained in the food or water. The degree of absorption generally depends upon the hydrophilic (easily soluble in water) or lipophilic (easily soluble in organic solvents or fats) nature of the ingested chemical. Lipophilic compounds (e.g., organic solvents) are usually well absorbed, since the chemical can easily diffuse across the membranes of the cells lining the GI tract. Hydrophilic compounds (e.g., metal ions) cannot cross the cell lining in this way and must be "carried" across by a transport system(s) in the cells. The extent to which the transport occurs depends upon the efficiency of the transport system and upon the resemblance of the chemical to normally transported compounds.

If the ingested chemical is a weak organic acid or base, it will tend to be absorbed by diffusion in the part of the GI tract in which it exists in its most lipid-soluble (least ionized or polar) form. Since gastric juice in the stomach is acidic and the intestinal contents are nearly neutral, the polarity of a chemical can differ markedly in these two areas of the GI tract. A weak organic acid is in its least polar form while in the stomach and therefore tends to be absorbed through the stomach. A weak organic base is in its least polar form while in the intestine and therefore tends to be absorbed through the intestine. Some caustics can cause acute reactions within the GI tract.

Another important determinant of absorption from the GI tract is the interaction of the chemical with gastric or intestinal contents. Many chemicals tend to bind to food, and so a chemical ingested in food is often not absorbed as efficiently as when it is ingested in water. Additionally, some chemicals may not be stable in the strongly acidic environment of the stomach and others may be altered by digestive enzymes or intestinal bacteria to yield different chemicals with altered toxicological properties. For example, intestinal bacteria can reduce aromatic nitro groups to aromatic amines, which may be carcinogenic (ICAIR 1985, pp. 4-1, 4-3). Irrespective of the route of absorption, once the chemical enters the bloodstream, it is then delivered to the target organ.

The ingestion route of exposure is seldom a factor in industrial situations, with the exception of the inadvertent incident. For example, workers eating lunch in a battery factory might ingest lead with their sandwiches (Beaulieu and Beaulieu 1985, p. 12). Ingestion gains importance with long-term intake of contaminants in water supplies.

Inhalation. Inhalation brings chemicals into contact with the lungs. Most inhaled chemicals are gases (e.g., carbon monoxide) or vapors of volatile liquids (e.g., trichloroethylene). Absorption in the lung is usually great because the surface area is large and blood vessels are in close proximity to the exposed surface area. Gases cross the cell membranes of the lung via simple diffusion, with the rate of absorption dependent upon the solubility of the toxic agent in blood. If the gas has a low solubility (e.g., ethylene), the rate of absorption is limited by the rate of blood flow through the lung, whereas the absorption of readily soluble gases (e.g., chloroform) is limited only by the rate and depth of respiration.

Chemicals may also be inhaled in solid or liquid form as dusts or aerosols. Liquid aerosols, if lipid-soluble, will readily cross the cell membranes by passive diffusion. The absorption of solid particulate matter is highly dependent upon the size and chemical nature of the particles. The rate of absorption of particulates from the alveoli[4] is determined by the compound's solubility in lung fluids, with poorly soluble compounds being absorbed at a slower rate than readily soluble compounds. Small insoluble particles may remain in the alveoli indefinitely. Larger particles (2 to 5 µm) are deposited in the trachea or bronchial (upper) regions of the lungs where they may be cleared by coughing or sneezing or they may be swallowed and deposited in the GI tract. Particles of 5 µm or larger are usually deposited in the nasal passages or the pharynx where they are subsequently expelled or swallowed (ICAIR 1985, p. 4-3). A chronic effect on the lung can be caused if the defense mechanisms are overwhelmed with particles from smoke, coal dust, etc.

Inhalation of air contaminants is probably the most important route of entry of chemicals to the body in industrial situations. A worker exposed to 1000 parts per million (ppm) of toluene vapor, over an 8-hr work shift, could be expected to show dramatic symptoms of eye and respiratory irritation and depression of the central nervous system (CNS). This response to toluene demonstrates local effects (at the point of entry — eye, lung) and systemic effects where the chemical was absorbed into the bloodstream and affected the CNS.

Some chemicals do not provide "warning properties" in the gaseous or vapor state. For example, carbon monoxide (CO) is odorless and colorless and can inflict serious toxic effects to the unsuspecting victim. Other chemicals may have the property of desensitizing the receptor. For example, hydrogen sulfide (H_2S) has the prominent "rotten egg" odor at low concentrations. However, at high concentrations the olfactory senses become paralyzed and the exposed individual can be quickly overcome with the toxic effect (Beaulieu and Beaulieu 1985, p. 14). High concentrations of H_2S can also cause respiratory arrest.

Long-term chronic health effects may be experienced by humans in various situations. For example, chronic bronchitis has been convincingly linked to long-term inhalation of sulfur dioxide, one of the more prominent urban air pollutants (Hodgson and Levi 1987, pp. 189–190). Emphysema, asbestosis, silicosis, and berylliosis have all been associated with exposure to dusts and/or fumes.

Dermal Absorption. Absorption of toxicants through the epidermal layer of the skin, and into the bloodstream, is hindered by the densely packed layer of rough, keratinized[5] epidermal cells. Absorption of chemicals occurs much more readily through scratched or broken skin. There are significant differences in skin structure from one region of the body to another (palms of hands vs. facial skin), and these differences further influence dermal absorption.

---

[4] Tiny cavities at the terminal end of the bronchiole, in the lungs, where the exchange of oxygen and carbon dioxide occurs.
[5] The layer of keratin, a tough fibrous protein containing sulfur and forming the outer layer of epidermal structures, such as hair, nails, horns, and hoofs.

Absorption of chemicals by the skin is roughly proportional to their lipid solubility and can be enhanced by application of the chemical in an oily vehicle and rubbing the resulting preparation into the skin. Some lipid-soluble compounds can be absorbed by the skin in quantities sufficient to produce systemic effects. For example, carbon tetrachloride can be absorbed by the skin in amounts large enough to produce liver injury (ICAIR 1985, p. 4-3). The NIOSH "Pocket Guide to Chemical Hazards" and the ACGIH handbook of TLVs and BEIs provide guidance regarding dermal exposure to hazardous materials (*see also*: HHS 1985, p. 2-2).

Toxic Actions

Toxic chemicals can be categorized according to their physiological effect upon the exposed species. The categories often overlap, but can be (somewhat simplistically) separated into groups of irritants, asphyxiants, CNS depressants, and systemic toxicants.

Irritants. Chemicals that cause effects such as pain, erythema, and swelling of the skin, eyes, respiratory tract, or GI tract are considered irritants, a local effect at the point of entry to the body. An example is sodium hydroxide (caustic) dust on perspiration-moist skin. The pH of the fluid is quickly increased above normal resulting in irritation. Mechanical friction, such as the rubbing of shirt cuffs or collar, compounds the irritating effect. The effect may be as simple as a mild stinging sensation to the more serious blistering of the skin. Ammonia vapors or spray can irritate the mucous membranes of the respiratory tract, causing tearing and stinging in the nasal passages and throat.

Asphyxiants. Chemical asphyxiants are those that deny oxygen to cells of the host organism, thereby slowing or halting metabolism. Simple, or mechanical, asphyxiants displace the available oxygen in an air space to the point of producing an atmosphere unable to support life (less than 16% oxygen). Oxygen starvation may occur in a confined space where methane gas (CH_4) displaces oxygen to the extent that the oxygen content of the atmosphere falls to less than 16%. Conversely, carbon monoxide (CO) is a gas that chemically ties up the hemoglobin in blood after inhalation. With hemoglobin unable to transport oxygen to cells and carbon dioxide from the cells, the tissues cannot maintain natural metabolic functions, and death occurs.

Central Nervous System (CNS) Depressants. Inhalation of most organic solvent vapors and anesthetic gases, or the introduction of narcotics to the body in the form of alcohol or depressant drugs, causes a deadening of the nervous system. A worker who inhales trichloroethylene vapor during a workshift might not have the neuromuscular coordination to safely drive an automobile. The appearance of inebriation can be mistaken for the effects of elevated blood alcohol concentration.

Systemic Toxicants. Systemic toxicants are chemical compounds that exhibit their effect dramatically upon a specific organ system and possibly far from the site of entry. There is considerable overlap between the systemic toxicants and the other categories. For example, the organic solvent carbon tetrachloride (CCl_4) is definitely a CNS depressant as well as an irritant and can cause irreversible liver or kidney damage.

Mercury vapor does not seem to produce irritation upon inhalation, but causes serious impairment to nerve endings. Chronic inhalation of mercury vapor can result in serious disease of the nervous system, including insanity (*see also:* Manahan 1994, p. 677)

An agent that has the potential to induce the abnormal, excessive, and uncoordinated proliferation of certain cell types, or the abnormal division of cells, is termed a carcinogen or potential carcinogen. Inhalation of asbestos fibers has been firmly linked to the production of lung cancer and mesothelioma (cancer of the linings of lung tissues).

A chemical that causes mutations or changes in the genetic codes of the DNA in chromosomes is called a mutagen. Formaldehyde vapor causes these changes in the bacterial organisms *Salmonella* sp. This characteristic is the basis for the "Ames test," a bacterial procedure used for indication of mutagenicity of suspect substances. Mutagenic toxins may affect future generations.

A teratogen is a toxicant that produces physical defects in unborn offspring. A suspect substance may be administered to a test animal to determine if it will cause congenital abnormalities in a fetus produced by the test animal (Beaulieu and Beaulieu 1985, pp. 15–17). The birth defects of a teratogen are not passed to future generations (*see also:* Manahan 1994, p. 662).

Risk Assessment and Standards

The EPA and other regulatory agencies have, over the years, frequently opted for risk-based standards because of the court-imposed need to "show harm" when a particular standard is challenged. As noted above, CERCLA, in 1980, created ATSDR and tasked the EPA and the new agency with filling data gaps for 275 priority hazardous substances.[6] As noted in the "Introduction" to this chapter, this insistence upon a rational basis (i.e., a showing of harm) for environmental or exposure standards has caused the standards-setting process to be time consuming, laborious, and frustrating. In 1990 it became apparent that Congress was then steering the EPA back toward more reliance upon technology-based standards (*Environment Reporter,* 9 March, 1990, pp. 1840–1841). The 1990 Clean Air Act Amendments (CAAA) require that the EPA assign maximum achievable control technology (MACT) standards to the newly listed hazardous air pollutants. Yet Section 303 of the CAAA also establishes a Risk Assessment and Management Commission, which is to "… make a full investigation of the policy implications and appropriate uses of risk assessment and risk management in regulatory programs under various Federal laws to prevent cancer and other chronic human health effects which may result from exposure to hazardous substances" (42 USC 7412). Thus the search continues, on the part of Congress, for approaches to rationalize standards to protect human health, while continuing reliance upon control technologies.

The congressional focus upon technology-based standards is an expression of the frustration of that body with the slow pace of the standards-setting process, the endless arguments growing from the "how-clean-is-clean" issues, and the inherent flaws in biological research (conversion of test animal data to human exposure

---

[6] For an in-depth discussion of this effort, *see:* Johnson and DeRosa 1997.

application — more on this later in this chapter). Nevertheless, the courts can be expected to lend a sympathetic ear to pleas for rationality in standards. As data gaps are filled and as research and analytical techniques advance, risk-based standards will increasingly dominate the regulatory schemes.

Risk Assessment. The risk assessment process for evaluation of chemical hazards varies, in detail, according to the proclivities, experiences, focus, and/or mandates of the individual risk assessor, researcher, or regulatory agency. However, the general paradigm for risk assessment flows from the 1983 National Research Council (NRC) publication *Risk Assessment in the Federal Government.* The process usually consists of the following four steps:

- Hazard identification
- Dose-response evaluation
- Exposure assessment
- Risk characterization

Variations on the process for use in Superfund or RCRA site evaluations will be summarized in Chapter 10. EPA methodology for risk assessment is introduced in an unnumbered Technical Information Package titled "Risk Assessment," which can be accessed at <http://www.epa.gov/oiamount/tips/risktip.htm>. Much more detail, although in the Superfund site remediation context, is available in the referenced documents (EPA 1990; EPA 1992; *see also:* LaGoy 1999).

Hazard Identification. The first step in the process, if for establishing an exposure standard for an individual chemical, may take the form of a toxicological evaluation, wherein the answer is sought to the the question: "Does the chemical have an adverse effect?" This evaluation may be a "weight-of-evidence" process in which the available scientific data are examined to determine the nature and severity of actual or potential health hazards associated with exposure to the chemical. This step involves a critical evaluation and interpretation of toxicity data from epidemiological, clinical, animal, and *in vitro*[7] studies. Factors that should be considered during the toxicological evaluation include routes of exposure, types of effects, reliability of data, dose, mixture effects, and evidence of health end-points including developmental toxicity, mutagenicity, neurotoxicity, or reproductive effects. The toxicological evaluation should also identify any known quantitative indices of toxicity such as the *threshold level* or No Observable Adverse Effect Level (NOAEL), Lowest Observable Adverse Effect Level (LOAEL), carcinogenic risk factors, etc. (ICAIR 1985, p. 8-2; EPA Science Policy Council 1995; Schoeny et al. 1998, Chapter 9).

The dose-response relationship is the most fundamental concept in toxicology. The product of the dose-response evaluation is an estimate of the relationship between the dose of a chemical and the incidence of the adverse effect in the human population actually exposed, or in test organisms in the laboratory.

---

[7] Studies conducted in cells, tissues, or extracts from an organism, i.e., not in the living organism.

Dose-Response Evaluation. Once the toxicological evaluation indicates that a chemical is likely to cause a particular adverse effect, the next step is to determine the potency of the chemical. The dose-response curve describes the relationship that exists between degree of exposure to a chemical (dose) and the magnitude of the effect (response) in the exposed organism(s), usually laboratory animals. By definition, no response is seen in the absence of the chemical being evaluated. At low dose levels, response may not be evident, but as the amount of chemical exposure increases, the response becomes apparent and increases. Thus, a steep curve indicates a highly toxic chemical; a shallow curve indicates a less toxic substance. The toxicity values derived from this quantitative dose-response relationship, usually at very high exposure levels, are then extrapolated to estimate the incidence of adverse effects occurring in humans at much lower exposure levels. The EPA Integrated Risk Information System (IRIS) is a repository for data needed by the risk assessor in developing the dose-response relationship. EPA program offices also maintain program-specific databases, such as the Office of Solid Waste and Emergency Response (OSWER) Health Effects Assessment Summary Tables (HEAST). The EPA guidance provides a detailed discussion of the data requirements for the dose-response development (U.S. EPA Science Policy Council 1995).

Depending upon the mechanism by which the subject chemical acts, the dose-response curve may rise with or without a threshold. Figure 4.1 illustrates the NOAEL and LOAEL described earlier. The TD_{50} and TD_{100} points indicate the doses associated with 50 and 100% occurrence of the measured toxic effect (*see also: Chastain 1998*).

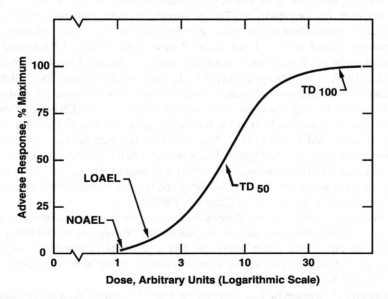

FIGURE 4.1 Hypothetical dose response curves. (Adapted from ICAIR Life Systems, Inc., *Toxicology Handbook,* prepared for the U.S. Environmental Protection Agency Office of Waste Programs Enforcement, Washington, D.C.)

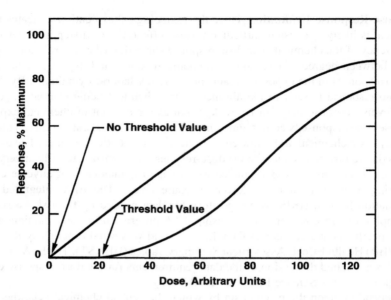

FIGURE 4.2 Hypothetical dose-response curves. (Adapted from ICAIR Life Systems, Inc., *Toxicology Handbook,* prepared for the U.S. Environmental Protection Agency Office of Waste Programs Enforcement, Washington, D.C.)

Figure 4.2 illustrates threshold and no threshold dose-response curves. In both cases, the response normally reaches a maximum after which the dose-response curve becomes flat or nearly so. The no-threshold curve coincides with a long-standing EPA assumption that damage to a single cell could trigger a chain reaction of mutations; therefore there is no "safe" dose of a carcinogen. EPA is said to be moderating this position because scientific studies indicate that exposure-caused damage to DNA is not always irreversible. In fact, some evidence shows that very high doses (i.e., maximum tolerated dose) and the methods used in dosing test animals may be biasing the results of cancer risk assessments (Chastain, 1998).

The dose-response evaluation for noncarcinogenic chemicals provides an estimation of the NOAEL or LOAEL. The NOAEL may then be assumed to be the basis for establishing an "Acceptable Daily Intake" (ADI) or Reference Dose (RfD). In practice, the NOAEL is adjusted by safety and uncertainty factors, which are an attempt to account for the "unknowns" involved (Assante-Duah 1993, pp. 94–102; Krieger et al. 1995, pp. 126–128; Chastain, 1998).

Mathematical models of the dose-response relationship for carcinogenic chemicals are used to derive estimates of the probability or range of probabilities that a carcinogenic effect will occur under the test conditions of exposure. Suggested readings providing examples of these models can be found in Assante-Duah 1993, pp. 91–92 and Krieger et al. 1995, Chapter 5.

Exposure Assessment. The assessor researches existing data and/or acquires specific exposure data to enable estimates of the magnitude of actual and/or potential human exposures, the frequency and duration of these exposures, the pathways by which humans are potentially exposed, and the numbers of humans who may be exposed. The data may include monitoring studies of chemical concentrations in exposure vehicles (ambient air, water supply, workplace environment, etc.). Modeling of the environmental fate and transport of contaminants may identify exposure links. Geographical locations and lifestyles of appropriate population subgroups must be considered. Intake[8] according to the routes of exposure (oral, inhalation, dermal) is determined or estimated. Uptake[8] across body barriers and other pharmacokinetics-related factors may be important (Schoeny et al. 1998, p. 206).

ATSDR cautions that "... at present, no single generally applicable procedure for exposure assessment exists, and, therefore, exposures to carcinogens must be assessed on a case-by-case or context-specific basis. While the need for, and reliance on, models and default assumptions is acknowledged, ATSDR strongly encourages the use of applicable empirical data (including ranges) in exposure assessments" (ATSDR 1993).

Risk Characterization. The final step in risk assessment, risk characterization, is the process of estimating the incidence of an adverse health effect under the conditions of exposure described in the exposure assessment. It is performed by integrating the information developed during the toxicity assessment (toxicological evaluation and dose-response evaluation) and the exposure assessment and may be a quantitative or qualitative (or both) evaluation. The degree of uncertainty and variability in all of the components of the assessments are evaluated and described. A variety of procedures have been developed and used for this final step in the risk assessment process (Assante-Duah 1993; EPA 1995; Schoeny 1998, pp. 205–211; Chastain 1998). The EPA is attempting to respond to pressures from Congress and the scientific community for new approaches to risk characterization, particularly with respect to "cumulative risk," i.e., exposure to multiple chemicals, mixtures or blends of chemicals, behavior of the mixtures under differing conditions, etc. The risk assessor must become familiar not only with the need to integrate quantitative and qualitative data, but with analytical techniques such as probabalistic risk assessment, the Monte Carlo[9] simulation technique, and others (Figure 4.3).

The EPA policy for conduct of risk assessment is set forth in a 1995 memorandum issued by Carol Browner, the EPA Administrator at that time. However, the "shelf life" of the document is probably limited by the range of pressures being applied upon the agency for improvements and increased rigor in the process.

The final risk assessment should include a summary of the risks associated with the exposure situation and such factors as the weight of evidence associated with each step of the process, the estimated uncertainty of the component parts, the

---

[8] Intake is the concentration or quantity of the subject agent that comes in direct contact with the body barriers. Uptake is the concentration or quantity moving across barriers, such as intestinal mucosa, alveoli, or epidermis.

[9] The EPA is preparing "Guidance for Conducting Health Risk Assessment of Chemical Mixtures," which has undergone peer review and was scheduled for release by the end of 1999. Other pertinent EPA publications are "Guiding Principles for Monte Carlo Analysis," EPA 630/R-97/001, March 1997, and "Use of Monte Carlo Simulation in Risk Assessment," EPA 903/F/94/001.

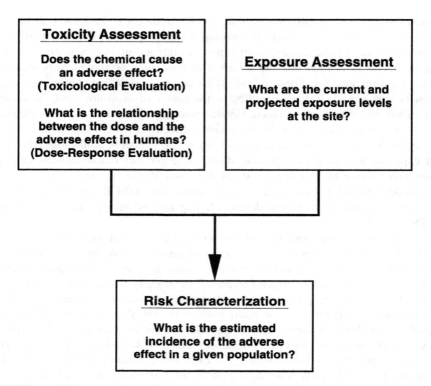

FIGURE 4.3 Risk assessment process at hazardous waste sites. (From the U.S. Environmental Protection Agency.)

distribution of risk across various sectors of the population, and the assumptions contained within the estimates.

The primary reason for interest in the details of a dose-response relationship for carcinogens is the need to estimate the risk to humans at low doses. Those responsible for promulgating risk-based standards want to know how small amounts of a chemical will affect lifetime disease incidence in humans. Typically, the only information is scant epidemiological data, together with results of animal experiments, both at high doses. Regulation, and the rationale thereof, would be much simpler if certain aspects of the dose-response relationship could be conclusively demonstrated. For example, if it could be conclusively demonstrated that there is a "threshold" dose below which there is no response, then exposure up to that threshold would evidently contribute no risk (Zeise et al. 1986, p. 1).

Thus, a risk-based standard may involve an exhaustive review of limited, questionable, or inappropriate exposure data; the need to extrapolate from observable effects at very high concentration exposure to very low concentration exposure criteria; the similar requirement to extrapolate from animal to human exposure criteria; and application under significantly different conditions than those prevailing in the data collection situation. This process, then, becomes the basis for establishing a standard at the predetermined risk level, i.e., 1 incidence per 100,000; 1,000,000;

10,000,000; etc. It is an imperfect process, vulnerable to assault, and frequently difficult to defend. It is clear that the technology-based standard is a more straightforward process and is therefore more appealing to those impatient with the slow pace of progress. It is similarly clear that risk-based standards are an imperative to those seeking improved rationality in the process (*see also*: EPA 1994; Zeise et al. 1986, pp. 43, 124–125; ICAIR 1985, Chapter 8; U.S. EPA 1989a, Chapters 4 to 7; EPA 1992; Hodgson and Levi 1987, pp. 281–283; Assante-Duah 1993; Krieger et al. 1995, pp. 123–133; Kester et al. 1995, Chapter 12; Douben 1998; Wickramanayake and Hinchee, Eds. 1998; Chastain, 1998; LaGoy 1999; Uliano, 2000; 64 FR 23833).

Other Hazards

Explosion and Fire. Prevention of fires and explosions is a major focus of RCRA and CERCLA and other environmental and workplace statutes. In fact, as noted in Chapter 1, fires and explosions were initially the primary concern of RCRA and the hazardous waste/materials management programs, and prevention thereof continues to be a major aspect of EPA, Department of Transportation, and Occupational Safety and Health Administration regulations and program guidance. Potential causes of explosions and fires at controlled and uncontrolled hazardous waste sites are numerous, including

- Chemical reactions that produce explosion, fire, or heat
- Ignition of explosive or flammable chemicals
- Ignition of materials due to oxygen enrichment
- Agitation of shock- or friction-sensitive compounds
- Sudden release of materials under pressure

Explosions and fires may arise spontaneously even at well-managed facilities. Such events are more likely to result from carelessness or poor practice on active sites or cleanup activities on abandoned sites. Examples include activities such as moving drums, mixing incompatible chemicals, or introducing an ignition source (such as electrical, electrostatic, or friction-generated spark) into an explosive or flammable environment. At hazardous waste sites, explosions and fires not only pose the obvious hazards of intense heat, open flame, smoke, and flying objects, but may cause the release of toxic chemicals into the environment. Such releases are a threat to workers on the site and to the general public living or working nearby (HHS 1985, p. 2-2). Regulated treatment, storage, and disposal sites are specifically designed and operated to prevent such incidents. A wide range of applicable fire prevention/protection standards have been promulgated by the American Society for Testing and Materials, the American National Standards Institute, The National Fire Protection Association, The American Petroleum Institute, and Underwriters' Laboratories. The regulatory agencies routinely adapt or excerpt from these standards (*see also*: Dawson and Mercer 1986, pp. 62–73; Meyer 1989, Chapter 13; Woodside 1999, Chapters 3, 7, and Appendices A to F).

Ionizing Radiation. Radioactive materials emit one or more of three types of harmful radiation: alpha particles, beta particles, and gamma rays, frequently identified by the Greek alphabet characters α, β, and γ. Alpha particles have limited penetration ability and are usually stopped by clothing and the outer layers of the skin. Alpha radiation poses little threat outside the body, but can be hazardous if alpha emitters are inhaled or ingested. Beta particles can cause harmful "beta burns" to the skin and damage the subsurface blood system. Beta emitters are also hazardous if inhaled or ingested. Gamma rays easily pass through clothing and human tissue and can cause serious permanent damage to the human body.

Several major health hazards may result from exposure to radiation, including burns or damage to internal organs, accumulation in the body until toxic levels are reached, malignancies, sterility, and/or harmful mutations. Acute exposure can result from improper handling of radioactive materials or improper disposal or storage in nonsecure facilities. Chronic exposure can potentially result from leaching of land-fills, volatilization of radioactive materials, or proximity of subjects to radiation sources (*see also*: Dawson and Mercer 1986, pp. 65–67; Corbitt 1989, pp. 9.87–9.111; Nebel and Wright 1993, pp. 489–490; Woodside 1999, Chapter 8).

Until recently, radioactive waste management and standards development have been regulated by statutes and agencies other than RCRA and the EPA and have not been considered a subset of regulated hazardous waste. The EPA has now promulgated regulations for the management of "mixed waste" to deal with wastes meeting both hazardous and radioactive waste definitions and having both characteristics. These statutes, regulations and standards will be discussed in Chapter 13.

Biomedical Hazards. As discussed in Chapter 12, the AIDS epidemic has brought the management of biomedical wastes sharply into the forefront. Wastes from health care, research, and biomedical manufacturing facilities may contain a variety of other infectious and/or pathogenic wastes.[10] Infectious wastes are those materials that contain disease-causing organisms or matter. Wastes that are infectious or contain infectious materials pose a hazard to handlers and the public if they are not isolated and/or disposed of in a manner that destroys the viability of the infectious matter. The EPA played a semi-active role in medical waste regulation from 1989 through 1991 based upon the authorities of RCRA Subpart J.[11] Following that period, the agency has issued rules and emission guidelines which apply to existing medical waste incinerators and has promulgated new source performance standards (NSPS) for newly constructed or modified hospital/medical/infectious waste incinerators (HMIWI). The active role in medical waste management has remained with state and local authorities. Background, evolution, current practice, and technologies of medical waste management and regulation are covered in Chapter 12.

Additional Hazards Associated with Hazardous Waste Management. Hazardous wastes and hazardous waste facilities may subject workers to a variety of

---

[10] There is inconsistency in the terminology used to define these wastes. The descriptors infectious, pathogenic, biomedical, biohazardous, toxic, and medically hazardous have all been used to describe infectious wastes. The EPA defines medical waste as any solid waste that is generated in the diagnosis, treatment, or immunization of human beings or animals in related research, biologicals production, or testing. An attempt is made in Chapter 12 to accommodate the definitional confusion.

[11] The short-lived Medical Waste Tracking Act (MWTA) of 1988, which was not renewed upon expiration in 1991 (*see:* Chapter 12).

other hazards, including physical hazards such as injury by heavy equipment, confined spaces, heat stress, engulfment, container handling, and electrical energy; exposure hazards such as oxygen deficiency, irritation, or corrosiveness; transportation incidents; and workplace violence. These hazards are the subject of Chapter 15, wherein their prevention and management will be explored in some detail. In most cases such exposures are limited to workers in direct contact or close proximity to the wastes. However, on- or off-site spills, uncontrolled releases, inadequate site security, or transportation accidents can subject the public to harmful exposures to these hazards (*see also*: Dawson and Mercer 1986, pp. 68–70; HHS 1985, p. 2-2; and Danby 1995, Chapter 9).

REGULATORY APPLICATION OF HEALTH STANDARDS AND CRITERIA

Technology-Based Standards

As discussed in the foregoing material, technology-based standards are best described as those grounded in treatment and/or control technologies, gradations of primitive to sophisticated processes, cost-effectiveness, economic feasibility, aesthetics, and political considerations. Some examples include

- The Clean Water Act requirements for definition and application of best practicable control technology currently available for classes and categories of point sources (other than publicly owned treatment works)
- The Clean Water Act requirements for definition and application of secondary treatment for publicly owned treatment works, and the inclusion, by definition, of oxidation ponds, lagoons, and trickling filters, as secondary treatment
- The RCRA treatment standards for land disposal restricted wastes including those expressed as specified technologies for destruction, treatment, or disposal
- The 1990 Clean Air Act Amendments requiring maximum achievable control technology (MACT) by sources of hazardous air pollutants

Risk-Based Standards

Standards and criteria derived from risk analyses of the nature outlined earlier in this chapter and based upon a predetermined level of risk to the receptor population are referred to as risk-based standards. Some examples follow:

- The Safe Drinking Water Act charges the EPA with promulgating primary drinking water standards containing maximum contaminant levels (MCLs) for public water supplies. The MCLs are to be established for each contaminant found in public water supplies that may have adverse human health effects, at levels having no known or anticipated adverse human health effect, with an adequate margin of safety.
- The RCRA land disposal restrictions also include a large number of standards that are health-related or risk-based.

- The Superfund (CERCLA) cleanup standards which require that remedial actions attain a level of control which renders impacted waters at least as clean as the MCLs of the Safe Drinking Water Act and the water quality criteria of the Clean Water Act.
- The Risk-Based Corrective Action (RBCA) statistical estimate representing an average dose for comparison with risk based remediation objectives conducted pursuant to RCRA Subtitle C.

RCRA Standards

As will be seen in subsequent chapters, RCRA embodies both technology-based and risk-based standards. The requirements for impermeable liners for land disposal facilities; for storage of hazardous wastes in nonreactive containers; for burning of hazardous waste fuels in high-temperature furnaces; for 99.99%, or 99.9999%, destruction and removal efficiencies (DRE) in thermal units; and some of the land disposal restrictions are technology-based standards.

If concentrations of the 40 toxicity characteristic wastes (40 CFR 261.24) are equal to or more than 100 times the National Interim Primary Drinking Water Standards, the waste is hazardous and must be managed as such. Such risk-based standards are most prevalent in permits for treatment, storage, and disposal facilities; in groundwater monitoring requirements for land disposal facilities; and in remediation requirements.

Congress and the EPA have attempted to craft the RCRA regulatory approach in risk-based rationales, but the large numbers and quantities of chemicals and mixtures involved, together with the varieties of generator/source operations, have made that approach exceedingly difficult. As a result, RCRA (the Act and the program) has focused upon regulatory mechanisms which, in large measure:

- Attempt to identify wastes which are hazardous to human health and the environment and capture them in a "cradle-to-grave" management system
- Create engineering controls, e.g., physical and space barriers that isolate the public from contact with the identified hazardous wastes, during generation, transportation, storage, treatment, and/or disposal
- Minimize the generation of hazardous wastes
- Encourage the re-use and recycling of hazardous wastes and the treatment to nonhazardous or reduced hazard condition
- Ensure secure disposal of wastes which cannot otherwise be safely managed

Standards Implementing the Land Disposal Restrictions

The Hazardous and Solid Waste Amendments of 1984 (HSWA) imposed land disposal restrictions (LDRs or "land ban") upon certain hazardous wastes. Section 3004 of HSWA restricts the land disposal of hazardous waste beyond specified dates unless the wastes are treated to meet *treatment* standards. The treatment standards can be either concentration levels for hazardous constituents that the waste must meet or treatment technologies that must be performed on the waste before it can be disposed.

In promulgating the standards, the EPA researched available health exposure data and treatment technologies to identify which proven, available treatment methods were most capable of minimizing the mobility or toxicity (or both) of the hazardous constituents. That technology was designated Best Demonstrated Available Technology (BDAT) for the particular waste (U.S. EPA 1998, Chapter 6). The LDR standards, which are found in 40 CFR 268, are thus based upon both risk and treatment technology. Application of the LDR standards is discussed, as appropriate, in Chapters 5 and 7 (*see also:* Glossary).

TOPICS FOR REVIEW OR DISCUSSION

1. What does a dose-response curve that passes through the origin indicate with respect to acceptable dose?
2. Identify four categories of physiological effects imposed by chemical constituents of hazardous wastes.
3. In the hazardous waste lexicon, what is meant by the term "toxicity hazard?"
4. Which of the routes of exposure is considered least likely to be a factor to workers on industrial sites? Why?
5. The risk assessment process for evaluation of a hazardous waste site usually consists of:
   a. _____
   b. _____
   c. _____
   d. _____
6. Carbon tetrachloride, a widely distributed pollutant, may cause damage to what human organs?
7. Identify some of the physiological effects on humans of exposure to mercury.
8. The current regulatory scheme for hazardous waste management generally relies upon risk-based or technology based standards. Briefly explain each. What are the arguments favoring each?
9. Which agencies and departments of the federal government classify hazardous materials and their constituents as "carcinogenic?" Construct a matrix showing which chemicals, by agency, are carcinogenic, suspect carcinogens, probable carcinogens, etc.

REFERENCES

American Conference of Governmental Industrial Hygienists. 1997. *Threshold Limit Values for Chemical Substances and Physical Agents and Biological Exposure Indices.* Cincinnati, OH.

Assante-Duah, D. Kofi. 1993. *Hazardous Waste Risk Assessment.* CRC Press, Boca Raton, FL

Beaulieu, Harry J. and Diane L. Beaulieu. 1985. *Toxicology,* National Environmental Health Association.

Chastain, James R., Jr. 1998. *Orientation to Health Risk Assessment for Practicing Engineers.* Practice Periodical of Hazardous, Toxic, and Radioactive Waste Management/January 1998, American Society of Civil Engineers, Reston, VA.

Corbitt, Robert A. 1989. *Standard Handbook of Environmental Engineering.* McGraw-Hill, NY.

Danby, John G. 1995. "Health and Safety Training for Hazardous Waste Activities," in *Accident Prevention Manual for Business and Industry — Environmental Management.* Gary R. Krieger, Ed., National Safety Council, Itasca, IL.

Dawson, Gaynor W. and Basil W. Mercer. 1986. *Hazardous Waste Management.* John Wiley & Sons, NY.

Douben, Peter E.T., Ed. 1998. "Perspectives on Pollution Risk," in *Pollution Risk Assessment and Management,* Peter E.T. Douben, Ed., John Wiley & Sons, NY.

Enger, Eldon D., J. Richard Kormelink, Bradley F. Smith, and Rodney J. Smith. 1989. *Environmental Science: The Study of Interrelationships.* Wm. C. Brown, Dubuque, IA.

Hodgson, Ernest and Patricia E. Levi. 1987. *A Textbook of Modern Toxicology.* Elsevier, NY.

ICAIR Life Systems, Inc. 1985. *Toxicology Handbook.* Prepared for EPA Office of Waste Programs Enforcement, Washington, D.C., TR-693-21A.

Johnson, Barry L. and Christopher T. DeRosa. 1997. "The Toxicologic Hazard of Superfund Hazardous Waste Sites," in *Reviews on Environmental Health, Volume 12,* No. 4, pp. 235–251, Freund, London.

Kamrin, Michael A. 1989. *Toxicology.* Lewis Publishers, Chelsea, MI.

Kester, Janet E., Holly A. Hattemer-Frey, Joseph W. Gordon, and Gary R. Krieger. 1995. "Risk Assessment," in *Accident Prevention Manual for Business and Industry — Environmental Management.* Gary R. Krieger, Ed., National Safety Council, Itasca, IL.

Krieger, Gary R., Mark J. Logsdon, Christopher P. Weis, and Joanna Moreno. 1995. "Basic Principles of Environmental Science," in *Accident Prevention Manual for Business and Industry — Environmental Management.* Gary R. Krieger, Ed., National Safety Council, Itasca, IL.

LaGoy, Peter K. 1999. "Risk Assessment in Remediation: Accurately Accounting for Uncertainty," in *Remediation,* John Wiley & Sons, NY.

Manahan, Stanley E. 1994. *Environmental Chemistry, Sixth Edition.* CRC Press, Boca Raton, FL.

Meyer, Eugene. 1989. *Chemistry of Hazardous Materials.* Prentice-Hall, Englewood Cliffs, NJ.

Munter, Florence, Stephen W. Bell, Robert Hollingsworth, Joseph W. Gordon, and Charles N. Lovinski. 1995. "Hazardous Wastes," in *Accident Prevention Manual for Business and Industry — Environmental Management.* Gary R. Krieger, Ed., National Safety Council, Itasca, IL.

National Research Council. 1983. *Risk Assessment in the Federal Government.*

Nebel, Bernard J. and Richard T. Wright. 1993. *Environmental Science.* Prentice-Hall, Englewood Cliffs, NJ.

Schoeny, Rita, Pavel Muller, and Judy L. Mumford. 1998. "Risk Assessment for Human Health Protection — Applications to Environmental Mixtures," in *Pollution Risk Assessment and Management.* Peter E.T. Douben, Ed., John Wiley & Sons, NY.

Sittig, Marshall. 1992. *Handbook of Toxic and Hazardous Chemicals and Carcinogens, Third Edition,* Noyes Publications, Park Ridge, NJ.

Uliano, Jr., Tony. 2000. "Environmental Health and Safety Risk Analysis," in *Hazardous Materials Management Desk Reference,* Chapter 10, Doye B. Cox, Editor-in-Chief, Adriane P. Borgias, Technical Editor, McGraw-Hill, NY.

U.S. Department of Health and Human Services. 1975 (earlier editions, U.S. Department of Health, Education, and Welfare). Registry of Toxic Effects of Chemical Substances. Superintendent of Documents, U.S. Government Printing Office, Washington, D.C.

U.S. Department of Health and Human Services. 1985. NIOSH/OSHA/USCG/EPA. Occupational Safety and Health Guidance Manual for Hazardous Waste Site Activities. Superintendent of Documents, U.S. Government Printing Office, Washington, D.C.

U.S. Department of Health and Human Services. 1990 (Draft, latest draft 1999). Health Assessment Guidance Manual. Agency for Toxic Substances and Disease Registry, Atlanta, GA.

U.S. Department of Health and Human Services. 1993. ATSDR Cancer Policy Framework. Agency for Toxic Substances and Disease Registry, Atlanta, GA.

U.S. Department of Health and Human Services. 1997. NIOSH Pocket Guide to Chemical Hazards. Public Health Service, Superintendent of Documents, U.S. Government Printing Office, Washington, D.C.

U.S. Environmental Protection Agency. 1989a. Risk Assessment Guidance for Superfund — Volume I — Human Health Evaluation Manual (Part A), Office of Emergency and Remedial Response, Washington, D.C., EPA 540/1-89/002.

U.S. Environmental Protection Agency. 1989b. Managing and Tracking Medical Wastes, Office of Solid Waste and Emergency Response, Washington, D.C., EPA 530-SW-89-022.

U.S. Environmental Protection Agency. 1992. Risk Assessment Guidance for Superfund, Volumes I and II, (RAGS), National Service Center for Environmental Publications, Cincinnati, OH, EPA 540/R/92/003.

U.S. Environmental Protection Agency. 1994. Monte Carlo Simulation in Risk Assessments, U.S. EPA Region 3 Hazardous Waste Management Division-3HW53, Philadelphia, PA, EPA 903/F/94/001.

U.S. Environmental Protection Agency. 1995a. Guidance For Risk Characterization. Science Policy Council, Washington, D.C.

U.S. Environmental Protection Agency. 1995b. Policy For Risk Characterization at the U.S. Environmental Protection Agency. Science Policy Council, Washington, D.C.

U.S. Environmental Protection Agency. 1997. Guiding Principles for Monte Carlo Analysis. National Service Center for Environmental Publications, Cincinnati, OH, EPA 630/R/97/001.

U.S. Environmental Protection Agency. 1998. RCRA Orientation Manual, Office of Solid Waste, Washington, D.C., EPA 530/R/98/004.

Wickramanayake, Goddage B. and Robert E. Hinchee, Eds. 1998. *Risk, Resource, and Regulatory Issues — Remediation of Chlorinated and Recalcitrant Compounds.* Battelle Press, Columbus, OH.

Woodside, Gayle. 1999. *Hazardous Materials and Hazardous Waste Management, Second Edition.* John Wiley & Sons, NY.

Zeise, Lauren, Richard Wilson, and Edmund A.C. Crouch. 1986. *The Dose-Response Relationships for Carcinogens: A Review.* John F. Kennedy School of Government, Harvard University, Cambridge, MA.

5 Hazardous Waste Sources/Generators

OBJECTIVES

At completion of this chapter, the student should:

- Have familiarity with some of the common industrial sources of hazardous waste and the RCRA approach to regulation of wastes from specific industries/processes.
- Understand the role that the generator plays in the "cradle-to-grave" management of hazardous wastes and the basic requirements RCRA imposes upon generators.
- Understand the RCRA focus on controls based upon the three categories of generators, i.e., nature and composition of a waste, environmental and health impacts of a waste, and/or quantity of waste produced.

INTRODUCTION

In the previous chapters we have shown that the increasing numbers of hazardous waste incidents during the 1970s brought about increasing public alarm and pressure upon Congress and the state legislatures to take decisive action to protect human health and the environment. We have also shown that Congress has, through original enactments and subsequent amendments, steadily strengthened and tightened the Resource Conservation and Recovery Act (RCRA) and the Comprehensive Environmental Response Compensation and Liability Act (CERCLA)[1] in its effort to achieve timely control and remediation of hazardous waste impacts. We have illustrated the wide variety of hazardous waste abuses and disposal practices that have helped to shape the statutory and regulatory structures.

We have shown the interrelationships of hazardous waste releases to the atmosphere, publicly owned wastewater treatment facilities, surface streams, the land surface, and to the earth's crust. We have shown the routes of movement through the environment and the mechanisms of human and environmental exposure. With this background overview, we may now begin to consider the generic approaches to management and control, as well as those embodied in the RCRA and CERCLA.

---

[1] CERCLA was being referred to as "Superfund" before it was written, formally named, and inacted, and the term is generally used to identify the program implemented by CERCLA and the amendments thereto.

Enactment of RCRA in 1976 enabled the EPA and the delegated (or "primacy") states to develop programs to implement the cradle-to-grave management of hazardous wastes. Remediation of abandoned hazardous waste sites, those for which "responsible parties" could not be found and those having responsible parties that are unable or unwilling to conduct cleanup operations, was not provided for until enactment of CERCLA in 1980. Although subsequent amendments of both Acts have blurred this historical distinction with some overlapping authorities, practitioners have come to think of hazardous wastes as being either RCRA (currently generated or released) or CERCLA (residual) wastes. In this chapter we will overview RCRA-specific management of hazardous waste generator activities. Discussion of the management of wastes contributed by site remediation is deferred to Chapter 11.

THE GENERATOR DEFINED

The "generator" is the first element of the RCRA cradle-to-grave concept, which includes generators, transporters, treatment plants, storage facilities, and disposal sites. The RCRA regulations define a generator as: ... any person, by site, whose act or process produces hazardous waste identified or listed in Part 261 of this chapter or whose act first causes a hazardous waste to become subject to regulation (40 CFR 260.10). In more practical terms, the generator is the creator of a hazardous waste who must analyze all solid wastes produced or use knowledge of the wastes to determine if they meet the RCRA Subtitle C definitions or listings of hazardous wastes. As indicated earlier, there are more than 20,000 large quantity generators reporting to the EPA and subject to the generator regulations. The RCRA definition is generally unambiguous with respect to conventional industrial sources, but can become less clear in the event of an accidental release. In site remediation activity, identification of the creator of the waste can become a highly contentious issue and/or the basis for major litigation (more on this in Chapter 11). Once a waste has been identified as a RCRA hazardous waste, it becomes subject to the Subtitle C regulations, and the generator assumes very significant responsibilities for the correct management thereof.

The Three Classifications of Generators

Congress and the EPA initially recognized that large numbers of generators, particularly small businesses, produce relatively minor quantities of hazardous wastes and, accordingly, created two categories of generators. The generator of more than 1000 kg of hazardous waste, per month, or more than 1 kg of acutely[2] hazardous waste, per month, was designated a "generator" (frequently spoken of as a "large quantity generator" or LQG) and was (and is) subject to the full content of the 40 CFR 262 regulation. Those generating less than 1000 kg of hazardous wastes or less than 1 kg

---

[2] Acutely hazardous wastes are wastes that the EPA has determined to be so dangerous that small amounts are regulated in a manner similar to larger amounts of other hazardous wastes. They are, specifically, F020-F023 and F026-F028 identified in 40 CFR 261.31 and the "p" wastes listed in 40 CFR 261.33.

of acutely[2] hazardous waste were classified as small quantity generators (SQGs) and were exempted from most of the generator requirements of the RCRA regulations.[3]

Because of concerns that wastes exempted from regulation by the SQG exclusion could be causing significant environmental harm, Congress amended the definition of SQGs in the Hazardous and Solid Waste Amendments of 1984 (HSWA). SQGs were redefined as producers of between 100 and 1000 kg of hazardous waste per month or less than 1 kg of acutely hazardous waste. As before, SQGs may accumulate less than 6000 kg of hazardous waste at any time. EPA reports indicated that there were approximately 236,000 SQGs in 1997 (U.S. EPA 1998, p. III-47). The SQG was made subject to new restrictions, which are summarized herein.

A new classification, the Conditionally Exempt Small Quantity Generator (CESQG), was defined as producing less than 100 kg of hazardous waste and less than 1 kg of acutely hazardous waste per month. CESQGs are limited to accumulation of less than 1000 kg of hazardous waste, less than 1 kg of acutely hazardous waste, or 100 kg of any residue from the cleanup of a spill of acutely hazardous waste at any time. EPA reports indicated that there were between 455,000 and 700,000 CESQGs in 1997 (U.S. EPA 1998, p. III-47). This category of generator is exempt from most generator requirements (*see also:* 40 CFR 261.5; 40 CFR 262, Subparts C and D; Ostler 1998, pp. 39–41; U.S. EPA 1995; U.S. EPA 1998, Chapter 3).

The RCRA Subtitle C regulations recognize three categories of generators

- *Large quantity generators* ("generators") (LQG) are defined as those facilities that generate more than 1000 kg of *hazardous waste* or more than 1 kg of *acutely hazardous waste* per month.
- *Small quantity generators* (SQGs) are defined as *producing* more than 100 kg, but less than 1000 kg of *hazardous waste* per month, or less than 1 kg of *acutely hazardous waste* per month; or as *accumulating* less than 6000 kg of hazardous waste at any one time or less than 1 kg of acutely hazardous waste at any one time.
- *Conditionally exempt small quantity generators* (CESQG) are those that generate less than 100 kg of *hazardous waste* per month or less than 1 kg of *acutely hazardous waste* per month; or *accumulate* less than 1000 kg of *hazardous waste* at any time or less than 1 kg of *acutely hazardous waste* at any one time.

WASTES GENERATED

In 1997, the latest year for which RCRA Biennial Report* data were available in mid-2000, 20,316 large quantity generators reported 40.7 million tons of hazardous waste generated. These data show an apparent decrease of 551 generators and 173

---

[3] 40 CFR 260 to 265, in 1980.
* EPA 1999b.

million tons when compared to the 1995 data. As noted earlier, this apparent decrease is attributable to the discontinued reporting of wastewaters containing hazardous wastes (U.S. EPA 1999b, p. ES-4). Prior to this discontinuity, overall hazardous waste generation had continued in the 200 to 300 million tons per year range through most of the 1980s and early 1990s.

The Toxic Release Inventory (TRI) ranks the total release quantities of TRI chemicals by Standard Industrial Code (SIC) 21 through 39, thereby providing some sense of the relative contributions of hazardous waste by type of *manufacturing* industry (Table 5.1). Again, as discussed in Chapter 3, the TRI *release* statistics are not comparable to RCRA hazardous waste generation quantities reported by the EPA. This distinction grows from the differences in reporting requirements for the TRI[4] and the biennial reporting requirements for LQGs.[5]

A few examples of basic industries and the types of hazardous wastes produced are listed in Appendix A to this chapter to illustrate the wide variety and complexity of the wastes. The reader should consult specific industry trade publications for details regarding industry-specific wastes produced. Those few examples are inadequate to suggest the numbers and kinds of hazardous chemical constituents in hazardous wastes that must be managed. There are approximately 900 listed wastes in 40 CFR 261 and countless more of characteristic wastes. The traditional intensity of industrial and business competition engenders the introduction of new products and thus new wastes to be managed. This historical burgeoning of waste generation, the health concerns and environmental degradation, the public and political pressures that arise, the ever increasing costs of waste management, and liability concerns have brought about intensified efforts to reduce quantities of wastes generated and to reuse and recycle wastes much more effectively. Hazardous waste minimization, reuse, and recycling are discussed in Chapter 8 (*see also:* Dawson and Mercer 1986, p. 119–129; Phifer and McTigue 1988, Chapter 3; Nebel and Wright 1993, Chapter 14; U.S. EPA 1999b, p. ES-4; U.S. EPA 1999c).

REGULATORY REQUIREMENTS

LQGs and SQGs are subject to regulations contained in 40 CFR 262. These regulations require them to:

- Identify and quantify wastes generated.
- Obtain an EPA identification number.
- Comply with accumulation and storage requirements (including requirements for training, contingency planning, and emergency arrangements).

---

[4] Reporting requirements for the TRI are derived from Section 313 of the Emergency Planning and Community Right-to-Know Act of 1986 and require reporting of releases of (currently) 654 *chemicals*. Instructions for TRI reporting are found in EPA document number EPA K-97-001 and are available on CD-ROM from EPA Regional Offices or the TRI homepage <http://www.epa.gov.opptintr/tri>.

[5] Biennial reporting is required of LQGs who ship any hazardous waste off-site to a treatment storage and disposal facility in the U.S. by 40 CFR 262.41(a) (*see:* section on Biennial Reports below). LQGs should also take note of § 262.41(b) regarding on-site disposal and § 262.56 regarding exports of hazardous waste.

TABLE 5.1
Quantities of TRI Releases and Transfers by Industry Type (1997)

SIC Codes	Industry	Releases	Transfers	Totals
20	Food products	92,040,698	1,527,792	93,568,490
21	Tobacco manufacturing	3,961,646	387,968	4,349,614
22	Textile mill products	18,822,616	613,643	19,436,259
23	Apparel	763,620	140,912	904,532
24	Lumber and wood	26,990,140	2,669,283	29,659,423
25	Furniture and fixtures	24,845,983	267,933	25,113,916
26	Paper products	228,783,236	4,747,148	233,530,384
27	Printing and publishing	24,521,063	122,715	24,643,778
28	Chemical products	742,647,731	54,849,079	797,496,810
29	Petroleum refining	66,054,902	3,268,549	69,323,451
30	Rubber and plastics	99,090,526	9,375,097	108,465,623
31	Leather products	2,703,714	2,030,230	4,733,944
32	Stone, glass, clay	35,217,660	8,003,691	43,221,351
33	Primary metals	405,945,561	288,716,526	694,662,087
34	Fabricated metals	66,338,048	29,198,597	95,536,645
35	Machinery, nonelectrical	17,935,728	4,465,584	22,401,312
36	Electrical	21,470,770	12,674,200	34,144,970
37	Transport equipment	91,538,326	10,519,302	102,057,628
38	Measuring, photographic	12,442,580	799,306	13,241,886
39	Miscellaneous	9,423,585	818,874	10,242,459
	Multiple SIC in 20-39	112,177,397	25,444,493	137,621,890
	No SIC in 20-39	12,738,725	457,907	13,196,632
Totals		2,116,454,255	461,098,829	2,577,553,084
Federal facilities		6,236,657	336,688	6,573,345

Note: In 1998, the EPA added six SIC codes to the reporting list: Metal Mining(10), Coal Mining (12), Electric Utilities (49), Chemical Wholesalers (5169), Petroleum Bulk Terminals (5171), RCRA/Solvent Recovery (4953/7389).

Source: EPA (1999a, p. 4–7).

- Properly prepare wastes prior to transport.
- Track the shipment and receipt of hazardous waste by use of the manifest system.
- Perform record keeping and reporting as required.

In consideration of the fact that SQGs produce a smaller portion of the total hazardous waste generated and the burdens of full compliance with the regulatory requirements, Congress authorized the EPA to ease the original requirements placed upon SQGs, provided that the requirements remained protective of human health and the environment. As noted earlier, CESQGs are subject to only minimal regulation, provided that they do not exceed generation limits for that category. Each requirement is discussed below, with differences in requirements for the three categories described as appropriate.

EPA ID Number

One of the ways by which the EPA and the primacy states monitor and track generator activity is the assignment of a unique identification number to each generator (and SQG), transporter, and operator of a treatment, storage and disposal (TSD) facility. Without this number, the generator and SQG are barred from treating, storing, disposing of, transporting, or offering for transportation any hazardous waste. Neither category of generator may offer its RCRA-defined hazardous waste to any transporter or TSD facility that does not also have an EPA ID number. Generators, SQGs, and transporters obtain ID numbers by "notifying" the EPA of hazardous waste activity, using EPA Form 8700-12. CESQG are not required to obtain ID numbers. *The EPA ID number is not a certification nor an endorsement of the assignee's hazardous waste activity. It is for identification purposes only (see also:* U.S. EPA 1998, p. III-48).

Pretransport Regulations

Pretransport regulations specify actions which the generator must take to ensure that hazardous wastes are packaged, labeled, marked, and (if appropriate) placarded prior to offering the wastes for transportation. The pretransport requirements (40 CFR 262, Subpart C) refer to elements of the Department of Transportation (DOT) regulations for transporting hazardous materials (49 CFR 172, 173, 178, and 179).[6] The DOT regulations require:

- Proper packaging to prevent leakage of hazardous waste, during both normal transport conditions and potentially dangerous situations, e.g., a drum of waste dropped from a truck bed or loading dock
- Labeling, marking, and placarding of the packaged waste to identify the characteristics and dangers associated with transporting the waste

A thorough examination of these detailed and exacting regulations would greatly exceed the scope of this text. The student or practitioner contemplating or having responsibilities for pretransport preparation of hazardous waste shipments must complete the training required by 49 CFR Subpart H. The general thrust of the requirements can be ascertained by examining the Hazardous Materials Table column headings (Figure 5.1) and the "Eight-Step Procedure" prepared by the DOT Transportation Safety Institute, which is provided as Appendix B to this chapter.

In brief, the marking requirements include the requirement for individual containers to display a "Hazardous Waste" marking of the format shown in Figure 5.2. The marking must include a proper DOT "shipping name" that uses the standardized

---

[6] The DOT regulations for transportation of hazardous materials were significantly modified in 1990. The modifications implement the HM 181 "Performance Oriented Packaging Standards," which bring U.S. hazardous materials shipping standards nearer to accord with international standards.

Symbols (1)	Hazardous materials descriptions and proper shipping names (2)	Hazard class or Division (3)	Identification Numbers (4)	Packing group (5)	Label(s) required (if not excepted) (6)	Special provisions (7)	Exceptions (8A)	Non-bulk packaging (8B)	Bulk packaging (8C)	Passenger aircraft or railcar (9A)	Cargo aircraft only (9B)	Vessel stowage (10A)	Other stowage provisions (10B)
	Accellerene, see p-Nitrosodimethylaniline.												
D	Accumulators, electric, see Batteries, wet etc.												
D	Accumulators, pressurized, pneumatic or hydraulic (containing non-flammable gas).	2.2	NA1956		NONFLAMMABLE GAS.		306	306	None	No Limit	No Limit	A	
	Acetal.	3	UN1088	II	FLAMMABLE LIQUID.	T7.	150	202	242	5 L	60 L	E	
	Acetaldehyde.	3	UN1089	I	FLAMMABLE LIQUID.	A3, B16, T20, T26, T29	None	201	243	Forbidden	30 L	E	
A	Acetaldehyde ammonia.	9	UN1841	III	CLASS 9.	T8.	155	204	241	200 kg	200 kg	A	
	Acetaldehyde oxime.	9	UN2332	III	FLAMMABLE LIQUID.		150	202	242	5 L	60 L	A	34
	Acetic acid, glacial or Acetic acid solution, more than 80 per cent acid, by mass.	8	UN2789	II	CORROSIVE.	A3, A6, A7, A10, B2, T8.	154	202	242	1 L	30 L	A	12, 21, 48
	Acetic acid solution, more than 10 per cent but not more than 80 per cent acid, by mass.	8	UN2790	II	CORROSIVE.	A3, A6, A7, A10, B2, T8.	154	202	242	1 L	30 L	A	112
	Acetic anhydride.	8	UN1715	II	CORROSIVE.	A3, A6, A7, A10, B2, T8.	154	202	242	1 L	30 L	A	40
	Acetone.	3	UN1090	II	FLAMMABLE LIQUID.	T8.	150	202	242	5 L	60 L	B	
	Acetone cyanohydrin, stabilized.	6.1	UN1541	I	POISON.	2, A3, B9, B14, B74, B76, B77, N34, T38, T43, T45.	None	227	244	Forbidden	30 L	D	25, 40, 49, M2
	Acetone oils.	3	UN1091	II	FLAMMABLE LIQUID.	T7, T30.	150	202	242	5 L	60 L	B	
	Acetonitrile, see Methyl cyanide.												
	Acetyl acetone peroxide with more than 9% by mass active oxygen.	Forbidden											
	Acetyl benzoyl peroxide, solid, or more than 40% in solution.	Forbidden											
	Acetyl bromide.	8	UN1716	II	CORROSIVE.	B2, T12, T26.	154	202	242	1 L	30 L	C	8, 40
	Acetyl chloride.	3	UN1717	II	FLAMMABLE LIQUID. CORROSIVE.	A3, A6, A7, N34, T18, T26.	None	202	243	1 L	5 L	B	40
	Acetyl cyclohexanesulfonyl peroxide, more than 82 per cent wetted with less than 12 per cent water.	Forbidden											
	Acetylene, dissolved.	2.1	UN1001		FLAMMABLE GAS.		None	303	None	Forbidden	15 kg	D	25, 40, 57
	Acetylene (liquefied).	Forbidden											
	Acetylene silver nitrate.	Forbidden											
	Acetylene tetrabromide, see Tetrabromoethane.												
	Acetyl iodide.	8	UN1898	II	CORROSIVE.	B2, T9.	154	202	242	1 L	30 L	C	8, 40
	Acetyl methyl carbinol.	3	UN2621	III	FLAMMABLE LIQUID.	B1, T1.	150	203	242	60 L	220 L	A	
	Acetyl peroxide, see Diacetyl peroxide, etc.												
	Acetyl peroxide, solid, or more than 25 percent in solution.	Forbidden											
	Acid butyl phosphate, see Butyl acid phosphate.												
	Acid, sludge see Sludge acid.												
	Acridine.	6.1	UN2713	III	KEEP AWAY FROM FOOD.		153	213	240	100 kg	200 kg	A	
	Acrolein dimer, stabilized.	3	UN2607	III	FLAMMABLE LIQUID.	T1.	150	203	242	60 L	220 L	A	40

FIGURE 5.1 DOT hazardous materials table (49 CFR 172.101).

HAZARDOUS WASTE

PENNSYLVANIA AND FEDERAL LAWS PROHIBIT IMPROPER DISPOSAL

If found contact the nearest police, public safety authority, U.S. Environmental Protection Agency at 800-424-8802 or the Pa. Department of Environmental Resources at 717-787-4343, if found within the Commonwealth of Pennsylvania (Ref. 40CFR: 262-32 (6) – Pa. Title 25: 75.262).

GENERATOR'S NAME ___ Nosteller Paint & Chemical CO ___ EPA No ___ PAD0007654356 ___

ADDRESS ___ 6057 Philadelphia Blvd. ___

CITY ___ Boyertown ___ STATE ___ PA ___ ZIP ___ 19245 ___

EPA/DOT WASTE DESCRIPTION ___ "RQ" WASTE FLAMMABLE LIQUID POISONOUS, N.O.S. ___ HAZARD CLASS ___ FLAMMABLE LIQUID ___

UN/NA NO ___ 1992 ___ EPA WASTE CODE # ___ F003 ___

DATE OF ACCUMULATION ___ 7/6/84 ___ MANIFEST DOCUMENT # ___ 00576 ___

HANDLE WITH CARE – THIS CONTAINER IS DANGEROUS AND CONTAINS HAZARDOUS OR TOXIC WASTE

| IT IS RECOMMENDED THAT IF THIS LABEL WILL BE AFFIXED TO ANY CONTAINERS WHICH ARE TO BE EXPOSED TO THE ELEMENTS FOR ANY SUSTAINED PERIOD OF TIME, THAT EACH LABEL BE PERMANENTLY COVERED WITH LABELGARD TAPE. | IN THE EVENT OF A SPILL OR RELEASE OF THIS HAZARDOUS WASTE, CONTACT THE U.S. COAST GUARD NATIONAL RESPONSE CENTER AT 800-424-8802 FOR INFORMATION AND ASSISTANCE. |

©1981 HAZARDOUS MATERIALS PUBLISHING CO., KUTZTOWN, PA 19530

FIGURE 5.2 Marking hazardous waste shipment. (From the Environmental Protection Agency.)

language of 49 CFR 172.101 and 172.102. The labels on individual containers must accurately display the correct hazard class as prescribed by Subpart E of Part 172. A hazard class label is shown in Figure 5.3. Bulk shipments, whether motorized or containerized, must display the correct placard. A placard is shown in Figure 5.4.

The labeling, marking, and placarding requirements grow from the need for emergency responders to have the best possible knowledge of the materials involved in any actual or potential release situation. Efforts are in progress to achieve international consistency of marking, labeling, and placarding conventions.

The pretransport regulations apply only to generators shipping hazardous waste off-site for treatment, storage, or disposal. They do not apply to *on-site* transportation. Accordingly, the generator and SQG should carefully examine the unique 40 CFR 260.10 definition of the term *on-site*, which is also provided in the Glossary

FIGURE 5.3 DOT hazard label.

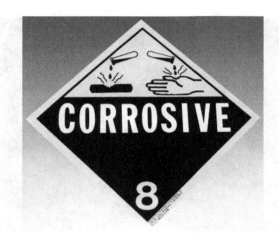

FIGURE 5.4 DOT hazard placard.

of this text. At 62 FR 6651, February 12, 1997, § 262.20 (f) was added to Subpart B, allowing transport "off-site" on a public highway within or along the border of contiguous property under the control of the same person, even if such contiguous property is divided by a public or private right-of-way. In those circumstances the generator or transporter must comply with the §§ 263.30 and 263.31 requirement for cleanup of hazardous waste discharges.

Accumulation of Waste

The regulatory material contained in 40 CFR 262.34, titled *accumulation time* is a "sleeper." The material greatly exceeds the apparent subject of accumulation time. The EPA chose to structure the content in the form — *generator may accumulate for 90 days without a permit, provided he/she complies with all that follows, including many references to other regulatory material.* It is a key portion of the generator regulations and requires much of the practitioner.

A generator may accumulate hazardous waste on-site for 90 days or less, provided the following accumulation-related requirements are met:

- Proper Management — The waste must be properly accumulated in containers, tanks, drip pads, or containment buildings. Containers must be kept closed and marked with the words "hazardous waste." Tanks and containers must be marked with the date on which accumulation began (Figure 5.5). The generator must ensure and document shipment of the waste off-site within the allowable 90-day period.

FIGURE 5.5 Dating accumulation container, satellite accumulation point, or temporary storage area. (From the Arizona Department of Environmental Quality.)

- Emergency Plan — A written contingency plan and procedures for managing spills or releases must be developed. Generators are required to have a written emergency plan, but SQGs are not.[7]
- Personnel Training — Facility personnel must be trained in the proper handling of hazardous waste. Generators are required to have an established training program.[8] SQGs are exempt from this requirement, but must ensure that employees handling hazardous wastes are familiar with proper procedures.

The 90-day accumulation period allows a generator to collect enough waste to make transportation more cost effective. If the generator accumulates hazardous waste on-site for more than 90 days, the generator becomes subject to the Subtitle C requirements for storage facilities, including the requirement for permitting. The regulations provide for a 1-time, 30-day extension under extenuating circumstances.

SQGs may store waste on-site for up to 180 days (or up to 270 days if the waste must be transported for 200 or more miles for off-site treatment, storage, or disposal), providing certain criteria are met. The on-site quantity of hazardous waste may not exceed 6000 kg at any time. The SQG exceeding the time or quantity limits becomes a storage facility and is subject to §§ 264 and 265 and permitting requirements. If a CESQG accumulates more than 1000 kg of hazardous waste, all of those wastes become subject to the SQG requirements. If the CESQG accumulates more than 1 kg of acutely hazardous waste, the acutely hazardous waste becomes subject to the full

---

[7] The small quantity generator is, however, required to meet minimal emergency planning requirements set forth in 262.34(d).

[8] The generator requirements for contingency planning and a training program are by reference, contained in 40 CFR 262.34, to 40 CFR 265, Subparts C and D, and to 40 CFR 265.16. Those new to the RCRA program, and subject to the generator regulations should read 40 CFR 262.34 and the referenced material in 40 CFR 265 very carefully.

regulation applicable to LQGs. Owners/operators of facilities that are SQGs or CESQGs should carefully read §§ 261.5, 264.34, and the sections referenced therein.

The Manifest

The Uniform Hazardous Waste Manifest (Form 8700-22, Figure 5.6) is the instrument that enables the tracking of, and accounting for, hazardous wastes in the cradle-to-grave system. Through the use of the manifest, generators, transporters,

Please print or type *(Form designed for use on elite (12 - pitch) typewriter)* *Form Approved. OMB No. 2050 - 0039 Expires 9 - 30 - 91*

UNIFORM HAZARDOUS WASTE MANIFEST

1 Generator's US EPA ID No. Manifest Document No.
2. Page 1 of
Information in the shaded areas is not required by Federal law

3. Generator's Name and Mailing Address
A. State Manifest Document Number
B. State Generator's ID
4. Generator's Phone ( )
5. Transporter 1 Company Name
6. US EPA ID Number
C. State Transporter's ID
D. Transporter's Phone
7. Transporter 2 Company Name
8. US EPA ID Number
E. State Transporter's ID
F. Transporter's Phone
9. Designated Facility Name and Site Address
10. US EPA ID Number
G. State Facility's ID
H. Facility's Phone

GENERATOR

11. US DOT Description *(Including Proper Shipping Name, Hazard Class, and ID Number)*	12. Containers No.	Type	13. Total Quantity	14. Unit Wt/Vol	I. Waste No.
a.					
b.					
c.					
d.					

J. Additional Descriptions for Materials Listed Above
K. Handling Codes for Wastes Listed Above

15. Special Handling Instructions and Additional Information

16. GENERATOR'S CERTIFICATION: I hereby declare that the contents of this consignment are fully and accurately described above by proper shipping name and are classified, packed, marked, and labeled, and are in all respects in proper condition for transport by highway according to applicable international and national government regulations.

If I am a large quantity generator, I certify that I have a program in place to reduce the volume and toxicity of waste generated to the degree I have determined to be economically practicable and that I have selected the practicable method of treatment, storage, or disposal currently available to me which minimizes the present and future threat to human health and the environment; **OR**, if I am a small quantity generator, I have made a good faith effort to minimize my waste generation and select the best waste management method that is available to me and that I can afford.

Printed/Typed Name | Signature | *Month Day Year*

TRANSPORTER

17. Transporter 1 Acknowledgement of Receipt of Materials
Printed/Typed Name | Signature | *Month Day Year*

18. Transporter 2 Acknowledgement of Receipt of Materials
Printed/Typed Name | Signature | *Month Day Year*

FACILITY

19. Discrepancy Indication Space

20. Facility Owner or Operator: Certification of receipt of hazardous materials covered by this manifest except as noted in item 19.
Printed/Typed Name | Signature | *Month Day Year*

EPA Form 8700 - 22 (Rev. 9 - 88) Previous editions are obsolete.

FIGURE 5.6 Sample uniform hazardous waste manifest form.

TSDFs,[9] and/or regulators can track the movement of hazardous waste from the point of generation to the point of ultimate treatment, storage, or disposal. RCRA manifests document the:

- Name, address, and EPA identification number of the generator, the transporter, and the facility where the waste is to be treated, stored, or disposed[10]
- Telephone number at which emergency response information may be accessed at all times (24 hours per day) while the shipment is enroute (*see:* 49 CFR 172, Subparts C and G)
- DOT description, including proper shipping name, of the waste being transported
- Quantities of waste being transported and the type of container

The EPA regulations pertaining to manifests do not cover some DOT requirements that pertain to "shipping papers," particularly with respect to emergency response requirements. The Uniform Hazardous Waste Manifest serves the purpose of a "shipping paper" for hazardous waste shipments. Thus, generators or generator representatives should become fully familiar with the 40 CFR 262 Subpart B and Appendix and 49 CFR 172 Subparts C and G prior to preparation of hazardous waste manifests.

The multiple-copy form is initially completed and signed by the hazardous waste generator. The generator retains Part 6 of the manifest, sends Part 5 to EPA or the appropriate state agency,[11] and provides the remaining parts of the manifest to the transporter. The transporter retains Part 4 of the manifest and gives the remaining parts of the manifest to the TSD facility upon arrival. The TSD facility retains Part 3 and sends Parts 1 and 2 to the generator and the regulatory agency (or agencies), respectively. Throughout this transition, the hazardous waste shipment is generally considered to be in the custody of the last signatory on the manifest.

If the generator does not receive Part 1 of the manifest form from the designated facility within 35 days from the date the waste was accepted by the initial transporter, the generator is required to initiate appropriate tracer activity. If the generator does not receive the Part 1 by the 45th day, the generator must file an *exception report* with the EPA regional office. The report must detail the efforts of the generator to locate the waste. SQGs that do not receive a copy of the signed manifest from the designated facility within 60 days must explain the exception on a copy of the original manifest and send it to the EPA Regional Administrator.

CESQGs are not subject to the EPA manifesting requirements; however, DOT requirements may apply to small hazardous waste shipments.[12] Access to and understanding of the DOT regulations cited above are necessary for all practitioners who may become involved in shipment of hazardous wastes.

---

[9] Treatment, storage, and disposal facility.
[10] It is good practice to also designate the name and address of an alternate TSDF. If for any reason the shipment is not accepted by the primary TSDF, the transporter must return the shipment to the shipper unless an alternate destination has been designated.
[11] Both the U.S. EPA and the state regulatory agency must receive copies in some situations.
[12] Some TSDFs may require partial or fully prepared manifests from SQGs and/or CESQGS.

The hazardous waste shipper, in signing the manifest form, certifies that:

- The shipment has been accurately described and is in proper condition for transport.
- The generator has a waste minimization program in place at its facility to reduce the volume and toxicity of hazardous waste to the degree economically practicable, as determined by the generator.
- The treatment, storage, or disposal method chosen by the generator is the most practicable method currently available that minimizes the risk to human health and the environment.

Biennial Reporting Requirements

Generators are subject to extensive record keeping and reporting requirements as set forth in 40 CFR 262, Subpart D. Generators who transfer hazardous waste off-site must submit a biennial report to the EPA Regional Administrator or state agency on EPA Forms 8700-13A and B[13] by March 1 of each even-numbered year. The EPA does not require SQGs to submit biennial reports; however, some state agencies require annual or biennial reporting. The federal biennial report covers generator activities during the previous year and includes

- EPA ID number, name, and address of the generator
- EPA ID number and name of each transporter used during the year
- EPA ID number, name, and address of each off-site TSDF and recycler to which waste was sent during the year
- Descriptions and quantities of each hazardous waste generated

Records Retention

Generators must keep copies of each signed manifest for 3 years from the date signed, a copy of each exception report filed, each annual or biennial report, and copies of analyses and related determinations made in accord with the generator regulations (40 CFR 262). In addition, copies of each signed manifest must be kept for 3 years or until a signed and dated copy is received from the designated facility. The manifest from the designated facility must be kept for at least 3 years from the date on which the hazardous waste was accepted by the initial transporter. The generator may be required to retain records beyond these specific limits where unresolved enforcement actions are in progress or upon request by the EPA Administrator.

Generators that treat, store, or dispose of their hazardous waste on-site must also "notify" the EPA of hazardous waste activity; obtain an EPA ID number; apply for a permit; and comply with the permit conditions. They too must comply with the federal biennial reporting requirement (U.S. EPA 1998, p. III-51).

---

[13] Some state agencies require submission of *annual* reports.

Exports and Imports of Hazardous Wastes

Export of hazardous waste from the U.S. to another country is prohibited unless:

- Notification of intent to export has been provided to EPA at least 60 days in advance of shipment.
- The exporter obtains written consent from the receiving country prior to shipment. This written consent must be attached to the manifest accompanying the shipment.
- The exporter has received a copy of the the EPA "Acknowledgement of Consent."
- The hazardous waste shipment conforms to the terms of the receiving country's consent (40 CFR 262.52).

Exporters should consult with state or EPA Regional officials regarding the present status of U.S. participation in the various international conventions, treaties, and other applicable trade agreements before engaging in hazardous waste export activity.

Any person who imports hazardous waste from another country into the U.S. must comply with the requirements of 40 CFR 262, i.e., the importer becomes the generator, for RCRA regulatory purposes, including responsibility for preparation of the manifest. Special instructions for importers completing the manifest for imported wastes are contained in Subpart F of 40 CFR 262 (see also: EPA 1998, pp. III-48-53; Ostler 1998, Chapter 3; Assante-Duah and Nagy 1998, Chapters 4, 5, and 6).

GENERATOR RESPONSIBILITIES FOR RESTRICTED WASTE MANAGEMENT

The land disposal restrictions (LDRs) and the implementing regulations are lengthy, detailed, and complex. In the interest of a concise overview, much of the following material on generator responsibilities is paraphrased from the 1998 RCRA Orientation Manual (EPA 530/R/98/004). The reader having or preparing for responsibilities for managing hazardous wastes that may be land-disposed should carefully study 40 CFR 268. The major elements of the LDR program are summarized in Chapters 4, 5, 7, and 11 of this text, in accord with the major topics of those chapters.

Generator Requirements

The generator must determine whether or not the waste generated is hazardous (either listed, characteristic, or both) by conducting waste analysis or by using knowledge of the waste. In order to prevent illegal dilution, the determination is made at the point of generation. If the generator determines that the waste is hazardous, the generator must then determine whether or not the EPA has established a treatment standard for the particular waste code. If so, the waste is restricted and the generator must manage it in compliance with all of the LDR requirements, whether on-site or off-site, and may not dispose of it on or beneath the land until it meets all applicable treatment standards. These standards may be either concentration levels for hazardous constituents that the waste must meet or treatment technologies that must be performed on the waste prior to disposal. As discussed in Chapter 4, the treatment standards are based on the performance of "Best Demonstrated Available Technology" (BDAT). The required "treatment" may confront the generator/SQG with the choice of obtaining a Part B permit or transporting the waste to a permitted facility for treatment. The delegated state agency or EPA Regional Office should be consulted if the §§ 264 and 265 requirements are in doubt.

The treatment standards are found in 40 CFR 268.40 and are expressed as a concentration or as a treatment technology. If the treatment specified for a waste is provided in the form of a concentration, the regulated entity may use any treatment technology that will achieve the concentration limit. If a treatment technology (a five-letter code found in § 268.42) is specified, that treatment technology must be used for treatment of the waste.

Treatment standards for most characteristic hazardous wastes require rendering the waste nonhazardous by removing ("de-characterizing") the constituent(s) that render the waste "characteristic." The generator or other regulated entity must examine the characteristic waste for *underlying hazardous constituents*. The underlying hazardous constituent(s) in a waste may not be the cause of the exhibited characteristic, but they are hazardous and must be treated to meet contaminant-specific concentrations. These concentrations are referred to as *universal treatment standards* (UTS). The UTS are listed in § 268.48. Characteristic hazardous wastes that have been de-characterized, and the underlying constituents treated to UTS concentrations, can be landfilled in a nonhazardous waste facility.

Several categories of exemptions, time extensions, and variances are available to the generator to enable disposal of LDR wastes under special circumstances. These are discussed in Chapters 7 and 11. The generator must comply with the record keeping requirements pertaining to operations subject to the LDR treatment standards (*see:* 40 CFR 268.7). The land disposal restrictions and the implementing regulations do not apply to conditionally exempt SQGs.

TOPICS FOR REVIEW OR DISCUSSION

1. Why are regulatory agencies (and their regulatory issue) so concerned with limiting the time over which hazardous wastes may be stored or accumulated?
2. What is the main purpose of the Hazardous Waste Manifest system?
3. How do the RCRA regulations apply the descriptors "hazardous waste" and "acutely hazardous waste" in making the distinctions between "conditionally exempt small quantity generator," "small quantity generator," and "generator?"
4. Why is it important to assign standardized shipping names for hazardous materials being shipped in commerce?
5. To what publication should one refer to determine the correct shipping name for a hazardous waste?
6. RCRA regulations (40 CFR 262.34) allow a small quantity generator to accumulate waste on-site for 180 days (or 270 days if the waste must be transported more than 200 mi to the treatment, storage, or disposal site).

What regulatory circumstance would cause the storage period to be automatically shortened?

7. The operator of a satellite accumulation point, as described by the RCRA regulations (40 CFR 262.34), accumulates U005 waste in excess of 1 quart and fails to begin accumulating the waste in another container. Is the facility now out of compliance? Explain.

8. A "small quantity generator" is not *required* to prepare and implement a contingency plan. What emergency planning is required?

REFERENCES

Assante-Duah, D. Kofi and Imre V. Nagy. 1998. *International Trade in Hazardous Waste.* E. & FN Spon-Routledge, NY.

Bloomer, Dorothy. 2000. Department of Transportation Hazardous Materials Regulations, in *Hazardous Materials Management — Desk Reference,* Doye B. Cox, Ed., The Academy of Certified Hazardous Materials Managers. McGraw-Hill, NY, Chapter 20.

Dawson, Gaynor W. and Basil W. Mercer. 1986. Hazardous Waste Management. John Wiley & Sons, NY.

Nebel, Bernard J. and Richard T. Wright. 1993. *Environmental Science the Way the World Works — Fourth Edition.* Prentice-Hall, Englewood Cliffs, NJ.

Ostler, Neal K. 1998. Generators of Hazardous Waste, in *Waste Management Concepts.* Neal K. Ostler and John T. Nielsen, Eds., Prentice-Hall, Inc., Upper Saddle River, NJ.

Phifer, Russell W. and William R. McTigue, Jr. 1988. *Handbook of Hazardous Waste Management for Small Quantity Generators.* Lewis Publishers, Chelsea, MI.

U.S. Department of Transportation. 1992. Intermodal Transportation of Hazardous Materials for Industry — Student Workbook. Transportation Safety Institute, Oklahoma City, OK.

U.S. Environmental Protection Agency. 1985. Does Your Business Produce Hazardous Waste? Office of Solid Waste and Emergency Response, Washington, D.C., EPA 530-SW-010.

U.S. Environmental Protection Agency. 1986. Understanding the Small Quantity Generator Hazardous Waste Rules: A Handbook for Small Business, Office of Solid Waste and Emergency Response, Washington, D.C., EPA 530-SW-86-019.

U.S. Environmental Protection Agency. 1995. Generation and Management of Conditionally Exempt Small Quantity Generator (CESQG) Waste, Office of Solid Waste and Emergency Response, Washington, D.C., EPA 530-R-95-017.

U.S. Environmental Protection Agency. 1998. RCRA Orientation Manual, Office of Solid Waste, Washington, D.C., EPA 530-R-98-004.

U.S. Environmental Protection Agency. 1999a. Toxics Release Inventory Reporting and the 1997 Public Data Release, Office of Pollution Prevention and Toxics, Washington, D.C., EPA 749-B-99-003.

U.S. Environmental Protection Agency. 1999b. The Biennial RCRA Hazardous Waste Report (Based on 1997 Data) Executive Summary, Solid Waste and Emergency Response, Washington, D.C., EPA 530-S-994-036.

U.S. Environmental Protection Agency. 1999c. RCRA, Superfund & EPCRA Hotline Training Module Introduction to Generators, Office of Solid Waste and Emergency Response, Washington, D.C., EPA 530-R-99-048.

APPENDIX A
Examples of Hazardous Wastes Produced by Basic Industries

Industry	Wastes produced
Chemical manufacturing	Spent solvents and still bottoms

Chemical manufacturing — Spent solvents and still bottoms

White spirits, kerosene, benzene, xylene, ethyl benzene, toluene, isopropanol, toluene diisocyanate, ethanol, acetone, methyl ethyl ketone, tetrahydrofuran, methylene chloride, 1,1,1-trichloroethane, trichloroethylene

Ignitable wastes not otherwise specified (NOS)

Strong acid/alkaline wastes

Ammonium hydroxide, hydrobromic acid, hydrochloric acid, potassium hydroxide, nitric acid, sulfuric acid, chromic acid, phosphoric acid

Other reactive wastes

Sodium permanganate, organic peroxides, sodium perchlorate, potassium perchlorate, potassium permanganate, hypochlorite, potassium sulfide, sodium sulfide

Emission control dusts and sludges

Spent catalysts

Ignitable paint wastes

Ethylene dichloride, benzene, toluene, ethyl benzene, methyl isobutyl ketone, methyl ethyl ketone, chlorobenzene

Ignitable wastes not otherwise specified (NOS)

Construction — Spent solvents

Methyl chloride, carbon tetrachloride, trichlorotrifluoroethane, toluene, xylene, kerosene, mineral spirits, acetone

Strong acid/alkaline wastes

Ammonium hydroxide, hydrobromic acid, hydrochloric acid, hydrofluoric acid, nitric acid, phosphoric acid, potassium hydroxide, sodium hydroxide, sulfuric acid

Metal manufacturing — Spent solvents and solvent still bottoms

Tetrachloroethylene, trichloroethylene, methylene chloride, 1,1,1-trichloroethane, carbon tetrachloride, toluene, benzene, trichlorofluoroethane, chloroform, trichlorofluoromethane, acetone, dichlorobenzene, xylene, kerosene, white spirits, butyl alcohol

Strong acid/alkaline wastes

Ammonium hydroxide, hydrobromic acid, hydrochloric acid, hydrofluoric acid, nitric acid, phosphoric acid, nitrates, potassium hydroxide, sodium hydroxide, sulfuric acid, perchloric acid, acetic acid

Spent plating wastes

Heavy metal wastewater sludges

Cyanide wastes

Ignitable wastes not otherwise specified (NOS)

Other reactive wastes

Acetyl chloride, chromic acid, sulfides, hypochlorites, organic peroxides, perchlorates, permanganates

Used oils

Paper industry — Halogenated solvents

Carbon tetrachloride, methylene chloride, tetrachloroethylene, trichloroethylene, 1,1,1,-trichloroethane, mixed spent halogenated solvents

APPENDIX A   *(Continued)*
Examples of Hazardous Wastes Produced by Basic Industries

Industry	Wastes produced
	Corrosive wastes
	Corrosive liquids, corrosive solids, ammonium hydroxide, hydrobromic acid, hydrochloric acid, hydrofluoric acid, nitric acid, phosphoric acid, potassium hydroxide, sodium hydroxide, sulfuric acid
	Paint wastes
	Combustible liquid, flammable liquid, ethylene dichloride, chlorobenzene, methyl ethyl ketone, paint waste with heavy metals
	Solvents
	Petroleum distillates

Source: EPA (1985).

APPENDIX B
DOT Eight-Step Procedure for Preparation of Hazardous Material Shipments

	Reference
1. Determine proper shipping name, hazardous class/division, ID number, and packing group.	172.101(2), (3), (4), (5)
2. Is this material regulated by 49 CFR?	172.101(2), (1) and Appendix to 172.101.
• As a hazardous material?	Col. (1), Col. (7)
• As a hazardous substance?	
• By highway mode?	
• As a poison inhalation hazard?	
3. Determine proper packaging.	
• Determine if an exception is authorized for the particular hazardous material.	172.101(8)(A) and reference to sections indicated
• If no exception is authorized, determine the specific packaging requirements.	172.101(B), (C) and reference to sections listed
• Determine the maximum net quantity of the hazardous material that may be shipped in one package by passenger-carrying and/or cargo-only aircraft as appropriate.	172.101(9)(A), (B)
• Ensure that completed package meets general packaging requirements.	173.24, 173.24a, 173.24b
• Determine special provisions.	172.101(7)
4. Mark the package.	Subpart D of 172 commencing at 172.300
5. Label the package.	
• With appropriate table label(s) unless excepted	172.101(6)
• With appropriate additional or multiple labeling requirements.	172.402, 404, 406
6. Prepare shipping papers with shipper's certification and signature.	172.200, 201, 202, 203, 204
7. Provide emergency response information.	Subpart G of 172 commencing at 172.600
8. Provide a placard as appropriate.	Subpart F of 172 commencing at 172.500

6 Transportation of Hazardous Wastes

OBJECTIVES

At completion of this chapter, the student should:

- Understand the advantages/disadvantages of the modes of transportation of hazardous wastes.
- Be familiar with the requirements for action by transporters of a release during transportation.
- Understand the general nature of the regulations imposed upon transport of hazardous wastes by RCRA and DOT regulations and their relationship to each other.
- Be aware of the regulatory complexities associated with transfrontier movement of hazardous wastes.

INTRODUCTION

Activity associated with transportation of hazardous wastes from the generator or source to intermediate destinations and to final disposition has been fraught with mismanagement of the wastes, especially during the early years of the Resource Conservation and Recovery Act (RCRA). During those years, and before, there were transportation incidents which frequently involved major threats to the environment and public safety. In pre-RCRA times, small locally based refuse haulers provided immediate and cheap removal of hazardous waste accumulations on a "no questions asked" basis. The most marginal of trucking operations could survive by removing unwanted wastes and disposing of them with abandon.

In the late 1970s/early 1980s period, as the RCRA became viable, many hazardous waste generators sought to avoid the new financial burdens of lawful waste management by hasty disposal of accumulated wastes. During this period, "midnight dumping" schemes became common. Truckers outfitted tankers with dumping valves so that liquid wastes could be dumped "on-the-run." Trailers loaded with drums of wastes were simply abandoned in random locations. Rural areas and deserts were littered and stained with all manner of hazardous wastes.

As RCRA regulations were implemented and the manifest system began to function, these practices were brought under control of the respective authorities.

As the regulatory agencies gained recognition and experience, most of the marginal transporters were "weeded-out" and transportation became a vital link in the cradle-to-grave management strategy. Although illegal transportation activities have continued to require the attention of law enforcement agencies, much of the regulatory focus has shifted to accident prevention, emergency response activity, surveillance of import-export activity, and tracking of wastes from source to ultimate disposition.

In the previous chapter, we covered the *pretransport* requirements with which *generators* of hazardous wastes must comply. In this chapter, we will overview basic hazardous waste transportation operations and cover the pertinent regulatory structures of the Environmental Protection Agency (EPA) and the Department of Transportation (DOT), as they pertain to *transporters* of hazardous wastes and hazardous materials. Statistical data on transportation of hazardous *waste* are difficult to obtain. In the following discussions it is frequently necessary to generalize in terms of hazardous *materials* that, in the DOT lexicon, include hazardous *wastes* (*see:* Glossary concerning these terms; Fox 2000, pp. 54, 67, 113; Eckmyre 2000, pp. 197ff).

MODES AND SCOPE OF HAZARDOUS WASTE TRANSPORTATION

A 1981 report prepared for the EPA estimated that 96% of the 264 million tons of hazardous wastes generated each year were disposed of at the site where they were generated and that most of the hazardous waste shipped off-site was transported by truck (Westat, Inc. 1981).[1] These shipments were usually over routes of 100 mi or less (ICF, Inc. 1984, p. 2). By 1989, the National Solid Wastes Management Association (NSWMA) stated that trucks traveling over public highways move 98% of the hazardous waste that is treated off-site. Rail freight moved the remainder (NSWMA 1989, p. 11). By 1993, the EPA counted 20,800 transporters of hazardous waste (EPA 1993) and by 1997 the transporter count had declined to 18,029 (EPA 1999). In 1997 trucks moved 98.6% of hazardous waste shipments, but only 80% of the tons moved. Rail shipments carried 1.4% of the shipments, but nearly 20% of the tons moved (U.S. DOT 1998, Table 2).

The decline in numbers of transporters is consistent with the apparent decline in hazardous waste treated off-site during the same general period. The EPA discounts the possibility that the decline in numbers of transporters reporting is due to the exclusion of wastewaters from the 1997 biennial reporting (discussed in Chapter 5), since most wastewaters are treated on-site. Shipments of hazardous waste by inland waterways and by air is infrequent and is not considered here.

Another important perspective can be gained from the statistics for hazardous *materials* transportation. A 1986 report stated that rail transportation moved about 8% of hazardous materials shipped, but 57% of the *ton-miles* of hazardous materials shipped (U.S. Office of Technology Assessment 1986, p. 46). By 1996, trucks moved 94% of the hazardous materials shipments, but only 35% of the tons moved. Rail shipments carried only 0.5% of the shipments, but 10.5% of the tons moved. Pipeline,

---

[1] Reliable current statistics on quantities treated on-site are difficult to obtain. In general, ever-tightening regulatory control, liability concerns, and availability of commercial treatment options have tended to encourage shipment of wastes off-site for treatment. Conversely, accelerating Superfund and RCRA site remediation activities involve more on-site treatment of hazardous wastes.

water, and air shipments accounted for the remainders of hazardous *materials* moved (U.S. DOT 1998, Table 2). The student or practitioner should have clearly in mind the fact that most "hazardous materials" become hazardous *wastes* when released to the environment. At that occurrence, the hazardous waste regulations of RCRA, CERCLA, and the state and local jurisdictions apply.[2]

Enactment of the Hazardous and Solid Waste Amendments of 1984 (HSWA) brought more than 100,000 new small quantity generators (SQGs) under regulation. Most of the SQGs have had no alternative to shipment of their hazardous wastes off-site for disposition. Thus, HSWA may have instigated some increase of small shipments and increased mixing of wastes. The addition of 25 new chemical constituents to Table 1, 40 CFR 261.24, in 1990 is said to have brought 17,000 new generators under RCRA regulation. In 1993, the EPA counted 266,000 generators of which approximately 240,000 were SQGs (EPA 1993). Although the 1997 generator count has declined slightly, very large numbers of SQGs have no options other than to ship hazardous wastes off-site for ultimate disposition; very large numbers of shipments involve small quantities and originate at small operations not accessible by rail; transportation of hazardous wastes is a major waste management activity and a major source of potential incident/release/exposure concerns.

Highway Shipment of Hazardous Wastes

As noted above, most hazardous waste transportation is accomplished via truck. Since implementation of RCRA regulations, most waste haulers generally fit one of three categories:

- Generators transporting their wastes to treatment, storage, and disposal (TSD) facilities
- Contract haulers collecting wastes from generators and transporting the wastes to TSD facilities
- TSD facilities collecting wastes from generators and transporting the wastes to their facilities

The highway transport mode is regarded as the most versatile. Tank trucks can access most industrial sites and TSD facilities, while rail shipping requires expensive sidings and is suitable only for very large quantity shipments. Cargo tanks are the main carriers of bulk hazardous *materials* (U.S. DOT 1998, Table 2); however, large quantities of hazardous *wastes* are shipped in 55-gal drums carried in nonbulk and less-than-truckload (LTL) shipments.

Because of the huge liabilities that can accrue from accidental release, improper handling, and illegal disposal, generators are said to prefer maintaining control of their wastes by transporting them in company-owned/operated trucks. Alternatively, generators may feel more confident that their wastes will reach the

---

[2] As discussed later in this chapter, DOT reporting requirements of 49 CFR 171.15–171.16 are also activated.

designated TSD facility if picked up and transported by that facility's trucking operation (Wentz 1989, p. 268). These considerations probably have much to do with the decline in numbers of small locally based waste-hauling firms.

Cargo tanks are usually made of steel or aluminum alloy, but can be constructed of other materials such as titanium, nickel, or stainless steel. They range in capacity from about 2000 to more than 9000 gal, depending upon road weight laws and the properties of the materials to be transported. Road weight laws usually limit motor vehicle weights to 80,000 lb gross; however, some states allow greater gross weights. The DOT specifications for cargo tanks used in bulk shipment of the common types of hazardous materials carried and example cargos are listed in Table 6.1. Figure 6.1 shows a DOT specification MC-306 tank trailer used for hauling combustible and flammable wastes. Figure 6.2 is an example of an MC-310 tank trailer for hauling corrosive wastes. These specification cargo tanks have been superseded by new specifications DOT 406, 407, and 412. However, the earlier specifications can continue in use after required modifications. The user must stay current with respect to 49 CFR 178 Continuing Qualifications requirements and schedules.

Railway Shipment of Hazardous Wastes

As noted earlier, rail shipments account for about 10.5% of the tonnage of hazardous *materials* transported annually, with about 4300 carloads shipped daily. The portion of these shipments that are hazardous *wastes* is included in the "other" category and

TABLE 6.1
Cargo Tank Table

Cargo Tank Specification Number	Types of Commodities Carried	Examples
MC-306, DOT 406 (MC-300, 301, 302, 303, 305)	Combustible and flammable liquids of low vapor pressure	Fuel oil Gasoline
MC-307, DOT 407 (MC-304)	Flammable liquids, Poison B materials with moderate vapor pressure	Toluene Diisocyanate
MC-312, DOT 412 (MC-310, 311)	Corrosives	Hydrochloric acid, caustic solution
MC-331, DOT 431 (MC-330)	Liquefied compressed gases	Chlorine, anhydrous ammonia, propane, butane
MC-338, DOT 438	Refrigerated liquified gases	Oxygen, methane

Note: The number in parentheses designates older versions of the specification; the older versions may continue in service until required phase-out, but all newly constructed cargo tanks must meet current specifications.

Source: 49 CFR 172.101 and 178.315-178.343.

FIGURE 6.1 DOT specification MC-306 tank trailer for transporting combustible and flammable materials.

FIGURE 6.2 DOT specification MC-310 tank trailer for transporting corrosive materials.

is estimated by DOT at 1.4% of shipments and 20% of the hazardous materials tonnage. Rail tank car specifications for transportation of pressurized hazardous materials are DOT 105, 112, and 114; for unpressurized shipments the numbers are DOT 103, 104, 111, and 115 (49 CFR173, Subpart F). Capacities for tank cars carrying hazardous materials are limited to 34,500 gal or 263,000 lb gross weight (49 CFR 179).

Accidents/Incidents Involving Hazardous Waste Shipments

Accident and transportation release statistics from the late 1970s and early 1980s provide insight regarding the relative hazards posed by the highway and rail modes

TABLE 6.2
Transportation Incidents Involving Hazardous Wastes — Rail and Highway

	Incident		Accidents/Derailments		Deaths	
Year	Hazwaste	Hazmat	Hazwaste	Hazmat	Hazwaste	Hazmat
1989	149	7,558	15	342	0	8
1990	194	8,883	10	299	0	8
1991	202	9,110	13	303	0	10
1992	413	9,351	17	284	0	15
1993	575	12,815	7	264	0	15
1994	547	16,087	NA	295	0	11
1995	676	14,743	NA	294	0	7
1996	458	13,950	NA	332	1	120
1997	419	13,995	NA	310	0	12
1998	421	15,343	NA	316	0	13

Note: "Incidents" do not equate with "accidents." A release incident can occur without an accident and, conversely, an accident can occur without a release; NA = not available.

Source: 1989–1993 data, Chemical Waste Transportation Institute; 1994–1998 data, U.S. DOT 1999.

of hazardous *materials* transportation. These data indicate that highway transport experienced 12 times the number of incidents involving hazardous materials, 4 times the number of fatalities, and 2 times the number of injuries as occurred in rail transport. However, rail accidents released approximately 50% greater quantities than did highway accidents involving hazardous *materials* (Blackman 1985, Chapter 2). Total transportation incidents, involving hazardous wastes, show significant increases from 1989 to 1995, followed by a reversal of the trend (Table 6.2). However, numbers of hazardous waste incidents involving accidents or derailments show no particular trends. One transportation incident-related death was reported during the period. Hazardous *materials* incidents are shown for comparison.

Hazardous materials transportation incidents tend to be spectacular, dangerous, freakish, and unpredictable (Figure 6.3). Rail accidents, as noted, involve containers

FIGURE 6.3 Highway transportation "incident."

FIGURE 6.4 Rail transportation "incident." (Courtesy of Envirosafe Services of Ohio, Inc. (ESOI).)

of up to 34,500 gal or 130 tons vs. the 9000-gal/40-ton limits for highway transportation. The greater quantity per container, chemical incompatibilities between rail tank car shipments, and the difficult accessibility encountered in rural locations leads to unmanageable fires which are frequently allowed to "burn themselves out" (Figures 6.4 and 6.5).

Incidents involving truck shipment of hazardous materials, when they occur in urban areas, are more likely to endanger human lives and property (Figure 6.6). Fires in populated areas typically must be controlled expeditiously in order to limit

FIGURE 6.5 Rail transportation "incident" morning after. (Courtesy of Envirosafe Services of Ohio, Inc. (ESOI).)

FIGURE 6.6 Highway transportation "incident" illustrating threat to urban areas.

exposure and property damage. (For a detailed and robust treatise on risk and risk assessment of hazardous materials transportation, by modes, *see:* Nicolet-Monnier and Gheorghe 1996, Chapters 3, 4, and 5).

REGULATORY STRUCTURES

In general, the DOT regulations deal with container and equipment specifications, packaging, categorization of wastes, and the determination of proper shipping descriptions. The EPA regulations provide the tracking mechanisms that are intended to maintain the cradle-to-grave management system.

Department of Transportation Regulations

As noted in Chapter 5, the Department of Transportation regulations dealing with transportation of hazardous materials are found at 49 CFR 171 thru 179 and are referred to as the HM 181,[3] "Performance-Oriented Packaging Standards." The content and detail of these regulations greatly exceed the scope of these chapters, but the general thrust can be understood by examining the column headings of the 49 CFR 172.101 Hazardous Materials Table (see Figure 5.1) and Appendix B of

---

[3] The Department of Transportation, as do other agencies, assigns a "docket number" to new regulatory proposals. The proposed regulations are referred to by the docket number throughout the promulgation process. Upon final publication of the rule package, DOT continues to refer to the implemented program by that number. Thus, the implementation of the Performance-Oriented Packaging Standards continues to be referred to as HM 181.

Chapter 5. The elements pertaining to transporters of hazardous materials focus on emergency response information and requirements, training of the "hazmat employee,"[4] and specialized training for drivers.

The transporter must maintain the emergency response information contained on the manifest in a manner that ensures that it is immediately accessible to emergency responders. For example, drivers of cargo tank vehicles must keep the manifest on the seat adjacent to the driver's seat or in the "pocket" of the door on the driver's side of the cab. Similar requirements apply to train crews and bridge personnel on vessels. If the transporter makes use of a transfer facility, the emergency response information must be maintained in a location that is immediately accessible to the personnel operating the facility.

Transportation Incidents Involving Hazardous Wastes

DOT immediate notification requirements for hazardous materials incidents are applicable to discharges of hazardous wastes. Notice is given by immediate telephonic report to the National Response Center (NRC), operated by the U.S. Coast Guard (800-424-8802). Specifically, the National Response Center must be notified when:

- A person is killed or injured to the extent that hospitalization is required, or
- Estimated carrier and/or property damage exceeds $50,000, or
- There is an evacuation of the general public lasting for 1 hour or more, or
- One or more transportation arteries or facilities are closed or shut down for 1 hr or more, or
- The operational flight plan of an aircraft is altered (49 CFR 171.15).
- Fire, breakage, spillage, or suspected radioactive contamination involving shipment of radioactive material (*see also:* 49 CFR 174.45, 176.48, and 177.807), or
- Fire, breakage, or spillage, or suspected contamination occurs involving shipment of infectious substances (etiologic agents), or
- There has been a release of a marine pollutant in a quantity exceeding 450 L (119 gal) for liquids or 400 kg (882 lb) for solids, or
- A situation exists such that in the judgment of the carrier, it should be reported to the NRC even though it does not meet the above criteria (49 CFR 171.15). Additionally, persons in charge of facilities (including transport vehicles, vessels and aircraft) must report any release of a hazardous *substance* in a quantity equal to or greater than its reportable quantity (40 CFR 302.6).[5]

A detailed written report is also required for all incidents for which a telephonic notice has been made, as well as for any time there is an unintentional release of a hazardous material during transportation (including loading, unloading, and tempo-

---

[4] Similarly, the training requirement of 49 CFR 172, Subpart H, is referred to as "the HM 126F training."
[5] More on reporting required by CERCLA may be found in Chapter 10.

rary storage related to transportation). The carrier must submit this report on DOT Form 5800.1 within 30 days of the date of discovery of the incident (49 CFR 171.16).

The HM 126F "hazmat employee" training requirement is prescribed in 49 CFR 172, Subpart H. The training is required for any employee who performs any function having to do with the safety of a hazardous material shipment (*see:* Glossary or 49 CFR 171.8 for the definition of "hazmat employee" and "hazmat employer"). The required training consists of three categories, which are

1. *General awareness/familiarization training* — the hazards associated with hazardous materials transportation, the hazard classes of HM 181, and hazard communication requirements
2. *Function-specific training* — the packaging, labeling, marking, and placarding of hazardous materials shipments, i.e., the Performance Oriented Packaging Standards
3. *Safety training* — the emergency response, personnel protective clothing and equipment, and methods and procedures for avoiding accidents and exposure

The standards also include driver training requirements and specialized training for drivers of vehicles transporting explosives, radioactive materials, or cryogenic gases. The hazmat employee must repeat the training at 2-year intervals; drivers must be trained annually (*see also:* Munter et al. 1995, pp. 207–208; Eckmyre 2000, pp. 197–208).

RCRA Regulations for Hazardous Waste Transporters

The RCRA transporter regulations (40 CFR 263) define "transporter," provide the tracking mechanisms that are intended to maintain the cradle-to-grave management systems for hazardous waste management, and impose cleanup and reporting requirements that apply in the event of the discharge of hazardous waste(s) during transport.

The transporter is defined as any person engaged in the off-site transportation of hazardous waste within the U.S. if such transportation requires a manifest.[6] This definition covers transportation by air, highway, rail, or water. The transporter regulations do not apply either to the on-site transportation of hazardous waste by generators who have their own treatment or disposal facilities, nor to TSD facilities transporting wastes within a facility (40 CFR 260.10 and Glossary). Both generator and treatment, storage, and disposal facilities (TSDF) owners and operators must heed the specific definition of the term "on-site" (*see:* Glossary, this text; 40 CFR 260.10.) As noted in Chapter 5, § 260.20(f) was added to Subpart B, allowing transportation of hazardous waste on a public highway within or along the border of contiguous property under control of the same person, even if such contiguous property is divided by a public or private right-of-way (62 FR 6651, February 12

---

[6] In 1998, the DOT extended application of the Hazardous Materials Regulations to include intrastate transportation of all hazardous materials in commerce.

1997). In those circumstances, the generator or transporter must comply with the §§ 263.30 and 263.31 requirements for cleanup of hazardous waste discharges.

Under some circumstances, transporters can become subject to the generator regulations by importing hazardous waste into the U.S. or by mixing hazardous wastes of different DOT shipping descriptions by placing them into a single container (40 CFR 263.10). As discussed later in this chapter, the transporter of hazardous waste is also responsible for cleanup of a discharge of hazardous wastes or commercial chemical product that occurred during transport (40 CFR 263.30).

A transporter may store hazardous wastes at a transfer station for up to 10 days without being subject to RCRA regulations other than those applicable to transporters. If the storage time exceeds 10 days, the transporter becomes a storage facility and must comply with the regulations pertaining to such a facility, including the requirements for obtaining a permit.

Transporters must comply with RCRA Subtitle C regulations that require:

- Obtaining an EPA Identification Number
- Complying with the manifest system
- Handling hazardous waste discharges

EPA ID Number

As discussed in the previous chapter, an EPA ID number is essential to the EPA and the primacy states in tracking transporter activity. Without this unique number, the transporter is forbidden to handle hazardous waste. Moreover, a transporter may not accept hazardous waste from an SQG or Generator nor transfer hazardous waste to a TSD facility unless they have EPA ID numbers.[7] A transporter obtains an ID number by "notifying" the EPA of hazardous waste activity, using the standard EPA notification form (EPA 8700-12).

The Manifest

The RCRA Subtitle C regulations prohibit transporters from accepting hazardous waste shipments from shippers without a manifest.[8] The function of the manifest system was described in the previous chapter. The transporter who accepts manifested hazardous wastes is required to sign and date the manifest and return a signed copy to the generator or previous transporter. The transporter is responsible for the shipment until the manifest is signed by the owner or operator of the receiving facility.

The transporter is required to deliver the entire quantity of waste accepted from either the generator or another transporter to the facility listed on the manifest or to the alternate facility if one is listed on the manifest. If the waste cannot be delivered as the manifest directs, the transporter must inform the generator and receive further instructions. The transporter must have the owner or operator of the TSD sign and date the manifest at the time of delivery to the TSD. The transporter retains Copy

---

[7] Transporters may, however, accept hazardous waste shipments from CESQGs.
[8] An exception is the provision for delivery of reclaimed wastes from SQGs [see: 40 CFR 263.20(h)].

4 of the manifest and gives the remaining three parts of the manifest to the TSD owner or operator. The transporter must retain a copy of the manifest for three years from the date the hazardous waste is accepted by the *initial transporter*.

Handling Hazardous Waste Discharges

In the event of a discharge of hazardous waste during transport, special requirements established by the EPA and DOT must be followed by the driver-transporter. A discharge of hazardous waste is defined as: "the accidental or intentional spilling, leaking, pumping, pouring, emitting, emptying, or dumping of hazardous waste into or on any land or water" (40 CFR 260.10).

EPA and DOT regulations pertaining to hazardous waste release events include provisions authorizing federal, state, or local government officials, acting within the scope of their official duties, to permit the immediate removal of hazardous wastes by transporters who do not have EPA ID numbers and without a manifest. Within 15 days following the incident, the transporter must obtain a temporary ID number and file a report, including a manifest, with DOT.[9] The EPA has also exempted all persons involved in treatment or containment actions taken during an immediate response to the discharge of hazardous wastes or materials from permitting requirements. All regulations for the final disposition of wastes must be followed after the emergency has been concluded.

EPA regulations similarly require transporters to clean up any discharges that occur during transport or take actions required or approved by appropriate government officials to mitigate human health or environmental hazards (40 CFR 263, Subpart C). Such cleanups are characteristically hazardous, not only to those doing the cleanup work, but to nearby residents, other users of the transportation system, and to the environment. It is rarely possible to achieve a totally satisfactory cleanup. Liquid wastes, liquid-borne solid wastes, and water from fire-fighting operations is often dispersed through storm drains and the soil, to the extent that it cannot be retrieved. Atmospheric releases are rarely controlled in a timely manner (*see also:* U.S. EPA 1998, Chapter 4).

Great advances have been made in response techniques and in cleanup procedures, but much more effort is needed in prevention of transportation-related hazardous material/waste releases. Figures 6.7 through 6.10 provide some sequences of transportation-related cleanup operations. (*See also*: Roberts 1985; Wentz 1989, pp. 263–267; Munter et al. 1995, Chapter 8; Eckmyre 2000, pp. 170–208).

Import/Export Activity

International movement of hazardous wastes is a matter of growing interest, complexity, and concern to responsible officials in the U.S., Mexico, Canada, and other nations. The transporter or shipper involved in or contemplating involvement in import/export of hazardous wastes should thoroughly examine 40 CFR 262, Subparts E and F. If transporting wastes to the Organization for Economic Cooperation and

---

[9] Immediate telephonic reporting is required under the circumstances outlined earlier in this chapter.

FIGURE 6.7 Release from a transportation "incident." (From the Arizona Department of Environmental Quality.)

FIGURE 6.8 Cleanup following a transportation "incident." (From the Arizona Department of Environmental Quality.)

Development (OECD) member countries, the provisions of Subparts E and F do not apply, and the shipper and transporter must carefully adhere to 40 CFR 262, Subpart H. The text *International Trade in Hazardous Wastes* by Assante-Duah and Nagy (referenced at the end of this chapter) is an excellent source for in-depth understanding of the intricacies of hazardous waste import/export regulation and control. Assante-Duah and Nagy (1998, p. 71) list several "stimuli" for transfrontier movements of wastes:

FIGURE 6.9 Cleanup following a transportation "incident." (From the Arizona Department of Environmental Quality.)

FIGURE 6.10 Release from a transportation "incident." (From the Arizona Department of Environmental Quality.)

- Overall stringent environmental laws in most industrialized countries
- Tightening of specific laws, regulations, and policies concerning disposal of certain types of wastes
- Rising costs of hazardous waste disposal in the home country where wastes are generated
- Diminishing domestic capacity for disposal of certain types of wastes
- Potential future liability for any damages caused by wastes disposed of domestically
- Market opportunities elsewhere for materials which can be recovered, reclaimed, or recycled from wastes otherwise destined for "final" disposal
- General economic growth which may result in more total generation of wastes
- Economies of scales associated with treatment, storage, and disposal facilities (TSDFs)
- Existence of TSDFs which may serve several countries
- Existence of an appropriate TSDF in a foreign country which is closer than a similar facility in the home country
- Lack of environmentally sound TSDFs in some waste-producing countries

Documented shipments to and from Mexico and Canada are a small but growing fraction of the quantities generated and managed in each nation. Table 6.3 summarizes those quantities for the years 1987 through 1995. The EPA has not completed compilation of data for subsequent years.

Import-export of hazardous waste to and from the U.S. is affected by a number of international treaties, bilateral agreements, prohibitions, etc. The scope of this chapter does not permit detailed explanation of these instruments; however, brief summaries of the content and implementation of two of the most important follow.

Basel Convention. The Basel Convention on the Control of Transboundary Movements of Hazardous Waste and Their Disposal was signed initially by 33 countries in 1989 and by more than 90 countries by 1996. Parties to the Basel Convention may not trade wastes covered by the convention with "nonparties" in the absence of a bilateral or multilateral agreement to govern transboundary movements. The U.S. is not presently a signator of the Basel Convention; however, U.S importers and exporters are subject to the conditions of bilateral agreements with four countries that are parties to the Basel Convention. Bilateral agreements covering import and export of hazardous wastes have been negotiated with Mexico and Canada. The U.S has bilateral agreements for import of hazardous wastes with Malaysia and Costa Rica.

Transfrontier Shipments of Hazardous Waste for Recovery within the OECD. The OECD "Council Decision" reached in 1992 by the 29 member nations identifies an extensive array of wastes that are subject to a graduated system of controls when wastes destined for recovery are moved across the borders of member nations. The agreement is intended to control trade in recyclables and minimize the possibility of abandonment or dumping of hazardous wastes. Transborder movement

TABLE 6.3
Transborder Shipment of Hazardous Waste — U.S., Mexico, and Canada (Metric Tons)

Year	From Mexico*	To Mexico*	From Canada**	To Canada**
1987		10,710	43,203	129,476
1988	990	15,615	66,304	144,613
1989	1,940	28,101	103,707	154,304
1990	3,261	39,209	136,752	143,411
1991	5,795	57,091	223,079	135,161
1992	6,806	72,178	174,682	123,998
1993	11,146	71,593	229,648	173,416
1994	10,133	75,582***		115,134***
1995	8,510	104,408***		121,014***
1996	6,983			
1997	11,057			

Note: Canadian definition of "hazardous waste" includes recyclables, gases, and biomedical wastes not included in the U.S. definition. Mexico bans import of hazardous waste unless it is to be recycled.

Source: *Unpublished EPA databases; **Environment Canada manifest database; ***EPA compilation from 1995 U.S. EPA Hazardous Waste Export Annual Reports.

among the members for purposes of disposal is not permitted. The key word in the title and in implementation is *recovery*. The system creates:

- A "Green-list" which imposes no additional controls beyond normal international commercial shipments
- An "Amber-list" of wastes that either (1) move on a shipment-by-shipment basis requiring prior written notification and consent from the importing and transit countries or (2) move to a facility that is preapproved by the importing country to accept that waste type with prior written notification only (In both cases the waste must be accompanied by a tracking document under a legally binding contract, chain of contracts, or equivalent arrangements within a corporate entity.)
- A "Red-list" wherein the wastes are handled in the same manner as Amber-list wastes except that prior written consent from importing and transit countries is always required and no facilities are preapproved to accept the wastes

Needless to say, practitioners contemplating activity involving import or export of hazardous waste must study the pertinent EPA and DOT regulations with great care (*see:* 40 CFR 262, Subpart H; Krieger 1995, Chapter 4; U.S. EPA 1998, Chapter III; U.S. EPA 1998).

Significant effort on the part of U.S. (state and federal) and Mexican officials has been committed to improvement of tracking and accountability for hazardous wastes in the border areas. Indeed, the "increases" in U.S.-Mexican shipments indicated in Table 6.3 may reflect some combination of improving awareness, surveillance, inspection, documentation, and enforcement pertaining to hazardous waste shipments. Nevertheless, much of the waste generated on the Mexican side is never accounted for and an occasional U.S. shipper attempts illegal movement of wastes to Mexico.

Researchers of the UCLA School of Public Health have documented border area industrial waste management problems since 1989. They report that, the 1983 U.S.-Mexico Agreement[10] notwithstanding, EPA records for 1988 show that only 1% (7 of 748) of the maquilidora industries operating in northern Baja California and Sonora requested shipment of hazardous wastes to the U.S. (Perry et al. 1990). The Baja industries are estimated to generate 100,000 tons of hazardous waste per year. By the end of 1990, the Secretaria de Desarrollo Urbano y Ecologia (SEDUE)[11] found that only 14.5% of the maquilas "... recycle legally or send back their residues to the United States" (Castillo and Perry 1992). The researchers suggest a variety of options for improved coordination, tracking, and accountability of hazardous wastes in the border areas (Perry et al. 1990; *see also:* Perry and Klooster 1992, Chapters 4, 5, and Executive Summary).

Two recent federal criminal prosecutions[12] growing from the attempted illegal shipment of waste PCBs to Mexico confirm the necessity for vigorous monitoring and enforcement of waste management statutes in the border area. To this end, the EPA provides technical and enforcement training to the U.S. Customs Service personnel, and the two agencies conduct joint training exercises at Border Patrol facilities. In recent years, the county, state, and federal environmental and law enforcement agencies have greatly improved cooperation and coordination of border area investigative activity.

TOPICS FOR REVIEW OR DISCUSSION

1. Discuss situations in which hazardous chemicals in transit burn or release flammables in rural and urban areas. When might it be advisable to attempt control and why? When might it be best to let the fire "burn itself out" and why?
2. The DOT lists 11 circumstances that may require notification of the National Response Center regarding a hazardous *materials* incident. What are these circumstances?
3. Discuss the advantages and disadvantages of highway vs. rail shipment of hazardous materials/wastes.

[10] U.S.-Mexico Agreement of Cooperation for the Protection and Improvement of the Environment in the Border Area (1983), Annex III.
[11] The federal environmental agency of Mexico.
[12] *U.S. v. Daniel G. Rodriguez-Castro* and *U.S. v. Weaver Electric Company, Inc.;* EPA (1994b).

4. There are several missteps that can cause a transporter of RCRA hazardous wastes to become subject to generator or storage facility (or both) requirements. List the missteps.

REFERENCES

Assante-Duah, D. Kofi and Imre V. Nagy. 1998. *International Trade in Hazardous Waste.* E. & FN Spon-Routledge, NY.

Blackman, William C., Jr. 1985. Environmental Impacts of Policies Toward the Rail and Motor-Freight Industries in the United States. Doctoral dissertation, Graduate School of Public Affairs, University of Colorado, Boulder.

Castillo, Victor M. and Diane Perry. 1992. "Environmental Implications of the Free Trade Agreement in the Maquiladora Industry," in *Transboundary Resources Report,* Summer 1992, Environmental Committee of the Tijuana-San Diego Region/United Nations Association of San Diego County, Centro Cultural de Tijuana, Baja California, Mexico.

Chemical Waste Transportation Institution. 1994. Unpublished database. Washington, D.C.

Eckmyre, Alan A. 2000. "Release Reporting and Emergency Notification," in *Hazardous Materials Management Desk Reference,* McGraw-Hill, NY.

Fox, Malcom A. 2000. *Glossary for the Worldwide Transportation of Dangerous Goods and Hazardous Materials,* CRC Press, Boca Raton, FL.

ICF, Inc. 1984. Assessing the Costs Associated with Truck Transportation of Hazardous Wastes. U.S. Environmental Protection Agency, Office of Solid Waste, Washington, D.C.

Krieger, Gary R. and Ian Austin. 1995. "Legal and Legislative Framework," in *Accident Prevention Manual for Business and Industry — Environmental Management.* Gary R. Krieger, Ed., National Safety Council, Itasca, IL.

Munter, Florence, Stephen W. Bell, Robert Hollingsworth, Joseph W. Gordon, and Charles N. Lovinski. 1995. "Hazardous Wastes," in *Accident Prevention Manual for Business and Industry — Environmental Management,* Gary R. Krieger, Ed., National Safety Council, Itaska, IL.

National Solid Waste Management Association. 1989. *Managing Hazardous Waste: Fulfilling the Public Trust.* Washington, D.C.

Nicolet-Monnier, Michel and Adrian V. Gheorghe. 1996. *Quantitative Risk Assessment of Hazardous Materials Transport Systems.* Kluwer, Norwell, MA.

Perry, Diane M. and Daniel J. Klooster. 1992. *The Maquiladora Industry: Generation, Transportation and Disposal of Hazardous Waste at the California-Baja California, U.S.-Mexico Border: Second Maquiladora Report.* School of Public Health, University of California, Los Angeles.

Perry, Diane M., Roberto Sanchez, William H. Glaze, and Maria Mazari. 1990. "Binational Management of Hazardous Waste: The Maquiladora Industry at the U.S.-Mexico Border," *Environmental Management,* 14(4), pp. 441–450.

Roberts, Alan I. 1985. "Transport of Hazardous Waste," in *Transfrontier Movements of Hazardous Waste.* Organization for Economic Cooperation and Development, Paris.

U.S. Department of Transportation, Research and Special Programs Administration. 1998. Hazardous Materials Shipments. Office of Hazardous Materials Safety, Washington, D.C.

U.S. Department of Transportation, Research and Special Programs Administration. 2000. 1999 Hazardous Materials Incident Data. Office of Hazardous Materials Safety, Washington, D.C.

U.S. Environmental Protection Agency. 1993. Catalog of Hazardous Waste Database Reports. Solid Waste and Emergency Response, Washington, D.C.

U.S. Environmental Protection Agency. 1994a. The Biennial RCRA Hazardous Waste Report (Based on 1991 Data) Executive Summary. Solid Waste and Emergency Response, Washington, D.C.

U.S. Environmental Protection Agency. 1994b. Enforcement Accomplishments Report FY 1993. Office of Enforcement, Washington, D.C., EPA 300-R-94-003.

U.S. Environmental Protection Agency. 1998. RCRA Orientation Manual. Office of Solid Waste and Emergency Response, Washington, D.C., EPA 530-R-98-004.

U.S. Environmental Protection Agency. 1999. The National Biennial RCRA Hazardous Waste Report (Based on 1997 Data) Executive Summary. Solid Waste and Emergency Response, Washington, D.C., EPA 530-S-99-036.

U.S. Office of Technology Assessment. 1986. Transportation of Hazardous Materials. Superintendent of Documents, U.S. Government Printing Office, Washington, D.C.

Wentz, Charles A. 1989. *Hazardous Waste Management.* McGraw-Hill, NY.

Westat, Inc. 1984. National Survey of Hazardous Waste Generators and Treatment Storage and Disposal Facilities Regulated under RCRA in 1981. U.S. Environmental Protection Agency, Office of Solid Waste, Washington, D.C.

7 Treatment and Disposal Methods and Processes

OBJECTIVES

At the completion of this chapter, the student should:

- Have overview knowledge of historical and traditional methods of treatment and disposal of hazardous wastes, and the environmental impacts of each.
- Have knowledge of past and present practices of land treatment and disposal, the environmental impacts thereof, and the RCRA land disposal restrictions.
- Have overview knowledge of nonpoint-source water quality impacts of hazardous waste treatment and disposal operations.
- Understand the air quality implications, residue management, and waste destruction capabilities of burning vs. incineration and the RCRA approach to each.
- Understand some of the classic reuse and recycling processes as a basic management approach and as an introduction to Chapter 8.
- Understand the basic differences between treatment, immobilization, and destruction and the processes associated with each category.
- Be familiar with history and practice of ocean dumping and underground injection and with concerns regarding potential environmental impacts of each.

INTRODUCTION

In the two previous chapters, we overviewed first the generation and then the transportation of hazardous wastes. We now take up the technologies, practice, and regulatory requirements associated with the ultimate disposition of hazardous wastes. Where appropriate, we will follow the pattern of previous chapters by beginning a topic with a discussion of "generic" practice or technology and follow with the regulatory requirements of the Resource Conservation and Recovery Act (RCRA).

As before, the format for a generic discussion of treatment and disposal is shaped by the RCRA format that groups treatment, storage, and disposal functions together as the "final link in the cradle-to-grave hazardous waste management system." The rationale for grouping *treatment* and *disposal* together is fairly clear.

Some recollection of the early practices and "horror stories" of Chapter 1 should serve to refresh our understanding of the abuses that were associated with accumulation of hazardous wastes. It is clear throughout the Subtitle C regulations that Congress intended that accumulation of hazardous wastes be controlled very rigorously. Thus, the grouping of treatment, *storage*, and disposal facilities (TSD facilities or TSDFs) as the final link, and as the entities requiring operating permits, became a regulatory format. For instructional purposes, it has become an entire way of thinking about the final disposition of hazardous waste.

The original RCRA legislation establishes two categories of TSD facilities based upon **permit** status. Section 3005(a) of the Act specifies that TSD facilities must obtain a permit to operate. In recognition of the fact that several years would be required for the EPA to issue permits to all operating facilities, Congress included § 3005(e), which established "**interim status.**" TSD facilities that were in existence on November 19, 1980, and met certain conditions, were allowed to continue operating until their permit was issued or denied. Such facilities are said to have **interim status** and are regulated by 40 CFR 265. The second category consists of those facilities having **permits**. Permitted facilities are regulated by 40 CFR 264.

Both **interim status** and **permit** standards consist of two types of requirements:

- Administrative and nontechnical requirements which are nearly identical for interim status and permitted facilities
- Technical and unit-specific requirements which embody significant differences for interim status and permitted facilities

Large numbers of **interim status** facilities continue to operate legally without fully approved **permits**, and it is expected that this situation will prevail for several more years. The 40 CFR 264 "finally permitted" standards, which will eventually apply to all TSD facilities, are more stringent than the Part 265 "interim status" standards. However, they are only a blueprint for the permit writer who must develop "best engineering judgment" standards for the specific facility. In the following paragraphs, we will overview both the interim status and permitted facility standards pertaining to administrative and nontechnical requirements. We will point out various differences that exist between the two sets of requirements.

Treatment, storage, and disposal practice involves a large variety of units and technologies. Thus, the TSD regulations are far more extensive than for generators and transporters. We will attempt to overview only the most important generic topics and salient features of the regulations. The student or reader is encouraged to (and the practitioner must) explore the technical literature and the RCRA Subtitle C regulations for details.

ADMINISTRATIVE AND NONTECHNICAL REQUIREMENTS

The administrative and nontechnical requirements are intended to ensure that owners and operators establish the necessary procedures and plans to operate the TSD facility according to established practice and to handle any emergencies or accidents. The administrative and nontechnical requirements for interim status and permitted TSDFs are very similar. These requirements are found in Subparts A through E of 40 CFR Parts 264 and 265.

Subpart A — Facilities That Are Subject to the Regulations

In general, all owners or operators of facilities engaged in the treatment, storage, or disposal of hazardous wastes must comply with the 40 CFR 264/265 regulations unless they are specifically excluded. Exceptions include

- A farmer who disposes of waste pesticides from his own operations
- Facilities that qualify for a "permit-by-rule"[1]
- The owner or operator of a totally enclosed treatment facility
- The owner or operator of an elementary neutralization unit
- The owner or operator of a wastewater treatment unit that is subject to Clean Water Act pretreatment standards or a National Pollutant Discharge Elimination System (NPDES) permit
- A person who responds to or cleans up a hazardous waste spill or release
- Facilities that legitimately reuse, recycle, or reclaim hazardous waste
- Generators, including small quantity generators (SQGs), that accumulate wastes within the time periods specified in 40 CFR 262
- Facilities that treat, recycle, store, or dispose of wastes generated by conditionally exempt small quantity generators (CESQGs)[2]
- A transporter that stores manifested shipments for less than 10 days (40 CFR 265.1)

Interim status TSDFs may manage dioxin-containing wastes (F020-F023, F026, and F027) only if the requirements of 40 CFR 265.1(d) pertaining to an immediate threat to human health, public safety, property, or the environment from the known or suspected presence of military munitions, other explosive material, or devices are met.

Subpart B — General Facility Standards

As was covered in the previous chapters, all facilities handling hazardous wastes must obtain an EPA identification number.[3] Owners and operators of TSDFs must ensure that the wastes being handled are correctly identified and managed according to the regulations. They must ensure that facilities are secure and are operating properly. Personnel working in the facilities must be trained to perform their duties

---

[1] Facilities that have a permit issued under other environmental laws, i.e., ocean disposal, underground injection, publicly owned treatment works, that meet the requirements of 40 CFR 270.

[2] *See:* 40 CFR 261.5.

[3] It may be conceptually useful to understand that *anyone* can apply for, and obtain, an EPA ID number. Issuance of the ID by the EPA does not amount to a permit or certification. It is a means of identification that the holder will be called upon to provide should he/she ultimately engage in hazardous waste management activity.

FIGURE 7.1 Site security requirement for RCRA sites.

correctly, safely, and in compliance with all applicable laws, regulations, and codes. In order to satisfy these requirements, owners and operators must:

- *Conduct waste analyses* prior to initiating treatment, storage, or disposal in accord with a written waste analysis plan (WAP), which must be kept on site. The WAP must specify tests and test frequencies that will provide the owner or operator with sufficient information on the properties of the waste to enable management of the waste in accord with the applicable laws, regulations, and codes (*see:* 40 CFR 264, 265.12).
- *Install security measures* to prevent accidental or unauthorized entry of people or animals onto the active portions of the TSDF (Figure 7.1). The facility must be surrounded by a barrier (i.e., a fence) with controlled entry systems or 24-hr surveillance. Signs carrying the warning "Danger — Unauthorized Personnel Keep Out" must be posted at all entrances (Figure 7.2). Signs must be printed in English and also in other languages predominant in the area surrounding the facility. Precautions must be taken to avoid fires, explosions, generation of toxic gases, and any other events that would threaten human health, safety, and the environment.
- *Conduct inspections* according to a written inspection plan and schedule to assess the compliance status of the facility and to detect potential problems such as malfunction, deterioration, operator error, and leaks or discharges. Observations made during the course of the inspections must be recorded in the facility's operating log and kept on file for 3 years. All problems noted must be remedied.
- *Conduct training* to reduce the potential for mistakes that might threaten human health and the environment. The regulations specify that the employee "… must successfully complete classroom instruction or on-the-job training that teaches them to perform their duties in a way that

FIGURE 7.2 Sign requirement for RCRA sites.

ensures the facility's compliance with the regulations." In addition, the Occupational Safety and Health Administration (OSHA) requires TSD facilities to implement a hazard communication plan, medical surveillance program, and a health and safety plan.[4] Decontamination procedures (Chapter 15) must be in place and employees must receive a minimum of 24 hr of safety training. The training must be completed within 6 months from the date the facility becomes subject to the TSDF standards or 6 months from the date the employee begins work at the facility. New employees must work under supervision until the training is completed, and the training must be reviewed annually.

- *Properly manage ignitable, reactive, or incompatible wastes.* Ignitable or reactive wastes must be protected from sources of ignition or reaction or be treated to eliminate the possibility. Owners or operators must ensure that treatment, storage, or disposal of ignitable, reactive, or incompatible waste does not result in damage to the containment structure, and/or threaten human health or the environment. Separation of incompatible wastes must be maintained. Part 264, Appendix V, provides a list of some common potentially incompatible wastes. It may be necessary to test the wastes to determine compatibility.
- *Comply with location standards* to avoid siting a new facility (subject to Part 264) in a location where flood or seismic events could affect a waste

---

[4] The required OSHA training is detailed in Chapter 15.

management unit. Existing facilities, subject to Part 265, are not required to meet this standard. Interim status and permitted TSDFs are prohibited from placement of noncontainerized or bulk liquid hazardous wastes in salt domes, salt beds, or underground mines or caves. The Department of Energy Waste Isolation Pilot Project (*see:* Chapter 13) has been granted exclusion from this prohibition by Congress.

- *Prepare and comply with the construction quality assurance (CQA) program requirements* that are applicable to foundations, dikes, soil liners, geomembranes, leachate detection, collection, and removal systems, and final cover systems at permitted and interim status facilities. The CQA program ensures that all design criteria are met during the construction of a unit. A written CQA plan is required. The CQA officer (a registered professional engineer) must certify that the unit meets all design criteria and permit specifications before waste can be received by the unit. These construction standards are extensive and will be covered in detail in the permitting process (40 CFR 264, 265 Subpart B).

Subpart C — Preparedness and Prevention

Facilities must be designed, constructed, maintained, and operated to minimize the possibility of a fire, explosion, or any unplanned sudden or nonsudden release of hazardous waste or hazardous waste constituents which could threaten human health or the environment.[5] Facilities must be equipped with:

- *An internal communications or alarm system* that can provide immediate emergency instructions to facility personnel
- *A telephone or two-way radio* capable of summoning emergency assistance from local police, fire, and emergency response units
- *Portable fire extinguishers*, fire, spill control, and decontamination equipment
- *Water* at adequate volume and pressure to supply water hoses, foam-producing equipment, automatic sprinklers, or water spray systems

All communications and emergency equipment must be tested as necessary to ensure proper operation in time of emergency. All personnel must have immediate access to the internal alarm or emergency communication system.[6] Aisle space (Figure 7.3) must be maintained to allow unobstructed movement of personnel and equipment during an emergency.

Owners or operators of TSDFs must attempt to make arrangements to:

---

[5] *See:* 40 CFR 264, 265.32 for exceptions to these rules where the nature of the hazard(s) cause the rule to be unnecessary.

[6] If there is ever only one employee on the premises while the facility is operating, the employee must have access to a telephone or hand-held two-way radio, capable of summoning external emergency assistance [40 CFR 264, 265.34(b)].

FIGURE 7.3 Aisle space, drum stacking limitation.

- *Familiarize police, fire, and emergency response teams* with the facility, wastes handled and their properties, work stations, and access and evacuation routes.
- *Designate primary and alternate emergency response teams* where more than one jurisdiction might respond.
- *Familiarize local hospitals* with the properties of the hazardous wastes handled at the facility, and the types of injuries or illnesses which could result from events at the facility.

Subpart D — Contingency Plan and Emergency Procedures

A contingency plan must be in effect at each TSD facility and by reference [§ 262.34(a)(4)] at each generator facility. The plan must be designed to minimize hazards to human health or the environment from fires, explosions, or any release of hazardous waste constituents. The plan must be implemented immediately whenever there is a fire, explosion, or release which could threaten human health or the environment.

The contingency plan must:

- Describe the actions which personnel must take to implement the plan.
- Describe arrangements concluded with local police, fire, and hospital authorities, contractors and emergency response teams to coordinate emergency services.
- List names, addresses, and the telephone numbers of all persons qualified to act as emergency coordinator for the facility.
- List emergency equipment, communication and alarm systems, and the location of each item.
- Include an evacuation plan for facility personnel.

The contingency plan must be maintained at the facility and at all emergency response facilities that might be called upon to provide emergency services. It must be reviewed and updated whenever any item affecting the plan is changed. A key requirement is the designation of an emergency coordinator who is responsible for directing response measures and reducing the adverse impacts of hazardous waste releases. There must be at least one employee present on the premises or on call at all times to fill this role. The emergency coordinator must have the authority to commit the resources needed to implement the emergency/contingency plan.

Other regulatory programs related to hazardous waste management, releases of hazardous materials to the environment, or exposures of humans to toxic materials also require emergency response planning, and/or preparation of contingency plans. These planning requirements are becoming more numerous, and the specifications are becoming more complex and sophisticated. Risk assessment and risk management measures are being required in a wide range of situations. Owners or operators of TSD facilities and their emergency coordinators will increasingly find it necessary to devote time and resources to the contingency planning effort. Among the different regulatory programs, some planning requirements are similar, duplicative, overlapping, or redundant. Time and resource commitments, training, drills, and coordination requirements can be economized by combining the required plans in one document. In an effort to assist preparers of these multiple planning requirements, five agencies[7] have collaborated in preparing the "'Four' Agency Integrated Contingency Plan." The National Response Team (NRT), chaired by the U.S. EPA, has issued "The National Response Team's Integrated Contingency Plan Guidance." The guidance is intended to provide a mechanism for consolidating the multiple plans, that would otherwise be required, into one functional emergency response plan or integrated contingency plan (ICP). A copy of the guidance can be obtained by calling the EPCRA/RCRA/Superfund Hotline at 800-424-9346 or electronically at the home page of EPA's Chemical Emergency Preparedness and Prevention Office (http://www.epa.gov/swercepp/).

The ICP guidance document provides a suggested structure for the facility's ICP and a detailed cross-reference matrix of ICP elements on the vertical axis and the regulatory Chapters and Parts on the horizontal headings. Table 7.1 lists the components and regulatory references upon which the ICP is based. The suggested structure is organized into three main sections: an introductory section, a core plan, and a series of supporting annexes. The structure of the core plan and annexes in the ICP guidance is based on the structure of the National Interagency Incident Management System (NIIMS) Incident Command System (ICS). NIIMS ICS is a nationally recognized system that has been used by federal, state, and local response organizations in a variety of emergency situations. The planner should find this guidance document to be helpful in developing a facility-specific ICP, which will dovetail with established response management practices, thereby facilitating its usefulness during an emergency.

---

[7] Environmental Protection Agency; Department of Transportation, Coast Guard (USCG), and Research and Special Programs Administration (RSPA); Department of the Interior, Minerals Management Service (MMS); Department of Labor, Occupational Safety and Health Administration (OSHA).

TABLE 7.1
Integrated Contingency Planning Components

Agency	Statute	Plan Requirement	Reference
EPA	RCRA	Contingency Plan	40 CFR 262.34, 264 and 265, Subparts C and D, and 279.52
EPA	CWA	Spill Prevention Control, a Countermeasure Plan	40 CFR 112
EPA	CAAA	Risk Management Program for Chemical Accidental Release Prevention	40 CFR 68
DOI/MMS	CWA/OPA[a]	Facility Response Plan	30 CFR 254
DOT/USCG	NNWA[b]	Facility Response Plan	33 CFR 154, Subpart F
DOT/RSPA	HMTA	Pipeline Response Plan	49 CFR 194
DOL/OSHA	SARA[c]	HazWOper Emergency Plan	29 CFR 1910.120
DOL/OSHA	CAA	Chemical Process Safety Standard	29 CFR 1910.119
DOL/OSHA	OSHA	Emergency Action Plans	29 CFR 1910.38

[a] Oil Pollution Act.
[b] Navigation and Navigable Waters Act.
[c] Superfund Amendments and Reauthorization Act (SARA) Title III includes emergency planning by State Emergency Response Commissions (SERCs) and Local Emergency Planning Committees (LEPCs). RCRA facilities may be required or may wish to coordinate their contingency plans with those of the SERC or LEPC or both.

Subpart E — Manifest System, Record Keeping, and Reporting

The operation of the manifest system has been previously described. The TSDF owner or operator receiving the waste is responsible for ensuring that the waste described on the manifest is the same as the waste on the truck. The intent is to ensure that there are no significant discrepancies in the amount (e.g., an extra drum) or type of waste (e.g., acid waste instead of paint sludge) that was shipped by the generator. If a significant discrepancy is discovered, the TSDF must reconcile the difference with the generator or transporter. If the difference cannot be cleared up, the EPA must be notified within 15 days of the incident.

The owner or operator or his agent must sign and date all copies of the manifest to verify that the waste has reached the designated facility. The copy of the signed manifest must be placed in the TSDF files, and a copy must be sent to the generator within 30 days. If it is necessary to send the waste to another facility, the owner/operator/agent must initiate a new manifest. Subpart E includes extensive record keeping and reporting requirements (EPA 1998, Section III).

GENERAL TECHNICAL STANDARDS FOR INTERIM STATUS AND PERMITTED FACILITIES

The 40 CFR 265, Subpart F, groundwater monitoring standards for **interim status** TSDFs were designed to minimize the potential for environmental and public health threats resulting from hazardous waste treatment, storage, and disposal at existing facilities that are awaiting permitted status. Over time, and because of the extended periods during which some facilities have operated under interim status, these standards have taken on greater importance than was originally intended and are indeed minimally adequate to serve their intended purpose. In contrast, the Part 264 standards entitled "Releases from Solid Waste Management Units" are detailed and comprehensive. Moreover, permit writers have wide latitude for imposing monitoring requirements that can reasonably be expected to detect releases from land-based units and/or provide the data needed for remediation. Nevertheless, there are common technical and environmental performance elements of the **interim status** and **permitted** facility standards. Both sets of standards are designed to protect the *uppermost* aquifer, which is the water-bearing geologic formation nearest the ground surface. Any and all deeper aquifers that are hydraulically connected to the uppermost aquifer are considered to be part of that aquifer.

Part 265, Subpart F — Groundwater Monitoring

Owners and operators of surface impoundments, landfills, and land treatment facilities used to manage hazardous waste must meet minimum groundwater monitoring requirements. The **interim status** facility requirements in 40 CFR 265.91 call for a monitoring system consisting of at least one well upgradient from the facility and three downgradient wells. The upgradient well(s) must provide data on groundwater that is not influenced by leakage from the waste management unit. The downgradient wells must be placed to intercept any waste migrating from the unit should a release occur.

Figure 7.4 illustrates a "one-up-and-three-down" layout of monitoring wells for a land-disposal facility — the upper left well being the upgradient or background data well and the others being the downgradient wells. Figure 7.5 illustrates an important problem, i.e., the nearby stream at low stage may draw the contaminant

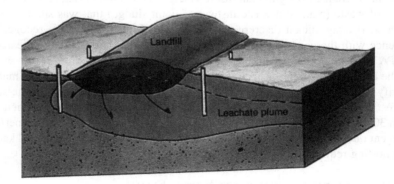

FIGURE 7.4 Groundwater monitoring well layout for a landfill disposal facility. (Adapted from Glenn R. Smart and David K. Cook, RCRA and CERCLA Groundwater Well Locations and Sampling Requirements, *Hazardous Materials Control*, 1(3), May/June 1988.)

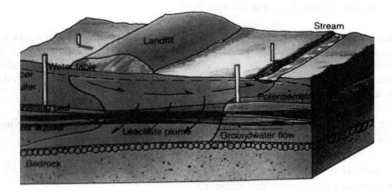

FIGURE 7.5 Natural drawdown interference with groundwater monitoring regime. (Adapted from Glenn R. Smart and David K. Cook, RCRA and CERCLA Groundwater Well Locations and Sampling Requirements, *Hazardous Materials Control,* 1(3), May/June 1988.)

plume into the base flow and away from the monitoring wells. Figure 7.6 illustrates a very common phenomenon — the creation of an artificial drawdown curve, which can alter the movement of the plume and generate misleading data from the monitoring well.

The one-up-and-three-down layout is generally understood to be a minimum pattern. As practitioners have gained in knowledge of plume behavior, older notions of vertical and transverse dispersion have given way to the understanding that

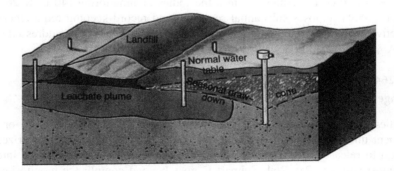

FIGURE 7.6 Artificial drawdown interference with groundwater monitoring regime. (Adapted from Glenn R. Smart and David K. Cook, RCRA and CERCLA Groundwater Well Locations and Sampling Requirements, *Hazardous Materials Control,* 1(3), May/June 1988.)

contamination releases may move in very narrow plumes. Recent tracer tests and detailed plume studies have established that:

> Because of weak dispersion, the degree of concentration heterogeneity diminishes very little down-gradient, requiring a more dense network of wells. In some plumes, the difference between detecting or missing a concentration zone orders of magnitude above a regulatory limit is the difference in positioning in depth of the critical well by only a meter or two (Ozbilgin et al., 1992).

The owner or operator must develop and follow a groundwater sampling and analysis plan which must include procedures and techniques for:

- Sample collection
- Sample preservation and shipment
- Analytical procedures
- Chain of custody control

Backgroundwater quality is determined by 1 year of quarterly monitoring of all well(s) for the 21 EPA Interim Primary Drinking Water Standards listed in Part 265, Appendix III; the six groundwater quality parameters of § 265.92(b)(2); and the four groundwater contamination parameters of § 265.92(b)(3). Thereafter the owner/operator must continue monitoring the wells for groundwater *quality* parameters at least annually and for groundwater *contamination* parameters at least semiannually.

Within 1 year, the owner/operator must prepare an outline of a more detailed ground-water *assessment* program that could be implemented to determine whether or not hazardous waste constituents have leached into the uppermost aquifer. The assessment program is implemented when/if there has been a statistically significant increase (SSI) in an indicator parameter. If a significant increase has occurred, the owner/operator must determine the rate and extent of the migration and the concentrations of the hazardous waste constituents in the plume. If no SSI is found to have occurred, the owner/operator resumes the indicator monitoring (40 CFR 265.93). Section 265.94 imposes substantial reporting and record-keeping requirements. If corrective action is required at an **interim status facility**, it will be addressed under RCRA § 3008(h), § 7003, or in the **permit** when issued.

Part 264, Subpart F — Releases from Solid Waste Management Units

Facilities with **permitted** landfills, surface impoundments, waste piles, or land treatment units must conduct groundwater monitoring to detect, characterize, and respond to releases of hazardous wastes or hazardous waste constituents into the uppermost aquifer.[8] Part 264, Subpart F, goes beyond compliance monitoring and establishes a three-stage program designed to detect and remediate any releases from regulated units:

---

[8] The practitioner considering compliance strategies should carefully examine the applicability provisions and the extensive groupings of waiver conditions and exemptions of 40 CFR 264.90.

authority and to the EPA or the state agency. The plat preserves a record of the exact location and dimensions of the hazardous waste activity for future reference. A notation must also be made on the deed to the property, notifying potential purchasers that the site was engaged in hazardous waste activity.

Following the closure, a 30-year post-closure period is established for facilities that do not "clean close" as described below. The post-closure care consists of at least the following:

- Groundwater monitoring and reporting
- Maintenance and monitoring of waste containment systems
- Continued site security

Facilities that leave hazardous waste in place at closure must prepare and submit a post-closure plan, which is similar to the closure plan. The closure and post-closure plans may be amended at any time and must be amended if there is any change of circumstances that affects the plan. The closure timetable and the post-closure care period may be lengthened or shortened by the EPA or the state agency.

Clean closure may be accomplished by the removal of all contaminants from impoundments and waste piles. At a minimum, owners and operators of surface impoundments and waste piles that wish to clean close must conduct soil analyses and groundwater monitoring to confirm that all wastes have been removed from the unit. The EPA and/or the state agency may establish additional clean-closure requirements on a case-by-case basis. (Figure 11.18 illustrates the extent of removal that may be necessary to "clean close" a former hazardous waste impoundment.) A successful demonstration of clean closure eliminates the requirement for post-closure care of the site (*see also:* U.S. EPA 1997b; U.S. EPA 1998, Chapter 5).

Subpart H — Financial Requirements

RCRA originally established financial requirements to assure that funds would be available to pay for closing a facility, for rendering post-closure care at disposal facilities, and to compensate third parties for bodily injury and property damage caused by accidents related to the operation of a TSDF. An obvious objective of Congress in establishing these requirements was the avoidance of necessity for cleanup under Superfund.

In the 1984 Hazardous and Solid Waste Amendments (HSWA), Congress mandated additional financial responsibility requirements, thereby emphasizing the importance of assured financial capability for completing needed remediation at TSDF sites. The Subpart H rules are detailed and extensive beyond reasonable summarization herein. The financial requirements for both are structured to achieve:

- Financial assurance for closure/post-closure
- Liability coverage for injury and property damage

Owners and operators must meet the financial assurance requirements by preparation of cost estimates for closure and, if required, post-closure. The cost estimates must reflect the actual and projected costs for the conduct of the activities described in the closure plan. Similarly, cost estimates for post-closure operations must be projected for the full post-closure period.

The owner or operator must then demonstrate to the EPA or the state agency the ability to pay the estimated amounts. The owner/operator may use any one or a combination of the following six mechanisms to comply with the financial assurance requirements:

- Trust fund
- Surety bond (two types)
  - Payment bond
  - Performance bond
- Letter of credit
- Closure/post-closure insurance
- Corporate guarantee
- Financial test

Liability insurance requirements include coverage of $3 million (annual aggregate $6 to $8 million) per sudden accidental occurrence such as a fire or explosion. Owners or operators must also maintain coverage of $4 million (annual aggregate $8 million) for nonsudden occurrences such as groundwater contamination. The liability coverage may be demonstrated using any of the six mechanisms allowed for assurance of closure and post-closure funds (*see also:* U.S. EPA 1998, Chapter 5).

HAZARDOUS WASTE TREATMENT

Hazardous waste treatment is a rapidly developing industry full of experimentation and innovation. This innovation is being driven by the need for effective and economical processes for reclaiming, treating, or destroying wastes rather than landfilling them without treatment. A hierarchy of general waste management options can be constructed as shown in Table 7.2. The most desirable option is source reduction through process modification (Combs 1989, p. XV-1). The less desirable options follow.

TABLE 7.2
Hazardous Waste Management Options and Priorities

- Source reduction (process modification)
- Separation and volume reduction
- Exchange/sale as raw materials
- Energy recovery
- Treatment
- Secure ultimate disposal (landfill)

Source: Combs (1989).

Source reduction approaches and waste exchanges will be discussed in Chapter 8. Overview discussions, examples, and regulatory requirements pertaining to separation and volume reduction practices, energy recovery, treatment and destruction methods, and secure ultimate disposal follow. The schematic of Figure 7.7 aligns types or categories of industrial wastes with the treatment processes and ultimate disposal usually applied.

The volume of a waste destined for treatment or disposal can often be reduced by physical processes such as adsorption, centrifugation, clarification,[9] evaporation, distillation, solvent extraction, or stripping. Figure 7.8 shows a typical centrifuge layout. Figure 7.9 illustrates simple gravity separation in cone-bottom tanks. These processes make use of differences in specific gravity or mass to separate harmless or nonhazardous components from the hazardous components. The nonhazardous component may then be routed to further treatment, disposal, or recycling, as appropriate. The hazardous component must be destroyed, rendered harmless, have toxicity reduced to acceptable levels, or be disposed of in a secure facility (*see also:* Glossary; Manahan 1994, pp. 582–586; Haas and Vamos 1995, Chapter 4; Woodside 1999, pp. 234–256; Watson 1999).

Chemical treatment processes may be used to alter chemical properties of wastes in order to facilitate or enable further treatment; to render the wastes nontoxic/nonhazardous for disposal; or to solidify or stabilize the wastes for ease of handling or reduced leachability or to render them nondegradable. Many varieties of chemical treatment have been devised by theory, stoichiometry, experimentation, accident, or combinations thereof. Consultant and commercial laboratory development of new waste-specific chemical treatments is a lucrative and rapidly growing global enterprise. The general categories of chemical treatment include acid/base neutralization, chemical precipitation, oxidation/reduction (redox), solidification/stabilization, electroysis, hydrolysis, chemical extraction and leaching (*see also:* Glossary; Manahan 1994, 587–594; Haas and Vamos 1995, Chapter 5; Woodside 1999, pp. 234–256; Watson 1999).

In situ biodegradation or "biological treatment" (not to be confused with bioremediation — see Chapter 11) of industrial wastewaters is usually accomplished in bioreactors and can effectively remove organic pollutants in wastewaters having low-to-moderate concentrations of simple organic compounds and lower concentrations of complex organics. The biota are generally less effective in attacking mineral or heavy metal constituents, but in carefully controlled conditions, can be usefully employed (*see:* Alexander 1999, Chapter 18).

Unlike physical treatment systems, biological treatment has the potential to transform organic pollutants into innocuous products rather than merely transferring the pollutant to another medium. Moreover, biotreatment is generally cheaper and enjoys a greater degree of public acceptance than some other forms of treatment, (e.g., incineration). Microorganisms can transform virtually any organic compound, whether man-made or naturally occurring, provided that environmental conditions (oxygen content, chemical composition, temperature, etc.) are correctly manipulated (Lewandowski and DeFilippi 1998, p. vii).

---

[9] Sedimentation and decantation may be aided by the addition of coagulants.

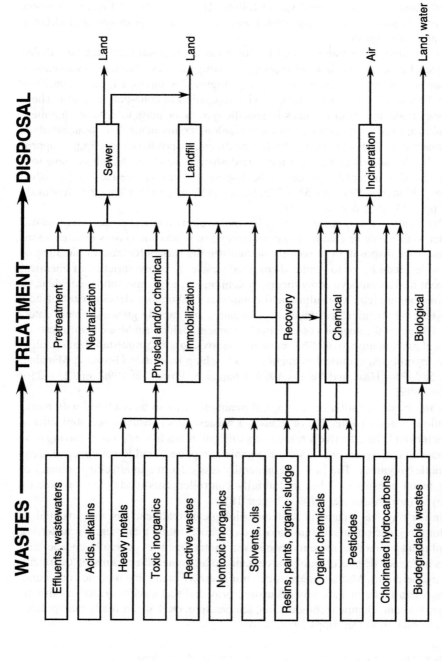

FIGURE 7.7 Treatment and disposal alternatives for industrial wastes. (Adapted from *Hazardous Waste Management*, © 1989, McGraw-Hill, New York. With permission from Charles A. Wentz.)

FIGURE 7.8 Centrifuge. Component of a hazardous waste solidification system.

FIGURE 7.9 Gravity separation cones. (From ROMIC Chemical Corporation, 2081 Bay Road, Palo Alto, CA 94303. With permission.)

Uncontrolled burning for energy recovery is a common method of hazardous waste management, and the practice can be a potential threat to human health and the environment. Burners mix hazardous wastes with fuel oil or other fuel mixtures and burn the mix in low temperature/low pressure boilers or other combustion units. Some flammable wastes continue to be burned in disregard of federal, state, and local regulations and ordinances. Such low-temperature burning does not destroy most hazardous components of the waste and, in fact, causes their dispersion in the atmosphere.

Legitimate (and regulated) burning of hazardous waste fuel can be a useful disposition of the waste and an economical energy source. Cement kilns and industrial furnaces, having adequate operating temperatures, dwell times, and emission controls, are allowed to burn some organic hazardous wastes. The EPA recently promulgated new regulations pertaining to combustion of hazardous wastes in boilers and industrial furnaces (BIFs), including cement kilns. These regulations and practices have been the source of great contention, as will be discussed.

As indicated earlier, treatment technologies for hazardous wastes are available in ever-increasing numbers. Some of these technologies and the commonly practiced recovery and disposal practices can be categorized as shown in Table 7.3. Brief descriptions of the more commonly used treatment systems follow.

Activated Carbon Adsorption

Organic substances may be removed from aqueous or gaseous waste streams by adsorption[10] of the chemical substances onto a carbon matrix. The carbon may be used in either granular or powdered form, depending upon the application and the process economics. The effectiveness of activated carbon in removing hazardous constituents from aqueous streams is directly proportional to the amount of surface area of the activated carbon. The carbon is highly porous, having total surface area in the range of 600 to 1000 m^2/g. Figure 7.10 diagrams a carbon adsorption system. The spent carbon is regenerated in ovens or by passing live steam through the carbon (Wentz 1989, pp. 172–173). A carbon regeneration system is diagrammed in Figure 7.11 (*see also*: Wilson and Thompson 1988; Voice 1989; Haas and Vamos 1995, pp. 105–106; Watson 1999, Chapter 2).

The adsorption behavior of an activated carbon often results from the carbon surfaces and their normal hydrophobic (water repelling) nature. Essentially any nonpolar molecule (such as most common hydrocarbons, trichloroethylene, trichloroethene, dichloroethane, polychlorobiphenyls, etc., which are common pollutants) will be adsorbed on an activated carbon (Watson 1999, Chapter 2) (*see also:* Woodside 1999, Chapter 12).

Other adsorbents in general use are manufactured or synthetic materials including silica gels, synthetic zeolites, and forms of cellulose. Natural adsorbents include coal, plant materials, and even sewage sludge (Haas and Vamos 1995, pp. 30–38). A recent development employs a cheap waste product, bagasse fly ash,[11] to adsorb hexavalent chromium from electroplating industry wastewater (Gupta et al., 1999).

Stripping

Air and steam stripping require mention in this introduction to hazardous waste treatment and disposal systems — not because of their particular effectiveness, but

---

[10] Adsorption is a yet incompletely explained physical, surface accumulation phenomenon that refers to the ability of certain solids to attract and collect organic substances from the surrounding medium. Granular activated carbon made from anthracite is widely used to adsorb organic components from liquid and gaseous waste streams.

[11] A waste product generated in the sugar industry.

TABLE 7.3
Hazardous Waste Treatment, Recovery, and Disposal Processes

1. Physical treatment processes
   a. Gas cleaning
      i.   Mechanical collection
      ii. Electrostatic precipitation
      iii. Fabric filter
      iv. Wet scrubbing
      v.   Activated carbon adsorption
      vi. Adsorption
   b. Liquids-solids separation
      i.   Centrifugation
      ii.   Clarification
      iii. Coagulation
      iv.   Filtration
      v.   Flocculation
      vi.   Flotation
      vii. Foaming
      viii. Sedimentation
      ix.   Thickening
   c. Removal of specific components
      i.   Adsorption
      ii.   Crystallization
      iii. Dialysis
      iv.   Distillation
      v.   Electrodialysis
      vi.   Evaporation
      vii. Leaching
      viii. Reverse osmosis
      ix.   Solvent extraction
      x.   Stripping
2. Chemical treatment processes
   a. Absorption
   b. Chemical oxidation
   c. Chemical precipitation
   d. Chemical reduction
   e. Wed oxidation
   f. Ion exchange
   g. Neutralization
   h. Chemical fixation and solidification
   i. Dehalogenation
3. Biological treatment processes
   a. Aerobic systems
   b. Anaerobic systems
   c. Activated sludge
   d. Spray irrigation
   e. Tricking filters

TABLE 7.3 *(Continued)*
Hazardous Waste Treatment, Recovery, and Disposal Processes

　　f.　Waste stabilization ponds
　　g.　Rotating bio contactors
　4.　Thermal treatment processes
　　a.　Incineration (see Table 7.4)
　　b.　Pyrolysis
　　c.　Vitrification
　5.　Ultimate disposal processes
　　a.　Deep-well disposal
　　b.　Dilution and dispersal
　　c.　Ocean dumping
　　d.　Sanitary landfill
　　e.　Land burial

---

because strippers have been employed in so very many of the early site remediation efforts. Stripping is most frequently used to remove volatile organics from wastewaters or contaminated groundwater. Strippers generally involve towers containing cascades, trays, or manufactured media, with induced-draft air or live steam passing upward and contaminated water cascading or trickling downward over optimized surface areas. A counter-current packed-tower air stripper is diagrammed in Figure 7.12. Steam stripping towers operate on a similar principle with live steam injected directly into the liquid waste.

In theory, the gas-liquid system reaches an equilibrium, based upon Henry's law,[12] and the volatile contaminants are preferentially removed and are carried out as a vapor with the exhaust air stream or the spent steam through the top of the unit. Some further treatment (carbon adsorption, incineration) must be applied to the exhaust vapors in order to capture and/or destroy the separated volatiles. A major problem with the early applications was the omission of this final stage and the uncontrolled release of the stripped volatiles to the atmosphere (*see also:* Haas and Vamos 1995, pp. 90–96; Woodside 1999, pp. 253–255).

Neutralization

Neutralization is a widely used chemical process in which the pH of an acidic, corrosive, or caustic wastewater or gas is adjusted to a more neutral range. Neutralization may be employed as a pretreatment step or final treatment process. Methods of neutralizing acidic wastes include

- Adding appropriate amounts of strong or weak base to the waste
- Passing acidic waste through limestone beds

---

[12] Henry's law: At constant temperature, the weight of gas absorbed by a given volume of a liquid is proportional to the pressure at which the gas is supplied, e.g., if a liter of water dissolves 5 g of a gas under 1 atm of pressure, it will dissolve 10 g of the same gas under 2 atm of pressure.

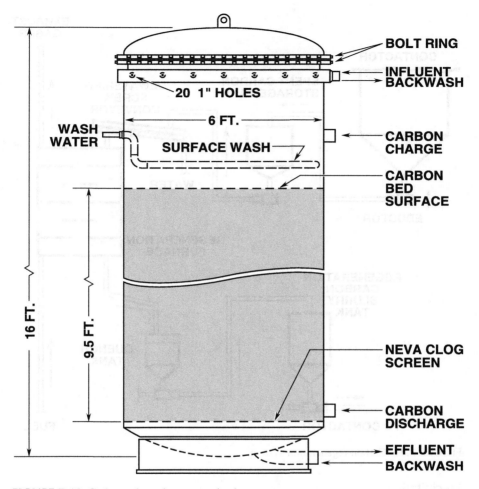

FIGURE 7.10 Carbon adsorption pressurized contactor.

- Mixing acidic waste with lime or dolomite lime slurries
- Mixing the acidic waste with a compatible alkaline waste[13]

Methods of neutralizing alkaline wastes include

- Adding appropriate amounts of strong or weak acid to the waste
- Adding compressed carbon dioxide gas to the waste
- Blowing flue gas through the waste
- Mixing the alkaline waste with a compatible acidic waste[13] (Hass and Vamos 1995, Chapter 5)

---

[13] Great care must be taken in determining compatibility of wastes to be mixed. For an exhaustive treatment of the neutralization processes, see Hass and Vamos 1995, Chapter 5.

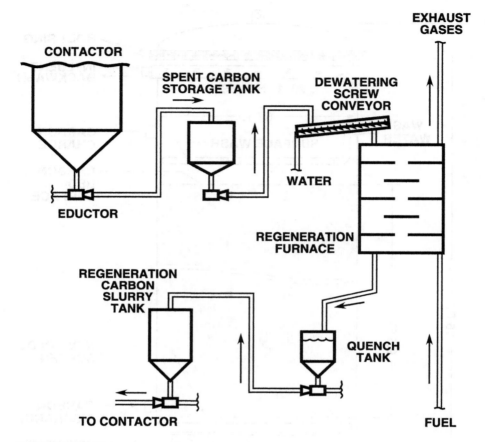

FIGURE 7.11 Carbon regeneration system.

Precipitation

If the solubility product of a metal hydroxide is suitable for precipitation, sodium hydroxide (caustic soda) or calcium hydroxide (lime slurry) can be used to treat liquid wastes containing heavy metals. The addition of a hydroxide ion precipitates the metals:

$$M^{+2} + 2(OH^-) \rightarrow M(OH)_2$$

The coagulation of the precipitated metals is both a physical and chemical process. The attraction of cations for anions causes the formation of a floc. It is frequently necessary to add a coagulant or flocculant to aid in separation of the precipitant from the remaining soluble phase. The mild turbulence in the stirred tank causes the small particles to collide forming a sludge with a concentration of 20 to 50% solids (DuPont 1988). This process must then be followed by solidification or other

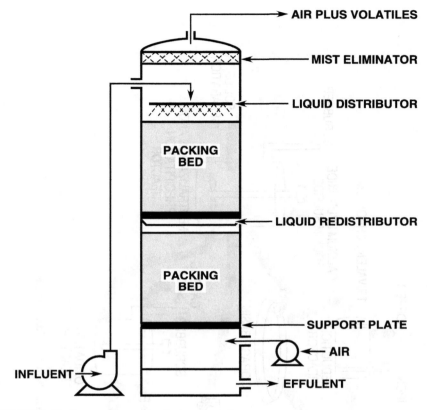

FIGURE 7.12 Counter-current packed tower stripper.

processes specific to the sludge formed, in order to render the sludge harmless to the environment.

Calcium hydroxide and sodium hydroxide are used by many industries to precipitate heavy metals; however, precipitation of chromium requires that all hexavalent chrome-containing ions be reduced to the trivalent state, since hexavalent chromium cannot be removed directly by hydroxide precipitation. Using sulfurous acid as a reducing agent:

$$3H_2SO_3 + H_2Cr_2O_7 \rightarrow Cr_2(SO_4)_3 + 4H_2O$$

$$Cr_2(SO_4)_3 + 3Ca(OH)_2 \rightarrow 2Cr(OH)_3\downarrow + 3CaSO_4$$

Green Precipitate

(George 2000, pp. 556–557). Cultured bacteria can also accomplish the necessary reduction of Cr^{+6} to Cr^{+3} under appropriate conditions (Alexander 1999, Chapter 18). Figure 7.13 illustrates a typical process in which an inorganic acid (perhaps a dilute

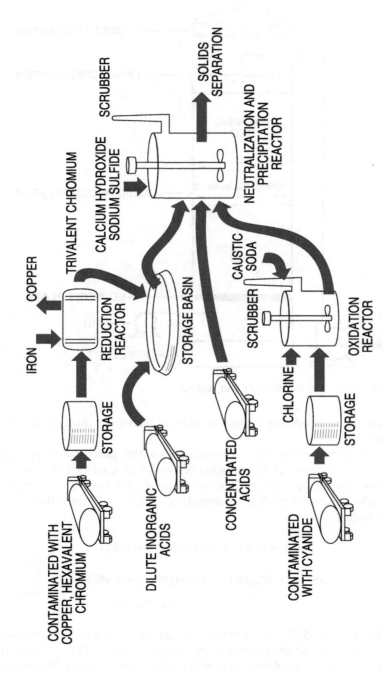

FIGURE 7.13 Chemical treatment: neutralization, precipitation, and chemical oxidation/reduction.

inorganic acid waste) is used as a reducing agent (*see also:* Haas and Vamos 1995, Chapter 5; Woodside 1999, pp. 245–249).

Stabilization and Solidification

Stabilization and solidification of liquid and semi-solid wastes are processes used to immobilize the hazardous constituents and provide physical structure to a waste, in order that it can be easily handled and land-disposed with minimized hazard to the land and groundwater. The processes that are used may involve some chemical reactions, but are primarily used to dewater and/or achieve physical encapsulation of the constituents.

Metals and nonmetals can be solidified with pozzolan[14] and lime after the waste has been precipitated. Metal hydroxides and calcium salts will combine with fly ash and lime in the presence of water to form a cementitious product. A typical formulation is

$$\text{Final Solid} = \text{Lime} + \text{Fly Ash} + \text{Waste} + \text{Water}$$

where lime is 5 to 15% by weight, fly ash is 50 to 65% by weight, waste is 8 to 19% by weight, and water is 10 to 60% of the original sludge by weight. For an organic sludge, a typical mixture ratio would be as above except having water at 10 to 20% by weight (DuPont 1988).

Figures 7.14, 7.15, and 7.16 illustrate a typical solidification process — liquid waste storage and blending, lime and fly ash storage and dispensing, followed by

FIGURE 7.14 Solidification process. Liquid waste storage and blending.

---

[14] An additive such as siliceous volcanic ash, or fly ash, originally used to improve the curing and strength properties of Portland cement concrete.

FIGURE 7.15 Solidification process. Lime and fly ash storage and dispensing.

FIGURE 7.16 Solidification process. Pugmill mixing and related equipment.

mixing in a pugmill. The mixed matrix is then spread in drying beds where it solidifies. The solidified material may then be loaded and transported to a land disposal facility.

Solidification and/or stabilization technologies that may be suitable for specific situations include

- Thermoplastic materials such as bitumen, asphalt, polyethylene, or polypropylene

- Thermosetting reactive polymers including reactive monomers urea-form-aldehyde, phenolics, polyesters, epoxides, and vinyls, which form a polymerized material when mixed with a catalyst
- Polymerization of spills of chemicals that are monomers or low-order polymers by adding a catalyst (Woodside 1999, Chapter 12).

Technologies for *in situ* solidification and/or stabilization are discussed in Chapter 11.

Oxidation and Reduction

The chemical processes of oxidation and reduction can be used to render hazardous wastes less hazardous or harmless. An *oxidation* reaction increases the valence of an ion with a loss of electrons. A *reducing* reaction decreases the valence with a gain of electrons. Reactions that involve both oxidation and reduction are known as *redox* reactions.

As shown earlier, hydroxide precipitation of hexavalent chromium cannot be accomplished without first being **reduced** to the comparatively innocuous trivalent chromium. The Cr^{+3} can then be precipitated as chromic hydroxide. Although the use of chemical reductants is commonly practiced in *ex situ* treatment of chromium-bearing waste, cultured organisms having a high degree of tolerance for Cr^{+6} can, in suitable conditions, affect the reduction (Alexander 1999, Chapter 18).

Cyanide-bearing wastewater, commonly generated by the metal-finishing industry, is typically **oxidized** with alkaline chlorine or hypochlorite solutions (the chlorine is **reduced**). In this process, the cyanide (the contaminant of interest) is initially oxidized to a less toxic cyanate and then to carbon dioxide and nitrogen in the following reactions:

$$NaCN + Cl_2 + 2NaOH \rightarrow NaCNO + 2NaCl + H_2O$$

$$2NaCNO + 3Cl_2 + 4NaOH \rightarrow 2CO_2 + N_2 + 6NaCl + 2H_2O$$

(Wentz 1989, p. 153). Oxidation of cyanide may also be accomplished with hydrogen peroxide, ozone, and electrolysis (Dawson and Mercer 1986, p. 333).

Biological Treatment

Organic waste constituents may be transformed, removed, and/or converted to inorganic byproducts by the use of aerobic, anaerobic, (or both) microorganisms. Biological treatment of municipal and industrial wastewaters (not to be confused with bioremediation; *see:* Chapter 11) is generally used for removal of organic pollutants from wastewater. Such systems employ the controlled use of micororganisms to destroy chemical compounds. It is effective with wastewaters having low-to-moderate concentrations of simple organic compounds and lower concentrations of complex organics.

Biological treatment of toxic organic components requires considerably more sophisticated operational control than is necessary with nontoxic wastewaters. The microorganisms used in biological treatment processes may be vulnerable to destruction by shock loading or rapid increases in the rate of feed. Acclimation and development of a functional population of biota may require considerable time, and the system is continuously subject to upset (adapted from Dawson and Mercer 1986, p. 335).

The biological treatment units, or bioreactors (Figure 7.17), used for treatment of hazardous waste components in industrial wastewaters are similar in configuration and operation to those used in municipal sewage treatment works. They include activated sludge, trickling filters,[15] biofilters, rotating biological contactors, aerated lagoons, oxidation ponds, and anaerobic sludge digesters (*see also:* Alexander 1999; George 2000, pp. 560–561; Haas and Vamos 1995, Chapter 6; Picardal et al., 1997; Govind et al., 1997).

Subpart Q — Chemical, Physical, and Biological Treatment

A list of some of the chemical, physical, and biological treatment processes which are regulated by 40 CFR 265, Subpart Q, is provided in Table 7.3. Some of the more commonly used processes are described in the preceding paragraphs. Examples of treatment processes not frequently used in treatment of RCRA wastes include distillation, reverse osmosis, ion exchange, and filtration. There are many different types of treatment processes, and the processes are frequently waste-specific. For these reasons, the EPA has not developed detailed regulations for any particular type of process or equipment. Instead, general requirements have been established in Part 265, Subpart Q, to assure safe containment of hazardous wastes.

The Subpart Q general requirements require owners/operators to:

- Avoid treating any waste that could cause equipment to rupture, leak, corrode, or otherwise fail.
- Equip continuous waste feed conveyances with a feed cut-off system.
- Comply with special requirements for ignitable or reactive wastes.
- Remove any waste characteristic before placing the waste in the process or equipment.
- Comply with special requirements for waste analyses in addition to general waste analysis requirements.
- Inspect discharge control, safety equipment, and monitoring data at least daily.
- Inspect construction materials of the treatment process for corrosion, leakage or erosion at least weekly.
- Remove all hazardous waste and residues from processes, equipment, and discharge confinement structures at closure (40 CFR 265.401-404).

---

[15] A misnomer: a "trickling filter" does no filtering. It is usually a round open-top tank filled with a media upon which biological films form. Wastewater is distributed over the media by rotary arms and "trickles" slowly over and through the mass while the microbes growing in and on the biofilm consume the food energy contained therein.

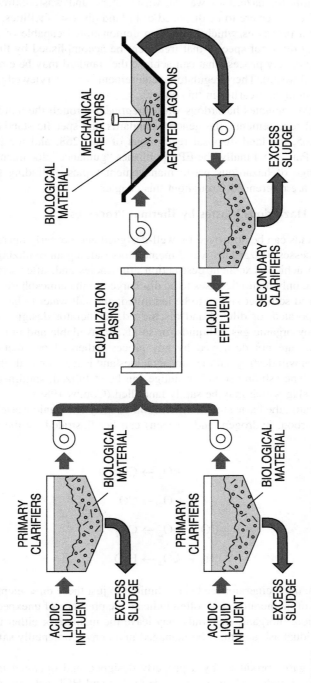

FIGURE 7.17 Biological treatment of industrial waste.

In the 40 CFR 268 Land Disposal Restrictions, the EPA establishes extensive treatment standards for hazardous wastes, wastewaters, and waste extracts which must be met if the wastes are to be disposed of in land disposal facilities. EPA also lists the treatment processes which have been demonstrated capable of achieving the standards, but does not specify that treatment be accomplished by the demonstrated process (i.e., any process that can achieve the standard may be employed to treat the regulated waste). These regulatory requirements are overviewed in the *Part 264, Subpart F* summary, earlier in this chapter.

Thus, the EPA regulates hazardous waste *treatment* through the administrative and nontechnical requirements, the general standards, the specific standards of 40 CFR 264 and 265, the land disposal restrictions of Part 268, and the permitting requirements of Part 270. Finally, the EPA publishes "guidance" documents dealing with a wide range of hazardous waste management topics, including treatment. Several of them are referenced throughout this chapter.

Destruction of Hazardous Wastes by Thermal Processes

Organic compounds can be destroyed by well-designed and properly operated high-temperature processes. Hazardous waste incinerators (adding an oxidizing agent to the process) can achieve excellent destruction efficiencies and, after scrubbing of the exhaust, leave only nontoxic gases to be discharged to the atmosphere; inorganic residues of ash and scrubber sludge to be landfilled; and salt water to be injected in deep wells, evaporated, or diluted and discharged. Incinerator designs which can effectively destroy organic gases, liquids, or solids are available and in use.

Heavy metals are not destroyed by any process (thermal or otherwise), but thermal processes will destroy sulfides and cyanides and leave all metals in the form of metal oxides. The ash and scrubber sludge can be stabilized, solidified, or converted to glassy slag which may be safely landfilled (Combs 1989).

In an incinerator, the basic stoichiometric combustion of organic waste materials (composed of carbon, hydrogen, and oxygen) can be illustrated by the following equations:

$$C + \tfrac{1}{2}O_2 \rightarrow CO$$

$$C + O_2 \rightarrow CO_2$$

$$CO + \tfrac{1}{2}O_2 \rightarrow CO_2$$

$$H_2 + \tfrac{1}{2}O_2 \rightarrow H_2O$$

In real terms, the incinerator feed is not limited to just these three elements. Gas chromatographs of incinerator gases often indicate the presence of unexpected products of combustion. Inorganic materials may leave the incinerator either in the flue gas or in the residual ash and must be managed in an environmentally safe manner (Brunner 1988).

Combustion gases produced by a properly designed and operated incinerator burning chlorinated hydrocarbons are CO_2, H_2O, N_2, and HCl. All except HCl are

---

TABLE 7.4
Incineration Processes

Multiple hearth
Fluidized bed
Recirculating fluidized bed
Liquid injection
Fume
Rotary kiln
Cement kiln
Large industrial boiler
Multiple chamber
Cyclonic
Auger combustor
Two stage (starved air)
Catalytic combustion
Oxygen enriched
Molten salt
Infrared (moving belt)

Source: Combs (1989).

---

completely nonhazardous. The HCl can be reacted with lime or caustic to produce nonhazardous salts, which can be landfilled (Combs 1989).

Excess air is supplied in order to ensure that the combustion reaction is driven to completion. Where insufficient air (oxygen) is supplied to the incinerator, the exhaust will contain "products of incomplete combustion" or PICs. The major operating parameters for destruction of organic wastes in incinerators are:

- Turbulence (a function of design)
- Excess air (nominally 25 to 100%)
- Destruction temperature (1200 to 3000 EF) (648 to 1694°C)
- Residence time (nominally 2 sec)

In practice, operating temperatures of 1600 to 2200°F are required to ensure destruction of organic wastes at 2-sec residence time.

As noted earlier, incineration of hazardous wastes is considered by many professionals and practitioners to be preferable to most treatment or destruction processes.[16] The drive to reduce or eliminate land disposal has sharpened the search for the ultimate incinerator. Several incinerator configurations and processes have been developed or are in development. Each variation is intended to meet a particular requirement, deal with a particular problem, or make use of an existing facility. Table 7.4 provides a list of the currently identified processes. Figure 7.18 diagrams

---

[16] Incineration of hazardous waste has its detractors. In the early 1990s, environmental activists mounted vigorous opposition to the siting and/or permitting of new hazardous waste incinerators. The issue surfaced in the 1992 presidential campaign, and shortly after taking office, the Clinton administration announced the "Hazardous Waste Minimization and Combustion Strategy" and suspended permitting of new incinerators.

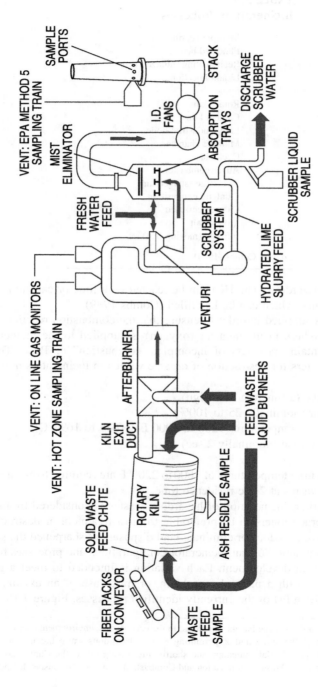

FIGURE 7.18 Rotary kiln incineration.

FIGURE 7.19 Rotary kiln incinerator layout. (ENSCO, El Dorado Facility, 309 American Circle, El Dorado, AR 71730. With permission.)

the rotary kiln incinerator, probably the most popular design in current use. Figure 7.19 is a view of a rotary kiln installation. Figure 7.20 diagrams a liquid injection incinerator, also a popular design.

As the land disposal restrictions were implemented, operators of boilers and industrial furnaces (BIFs), including cement kilns, many of which had been burning their own hazardous wastes, began operating as commercial burners. The practice was (and to some extent remains) fraught with uncertainty and contention. A central issue was the question of whether the wastes were legitimate hazardous waste fuels or if the BIFs were being used as incinerators, i.e., "sham recycling." In 1990, the EPA published standards for BIFs in 40 CFR 266, Subpart H, and began offering "interim status" to permit applicants while their permits were being processed. By July 1993, some 159 BIFs in 34 states and Puerto Rico had applied for permits. Further controversy arose over the standards, with incinerator operators protesting that BIFs were being given unfair competitive advantage in the less stringent standards. These and other issues, various petitions, legislative proposals, and related litigation were at a vigorous pitch when the EPA Administrator, on May 18, 1993, announced a "temporary capacity freeze" as the centerpiece of a "Hazardous Waste Minimization and Combustion Strategy." The freeze suspended new permitting, for 18 months, and the agency announced that it intended to propose new regulations for hazardous waste incinerators and industrial boilers and furnaces within 18 months to 2 years.

The "combustion" portion of this strategic effort[17] produced *proposed* Maximum Achievable Control Technology (MACT) combustion standards for incinerators, cement kilns, and lightweight aggregate kilns on April 19, 1996 and *final* MACT standards for the three categories on September 30, 1999 (*see:* Appendix A to this chapter). In a related development, on August 5, 1999, the EPA proposed a rule on the emission, transport, and disposal of cement kiln dust (CKD), a by-product of the cement production process. CKD had previously been excluded from the RCRA hazardous waste rules under the "Bevill Amendment" (*see:* Chapter 1 or Glossary). MACT standards for BIFs had not been proposed at the time of this writing, however, the health-based standards of 40 CFR 266, Subpart H, for BIFs continue in force. The "waste minimization" portion of the strategy is discussed in Chapter 8.

Many of the emerging technologies for hazardous waste treatment and/or destruction fall into the classification of "other thermal processes." Most employ oxidation (air/oxygen) as a final step, but this is not always necessary nor desirable. They are differentiated from incineration because they involve:

- Extensive electrical energy input (plasma arc pyrolysis, microwave discharge, advanced electrical reactor, *in situ* vitrification)
- Oxidation in the liquid phase (wet air and supercritical water oxidation)
- Pyrolysis or vaporization which may or may not be followed by incineration of the off-gases in a second stage (pyrolysis, calcination, thermal desorption)

If the gases produced in the first stages of these processes are not destroyed by incineration in the second stage, then some other treatment process must follow, e.g., carbon adsorption (Combs 1989).

The Dempsey and Oppelt paper, Incineration of Hazardous Waste: A Critical Update, in *Air and Waste*, Vol. 43, January 1993, is an outstanding overview and update of hazardous waste incineration technology. Most of the technical content of the paper remains valid in early 2000 (*see also*: Rappe et al. 1986, Sections I and II; EPA 1993a; Johnson and Cosmos 1989; Wentz 1989, Chapter 8; Barton et al. 1992; Haas and Vamos 1995, Chapter 7; Gouldin and Fisher 1997).

Subpart O — Hazardous Waste Incinerators

The Subpart O standards of Part 265 require **interim status** incinerators to meet general operating requirements that include

---

[17] Pertaining to approximately 1% of the annual hazardous waste generation in the U.S., by the pre-1997 RCRA definition of "hazardous waste;" or about 10% of the hazardous waste generated, based on the 1997 RCRA definition which excluded wastewaters (*see:* Chapter 2). A large portion of the hazardous waste subject to the 1999 MACT combustion rule is "wastewater." Thus, the total annual generation of hazardous waste remains minimally impacted by the combustion portion of the strategy, however, a high-profile effort has been made to assuage concerns for health risks in downwind areas.

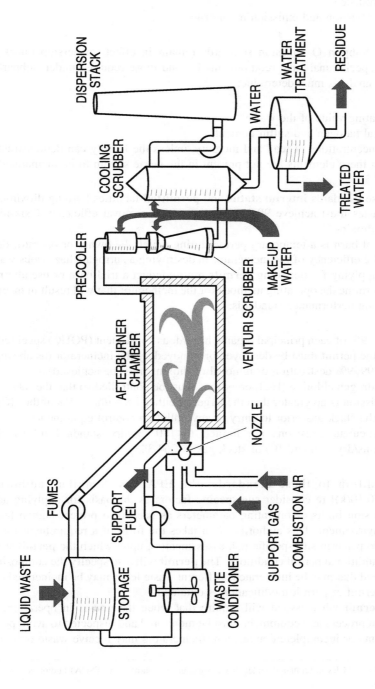

FIGURE 7.20 Liquid injection incinerator.

- Achieving normal steady-state combustion conditions before wastes are introduced
- Combustion and emission monitoring

While the Subpart O emission standards remain in effect, owners/operators are subject to special analytical requirements beyond those required under Subpart B. The waste analysis must determine:

- Heating value of the waste
- Total halogen and sulfur content
- Concentrations of lead and mercury, unless the facility can demonstrate that these elements are not present in the waste stream to be incinerated

Under these standards **interim status** and **permitted** facilities burning dioxin-containing wastes must achieve 99.9999% destruction removal efficiency ("six-nines DRE") of dioxins.[18]

The trial burn is a temporary period during which the owner or operator demonstrates the efficiency of the incinerator in destroying a surrogate hazardous waste. Facilities applying for operating **permits** must conduct a trial burn or use alternate data to determine the operating methods for the incinerator that will result in meeting the following performance standards:

- 99.99% of each principal organic hazardous constituent (POHC) specified in the permit must be destroyed or removed by the incinerator (as above, 99.9999% destruction or removal of dioxins must be achieved).
- Hydrogen chloride (HCl) emissions must be controlled so that the rate of emission is no greater than the larger of either 1.8 kg/hr or 1% of the HCl in the stack gas prior to entry to any pollution control equipment.
- Particulate emissions are limited to 180 mg/dry standard m^3 (0.08 grains/dry standard ft^3) of stack gas (§ 264.343).

In addition to these standards, the EPA has "omnibus authority" [40CFR270.10(k)] to consider site-specific factors as the basis for applying additional emission limits or operating parameters as needed to protect human health and the environment. This authority often takes the form of a requirement for the applicant to perform site-specific risk assessment(s), upon which the **permit** writer may base additional permit conditions. The **permit** will also specify the composition of waste feed that may be incinerated. Different waste feeds may be incinerated only if a new permit or permit modification is obtained.

The **permit** when issued will require that when the unit is in operation, the combustion process and equipment must be monitored and inspected to avoid potential accidents or incomplete combustion. Incinerators may receive waste only after

---

[18] Incineration of PCBs is regulated under Toxic Substances Control Act (TSCA) authorities (*see:* 40 CFR 761.70).

the destruction removal efficiency (DRE) has been achieved and the unit is complying with its operating requirements.

The newly promulgated Maximum Achievable Control Technology (MACT) standards[19] will eventually move the emission standards for combustion units to 40 CFR, Part 63 (Clean Air Act standards). **Interim status** and **permitted** combustion facilities in operation on September 30, 1999 have 3 years in which to achieve compliance with the MACT standards promulgated on that date. The new standards limit emissions of dioxides, furans, mercury, low-volatile and semi-volatile metals such as cadmium and lead, particulate matter, acid gases, and carbon monoxide. There are separate emission standards for each of three categories of combustion units. Appendix A summarizes numbers of existing incinerators, cement kilns, and lightweight aggregate kilns which are subject to the standards, the emission limits for each category, and the Federal Register page for each. The reader or practitioner may access the full documents at September 30, 1999, 64 FR 52827-52876 and 64 FR 52877-52926. The EPA has also produced an unnumbered question and answer publication that can be accessed at http://www.epa.gov/hwcmact.

Subpart P — Thermal Treatment of Hazardous waste

Incinerators, cement kilns, and lightweight aggregate kilns are considered a type of thermal treatment process, although the intent with each is destruction of the waste. As noted above, these units are regulated by the **permitted** facility standards of Part 264, the **interim status** standards of Part 265, and the MACT standards of Part 63.[19] Boilers and industrial furnaces are regulated [with certain exemptions — § 266.100(b)] by Part 266, Subpart H, and are subject to most of the same requirements as are incinerators. Less conventional methods of thermal treatment, such as molten salt combustion, calcination, wet air oxidation, and fluidized bed combustion, are regulated under Part 265, Subpart P. The EPA has never established final **permit** requirements for the thermal treatment units of Subpart P. Many of these Subpart P units, which do not fit the Parts 264 nor 266 unit descriptions have now met the miscellaneous unit standards of Part 264, Subpart X, and have been issued final **permits** (EPA 1997c).

ACCUMULATION AND STORAGE OF HAZARDOUS WASTE

As discussed earlier, the accumulation of hazardous waste has been one of the most troublesome of hazardous waste management issues for the regulatory agencies. Most of the early disasters and many of today's Superfund sites grew from the uncontrolled accumulation of hazardous wastes. Congress, in crafting the RCRA statutes, and the EPA, in the implementing regulations, have sought to impose rigorous controls and accountability upon all who accumulate and/or store hazardous wastes.

RCRA defines "storage" as the holding of hazardous waste for a temporary period, at the end of which the hazardous waste is treated, disposed of, or stored elsewhere. Throughout the Subtitle C regulations, the accumulation of hazardous waste beyond a prescribed period (in most situations 90 days) is considered to be

---

[19] *See:* Appendix A to this chapter.

FIGURE 7.21 Typical abandoned drum scene.

"storage." The owner or operator of a facility in which waste is to be held for more than 90 days must apply for a permit before commencing accumulation and must comply with the regulations pertaining to storage facilities.[20]

The 40 CFR 264/265 Subparts A through E, General Facility Standards, previously overviewed, contain the major provisions applicable to storage facilities. Subparts I and J, the standards for the use of containers and tanks, pertain to all RCRA facilities that use them.[21] Since the primary function of containers and tanks is "storage," we include our overview of those subparts here. The container and tank management requirements for **interim status** and **permitted** TSD facilities are very similar. The minor differences will be pointed out in the following sections.

Whether viewing the most modern and well-operated hazardous waste facility or the most outrageous of abandoned hazardous waste dump sites, Americans have come to think of the standard 55-gal drum when they think of hazardous waste (and vice versa). Figures 7.21 and 7.22 illustrate the point. The "drum" may be any one of several Department of Transportation (DOT) specified 55-gal containers, but they have collectively become the most frequently used (if not the standard) container for collection, storage, shipment, and disposal of liquid hazardous wastes.

Subpart I — Containers

Unless the owner/operator of a **permitted** or **interim status** RCRA storage facility can be certain that a hazardous waste container will never be shipped, selection of the proper drum or container for wastes that are to be stored should be made in

---

[20] Exceptions to this otherwise nearly universal rule are the exceptions for small quantity generators and conditionally exempt small quantity generators as noted in Chapter 5 and 40 CFR 261, Subpart A, and Part 262, Subpart C.

[21] Section 265.201 provides special requirements for SQGs that *accumulate* hazardous waste in tanks.

FIGURE 7.22 Shipment of liquid hazardous waste in 55-gal drum.

accord with the Chapter 49 DOT regulations. The user consults the 49 CFR 172.101 Hazardous Materials Table (see Figure 5.1). Column 7 is checked for any applicable sprecial provisions; then packaging authorizations of column 8 are consulted for the appropriate section of 49 CFR 173 where general and specific packaging and shipping requirements can be found. Figure 7.23 shows the location of the drum stamp information which must be in accord with the specification for the hazardous material to be contained.

FIGURE 7.23 Drum stamp information.

Containers selected for use only in storage and/or disposal of hazardous waste must be in good condition, clean, and free of rust, dents, and creases, prior to use. The regulations additionally require:

- Containers holding hazardous waste must always be closed, except when wastes are added or removed.
- Wastes in leaking or damaged containers must be recontainerized.
- Compatibility of the waste with the container (i.e., corrosive wastes should not be stored in metal containers) must be ensured.
- Hazardous wastes must not be placed in an unwashed container that previously held an incompatible waste or material.
- Handle containers properly to prevent ruptures and leaks.
- Prevent the mixture of incompatible wastes.
- Conduct inspections to assess container condition.

Containers holding ignitable or reactive waste must be located at least 15 m (50 ft) from the facility property line. Containers in storage areas must be placed to maintain aisle space to allow unobstructed movement of personnel, fire protection equipment, spill control equipment, and decontamination equipment to any area of facility operation (40 CFR 264 and 265.35). Containers holding liquids must not be stacked in a storage area (*see also:* Occupational Health and Safety Administration (OSHA) regulations and standards pertaining to container storage in 29 CFR 1910.106; 1910.120; 1915.173; 1926.65).

Container storage areas in **permitted** storage facilities must be structurally sound, free of cracks or gaps, and sufficiently impervious to contain leaks, spills, or accumulated precipitation. The base must be sloped or the system must be designed to facilitate drainage of spills to a collection point. There must be a secondary containment system with capacity to contain at least 10% of the volume of the containers, or 100% of the largest container, whichever is greater. Containers which hold no free liquids are exempt from this determination. **Interim status** storage facilities are not required to have secondary containment systems.

The 40 CFR 262 Subpart C, Pre-Transport Requirements for Hazardous Waste Generators, makes provision for generators to *accumulate* as much as 55 gal of hazardous waste, or 1 qt, of acutely hazardous waste in containers at or near any point of generation where wastes initially accumulate. This location, known as a *satellite accumulation point,* must be under the control of the operator of the process that is generating the waste. The generator facility is not required to have a permit or interim status, but must comply with §§ 265.171, 265.172, and 265.173(a), requiring containers to be in good condition, compatible with the wastes being stored, and to keep the container closed except when adding or removing the waste. The satellite container must be marked with the words "hazardous waste." If the hazardous waste or acutely hazardous waste container reach the specified limits, a new container must be started and marked with the

FIGURE 7.24 Hazardous waste storage tanks with secondary containment.

starting date. The accumulated wastes must be moved, within 3 days, to a temporary storage area or to a permitted or interim status hazardous waste management facility. The EPA has not, in mid-2000, placed limits upon the number of satellite accumulation points that a facility may operate. Owners/operators should read § 262.34(c) carefully and/or consult the state regulatory agency, as appropriate, before initiating satellite accumulation point operations.

Owners and operators of RCRA hazardous waste storage facilities are subject to the air emission standards of Parts 264 and 265, Subparts AA, BB, and CC.

Subpart J — Tanks

The Subpart J regulations apply to stationary tanks storing wastes that are hazardous under Subtitle C of RCRA (Figures 7.24 and 7.25). The regulations pertaining to underground storage tanks storing petroleum products (exempt from Subtitle C regulation) or hazardous *substances* are found in Subtitle I of RCRA.[22] Regulations governing hazardous waste tanks were substantially expanded under HSWA. There are five general operating requirements:

- *Tank assessment* must be completed to evaluate the structural integrity and compatibility with the wastes that the tank system is expected to hold. The assessment covers design standards, corrosion protection, tank tests, waste characteristics, and the age of the tank.

[22] Management of underground storage tanks for petroleum and hazardous substances is overviewed in Chapter 14.

FIGURE 7.25 Hazardous waste storage tanks with secondary containment.

- *Secondary containment and release detection* is required unless the tank does not contain free liquids and is located in a building with impermeable floors. Secondary containment systems must be designed, installed, and operated to prevent the migration of liquid from the tank system and to detect and collect any releases that do occur. Containment systems commonly used include liners, vaults, and double-walled tanks. Figures 7.26, 7.27, and 7.28 illustrate acceptable containment systems.
- *Operating and maintenance requirements* necessitate the management of tanks to avoid leaks, ruptures, spills, and corrosion. This includes using freeboard or a containment structure to prevent and contain escaping wastes. A shut-off or bypass system must be installed to prevent liquid from flowing into a leaking tank. Figure 7.29 illustrates an all-too-common practice that is certain to be the eventual cause of disaster.
- *Response to releases* must include immediate removal of the remaining contents of leaking tanks. The area surrounding the tank must be visually inspected for leaks and spills. Based on the inspection, further migration of the spilled waste must be stopped and contaminated soils and surface water must be disposed of in accord with RCRA requirements. All major leaks must be reported to the EPA or the state agency.
- *Closure and post-closure* requirements include the removal of all contaminated soils and other hazardous waste residues from the tank storage area

FIGURE 7.26 Secondary containment with additional containment for spillage from fill connection.

FIGURE 7.27 Invitation to disaster.

FIGURE 7.28 Secure landfill leachate detection and monitoring access gallery.

FIGURE 7.29 Land disposal liner sections being welded. (From GSE Lining Technology, Inc., 19103 Gundle Rd, Houston, TX 77073. With permission.)

at the time of closure. If decontamination is impossible, the tank storage area must be closed following the requirements for landfills.

RCRA facilities, whether subject to generator rules in **interim status** or finally **permitted**, are also subject to the Parts 264 and 265, Subparts AA and BB, covering atmospheric emissions from process vents and equipment leaks. Facilities that treat or store hazardous wastes in impoundments, tanks, or containers are subject to Parts 264 and 265, Subparts CC, which impose air emission standards on the treatment or storage units.

Subpart DD — Storage in Containment Buildings

In 1993 the EPA published final regulations pertaining to storage of hazardous wastes in containment buildings at 40 CFR 264 and 265, Subpart DD. All RCRA facilities which store bulk hazardous waste in "containment" buildings are subject to the rules. They provide extensive structural specifications and exacting operating requirements, all of which are designed to ensure that:

- Fugitive dusts are contained.
- Liquid wastes are collected and completely and securely contained.
- There are no cracks or other structural defects that could permit release of hazardous waste constituents.
- Tracking of wastes from the facility by persons or equipment is prevented.

The containment building design must be certified by a qualified, registered professional engineer prior to operation of the unit.

At closure of the unit, the owner or operator must remove or decontaminate all waste residues and all contaminated equipment, subsoils, and structures. If, after removing all residues and making all reasonable efforts to decontaminate the facility, not all contaminated subsoils can be practicably removed or decontaminated, the closure and post-closure requirements that apply to landfills become applicable to the facility.

LAND DISPOSAL OF HAZARDOUS WASTE

Hazardous waste disposal practice has, historically, followed the path of least resistance. As discussed in the introductory chapter, several factors have driven hazardous wastes onto and beneath the earth's surface. These factors include (1) the relatively low cost of land and land disposal procedures; (2) the environmental legislation of the 1970s and early 1980s which placed increasingly stringent controls on releases to the atmosphere and to "waters of the nation;" and (3) widely held beliefs to the effect that land disposal was safe and proper.

The tragic consequences of these practices are now upon us in the form of several thousand contaminated sites in various stages of remedial activity, some of which cannot be cleaned up; tens of thousands of "brownfields" that will receive minimal, if any, remediation; more than 6000 contaminated sites in Eastern Europe; untold thousands of sites elsewhere in the world; and contaminated groundwater, estuaries, and bays coinciding with irresponsible land disposal of hazardous wastes in countless sites around the world. Newly credentialed/assigned/designated professionals and practitioners must understand the calamitous extent of the problem, the magnitude of the task of correcting the existing insults, the absolute necessity of preventing new cases of ruined land and groundwater, and the desperate need for improved remediation technologies. In juxtaposition, there is similar need for political leadership, legislative, administrative, judiciary, and technological effort to avoid the imperatives that drive us to remedial overkill and squandered resources. The following few pages summarize and overview the principle technologies and regulatory processes in place in the U.S. at the turn of the century. A few of the promising technological innovations in hazardous waste treatment are cited.

In RCRA § 3004, Congress recognized nine types of land disposal units (LDUs) and defined land disposal as placement of hazardous wastes in any one of them:

- Landfills
- Surface impoundments
- Waste piles
- Injection wells
- Land treatment facilities
- Salt domes
- Salt bed formations
- Underground mines
- Underground caves

The EPA has promulgated unit-specific technical standards for four LDUs within the TSDF requirements of Parts 264 and 265. The four — landfills, waste piles, surface impoundments, and land treatment units — and the applicable subparts are discussed briefly in the following paragraphs. Although the design objectives of the four are unit-specific, the regulatory goal of minimizing the formation and migration of leachate is nearly identical for each. Salt dome/bed formations and underground caves/mines are regulated as Miscellaneous Units in Part 264, Subpart X. Underground injection wells are regulated in Parts 264 and 265, Subpart R and Subpart X (EPA 1997c).

Landfills

"Sanitary landfills" were developed for the disposal of municipal refuse, as an orderly alternative to the open dump. Early practice called for the working face to be compacted and maintained at a 30 degree slope, with 6 in. of daily cover on the working face and with 2 ft of top cover over the daily cells (Ehlers and Steel 1958, pp. 197–203).

Codes, specifications, and administrators concerned themselves with control of insects, rodents, odors, and blowing refuse. Few operators of sanitary landfills gave thought to the effects of hazardous waste. Sites receiving liquid wastes usually designated a discharge area away from the compacted cells in order to avoid erosion of the cells. Dry hazardous wastes were legally mixed with domestic refuse, compacted, buried, and forgotten.

With large numbers of these sites now crowding the National Priority List (NPL; for remediation under Superfund), regulators, designers, and operators of landfills, and "responsible parties," have become appropriately concerned that wastes be safely contained within secure landfills. Design and operating procedures have evolved to include elaborate safeguards against leakage and migration of leachates. Most sanitary landfills do not knowingly accept hazardous wastes, and landfill disposal of bulk liquids is banned by RCRA. Later in this chapter we will cover the RCRA Land Disposal Restrictions which are increasingly stringent with regard to all forms of land disposal.

FIGURE 7.30 Liner bedding for protection from puncture.

Secure landfills for hazardous waste disposal are now equipped with double liners, leakage detection, leachate monitoring and collection, and groundwater monitoring systems (see Figure 7.28). Synthetic liners are of minimum 30-mil thickness. Liner technology has improved greatly and continues to do so. Very large sections of liner fabric now minimize the numbers of joints. Adjacent sections are "welded" together to form leak-proof joints having a high degree of integrity (see Figure 7.29). Liners are protected by sand bedding or finer material devoid of sharp edges or points which might penetrate the liner fabric (Figure 7.30). Another layer of bedding protects the inner liner from damage by machinery working the waste (Figure 7.31). Some states allow one of the liners to be of natural clays. The completed liner must

FIGURE 7.31 Liner bedding for protection from puncture. (From GSE Lining Technology, Inc., 19103 Gundle Rd, Houston, TX 77073. With permission.)

FIGURE 7.32 Secure landfill leachate detection and monitoring access. (Courtesy of Safety-Kleen Corporation.)

demonstrate permeability of less than 10^{-6} cm/sec and must include a leachate collection system.

Leachate detection and collection systems are equipped with access to sumps via galleries, pumps, or other means of leachate sampling or removal (Figure 7.32). The double liner, leakage detection, and leachate collection systems are diagrammed in Figure 7.33.

Landfill caps are the subject of detailed guidance by the EPA. Figure 7.34 is a cross-sectional diagram of a typical cap design. The objective of the cap design is to protect the cells from erosion, to route potential run-on around and away from the cap, and to prevent buildup of generated gases within the landfill.

Groundwater monitoring schemes are designed to provide upgradient (background) water quality data and to detect downgradient differences in critical water quality parameters. As was discussed earlier in this chapter, RCRA requires upgradient and downgradient patterns of monitoring well placement to detect leakage from landfills (*see also:* EPA 1981, 1982, 1987a,b, 1988).

Subpart N — Landfills

Landfills have historically presented two general classes of problems. The first of these includes fires, explosions, production of toxic fumes, and related problems from the improper management of ignitable, reactive, and incompatible wastes. Owners or operators are required to analyze their wastes to provide enough information for proper management thereof. They must control the mixing of incompatible wastes in landfill cells. They may landfill ignitable and reactive wastes only when the wastes meet all applicable requirements of the land disposal restrictions of 40 CFR 268 and have been rendered unignitable or nonreactive (40 CFR 264,

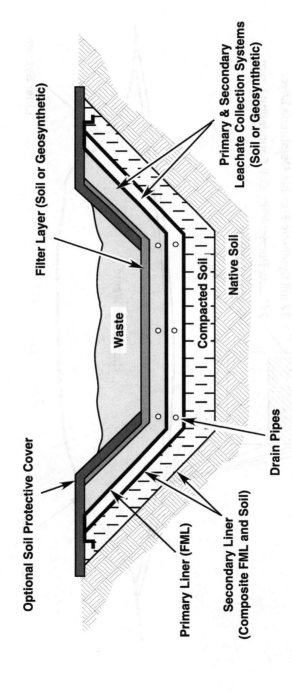

FIGURE 7.33 Cross-section of a secure landfill double-liner system.

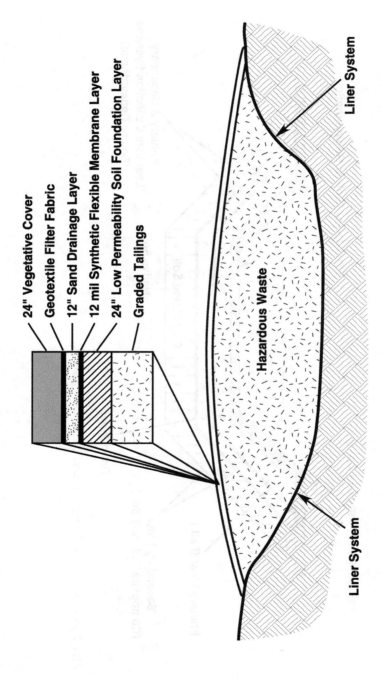

FIGURE 7.34 Land disposal site cap designed for maximum resistance to infiltration. (From U.S. Environmental Protection Agency.)

265.312). Placement of bulk or noncontainerized liquid hazardous waste or hazardous waste containing free liquids in any landfill is prohibited (§§ 264, 265.314).[23]

The second general class of landfill problems concerns the contamination of surface and groundwaters. RCRA hazardous waste landfills must have double liners, a leachate detection and removal system (LCRS) above each liner, and a leak detection system (§§ 264, 265.221). The LCRS must drain to a sump, where an approved "action leakage rate" (ALR) is measured.[24] The Parts 264 and 265 regulations require diversion of run-on away from the active face of the landfill; treatment of any liquid wastes or semisolid wastes so that they do not contain free liquids; and proper closure (including a cover) and post-closure care to control erosion and the infiltration of rainfall. Unless they are very small (such as ampules), containers placed in landfills must be at least 90% full or be crushed or otherwise reduced in volume, in order to prevent voids and collapse of final landfill covers. Groundwater monitoring, as described in Subpart F, is required, as is the collection of rainwater and other run-off from the active face of the landfill. Segregation of waste, such as acids, that would mobilize, solubilize, or dissolve other wastes or waste constituents, is required (EPA 1997d). Extensive monitoring and inspection requirements are imposed throughout the active life of the landfill and during the closure and post-closure periods (§§ 264.303, 265.304) (*see also:* EPA 1997d; EPA 1998, Chapter 5).

Surface Impoundments

Prior to imposition of the RCRA regulations, surface impoundments were very frequently used for "treatment" of wastewaters having hazardous components. Some were of flow-through design, but many were described by their owners/operators as evaporation ponds or treatment ponds. Most were unlined, thereby allowing infiltration and ultimate groundwater contamination. In the case of the flow-through ponds, some degree of treatment may have been achieved by sedimentation and/or solidification of solids, surface emission of volatile organics, and/or oxygen transfer at the surface. Evaporation ponds lost most of their VOC content to the atmosphere, and may have achieved some degree of treatment via oxygen exchange at the surface, and by settling of solids. Many industrial waste impoundments were created by earthen dams that conveniently failed as the impounded content approached capacity.

In approximately the eastern half of the U.S., annual rainfall exceeds annual evaporation, so that "evaporation" ponds in this region would have been losing liquid to the subsurface. Aerated impoundments, if properly managed, may have developed biota that were effective in attacking certain organic contaminants. However, the aeration accelerates the transfer of VOCs to the atmosphere, and the losses to the groundwater would have been similar to those discussed above. Impoundments are similar to other biological systems in ineffectiveness with heavy metals, most inorganic, and some organic components of hazardous wastewaters. In summary, it is questionable whether significant treatment beyond dispersion of pollutants into the

---

[23] In certain situations small containers (such as ampules placed in lab packs) holding free liquids may be landfilled (40 CFR 264, 265.316).

[24] The action leakage rate is the maximum design flow rate that the leak detection system can remove without the head on the bottom liner exceeding one foot (*see:* 40 CFR 264, 265.302).

atmosphere, surface streams, and the groundwater was achieved in and by these early facilities.

The EPA continues to recognize various configurations of surface impoundments as treatment units, and they continue to be used by some industries. The RCRA requirements for double liners, leachate detection and removal systems, and groundwater monitoring have brought about improvements in protection of surface and groundwaters. The double-liner system for an impoundment facility is diagrammed in Figure 7.35. Such units do not contain, retain, nor re-entrain VOCs released at the surface or stripped out by aeration. Moreover, sludges containing precipitated solids, expired aerobic biota, active anaerobic bacteria, inorganics including heavy metals, and remaining organics must be periodically removed and subjected to further management. RCRA guidance for design, construction, and operation of surface impoundments is discussed below (*see also:* EPA 1987a, b).

Subpart K — Surface Impoundments

Hazardous waste surface impoundment units, lateral expansions, and replacement units, upon which construction commenced after July 29, 1992, must have a double liner, a leachate collection and removal system (LCRS), and a leak detection system (§§ 264, 265.221). New, lateral expansion and replacement units upon which construction or reuse began between July 15, 1985 and July 29, 1992 were required to have at least one liner and a LCRS. There are conditions under which the EPA may approve alternate designs, but the double liner requirement applies to essentially all new and currently operating surface impoundments (*see:* 40 CFR 264.221(a) regarding the exceptions).

The double liner system consists of a top liner to prevent migration of hazardous constituents into the liner and a composite bottom liner consisting of a synthetic geomembrane and 3 ft of compacted soil material. The design must include an LCRS above both liners, which also serves as the leak detection system. The unit must be designed with a bottom slope of at least 1% be made of materials chemically resistant to the wastes to be placed in the unit, and be able to remove leakage of hazardous constituents at a minimum specified rate established by the **permit**. The required LCRS includes sumps serving each layer. Pumps must be of sufficient size to collect and remove liquids from the sump and prevent liquids from backing up into the drainage layer. The design must enable measurement and recording of liquid levels in the sumps and of liquids removed.

The requirements include preventing liquids from escaping due to overfilling or run-on and prevention of erosion of dikes and dams. Liners must meet permit specifications for materials and thickness. To ensure that a surface impoundment meets all technical criteria, the EPA requires a construction quality assurance (CQA) program, including a CQA plan. The plan prescribes methods of monitoring and testing of construction materials and their installation and the required documentation (§ 264.19).

The EPA requires two types of inspections in addition to the general inspection requirements of §§ 264, 265.15. Sections 264 and 265.226 require inspections of liners and covers for any problems after construction and continued weekly inspec-

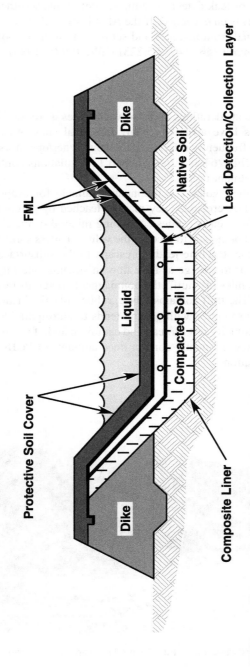

FIGURE 7.35 Cross-section of a liquid waste impoundment double-liner system.

tions after storms to monitor for evidence of deterioration, malfunctions, overtopping, sudden drops in liquid levels, etc. Another set of inspection requirements deals with monitoring of leak detection sumps to determine whether the allowable leakage rate (ALR) has been exceeded. If the ALR is exceeded, the owner/operator must notify the Regional Administrator and set in motion the response action plan incorporated in the **permit** (§§ 264, 265.223) (EPA 1997d, 1998).

Waste Piles

Essentially, the same considerations apply to waste piles as those discussed above. Hazardous waste piles have arisen on many industrial sites. As in the previous discussions, they were frequently referred to by their owner/operators as "treatment" piles. Indeed, the 40 CFR 264 and 265, Subpart L, regulations continue to refer to them as storage or treatment units.

Such piles make their volatile components available for evaporation and are subject to wind and water erosion. They may be leached by percolation of rainfall and run-on. As mentioned earlier, piles containing mineral or metal values may be leached with weak acid or caustic to recover the values. Unless carefully constructed over an impervious base, leachate tends to escape to the subsurface, there to contaminate groundwater or to emerge as base flow in streams (see Figure 1.20).

Figure 7.36 shows piles of "fluff," the nonmagnetic materials discarded in autoshredding operations. The fluff contains fabric, rubber, plastic, insulation, lead, and cadmium. It is usually saturated with oil and tends to auto-ignite. Samples of fluff usually fail the TCLP toxicity test for lead and cadmium and, if so, must be managed as hazardous waste; they frequently contain trace amounts of PCBs from electrical components and insulation.

FIGURE 7.36 Hazardous waste pile. "Fluff" from auto shredding operation.

FIGURE 7.37 Hazardous waste pile. "Dross" from aluminum salvage operation.

Aluminum dross, the ladle scum and dregs from aluminum salvage operations, is shown in Figure 7.37. The dross is allowed to accumulate in salvage yards where it is easily eroded by wind and run-on/run-off. Some dross fails characteristic tests for various metals and, if so, must be managed as hazardous waste.

Accepted practice has been the transfer of the contents of waste piles to landfills when pile size or management became a problem. Recent determinations of TCLP toxicity are requiring that dross and fluff piles be managed within acceptable RCRA management alternatives.[25] RCRA specifications for waste piles are similar to those for landfills and are discussed briefly below.

Subpart L — Waste Piles

Owners or operators of **permitted** waste piles that are located inside or under a structure; do not receive free liquid; are protected from surface water run-on; are designed and operated to control dispersal of the waste; and are managed to prevent generation of leachate may be exempt from groundwater monitoring requirements and the design and operation requirements of § 264.250(c). Otherwise, new units, lateral expansions, and replacement units require a double liner and leachate collection and removal system (LCRS) and in most cases require a second LCRS above the top liner (§§ 264.251, 265.254). Stormwater run-on and run-off controls are required and piles must be covered or managed to prevent dispersal of wastes by wind. Waste piles are subject to the same inspection and release response requirements including an approved allowable leak rate. Piles which are used as storage units must "clean close," and those used as treatment facilities must clean close or

---

[25] Some states, citing the variability in toxicity determinations on fluff and dross, have opted for "special waste" designations or other management options.

meet the closure requirements for landfills including post-closure care (EPA 1997d; EPA 1998).

Land Treatment

"Land treatment," "land application," and "land farming" are terms that have been used to label the practice of spreading hazardous wastes on the land surface. The practice ostensibly uses the interaction between plants and the soil surface to degrade or stabilize the waste. There have been successful applications of the practice in the treatment and disposal of hydrocarbon wastes from the petroleum industry and domestic sewage sludge. Chlorinated and other persistent compounds and wastes bearing heavy metals are not suitable for land applications for treatment or disposal. Sewage sludge containing heavy metals is similarly unsuitable. The practice requires very careful monitoring and controls to prevent contamination of surface and groundwaters.

Significant amounts of sewage sludge and petroleum industry wastes continue to be "land treated," but the practice is in disfavor among environmentalists and regulatory agencies. The EPA continues to legitimize the practice with the Subpart M regulations.

Subpart M — Land Treatment

Owners or operators of land treatment units (LTUs) must basically ensure that hazardous constituents placed in or on the treatment zone are degraded, transformed, or immobilized within the treatment zone. Whereas land *disposal* units generally require elaborate groundwater protection, land *treatment* units treat the waste within the matrix of the surface soil. If the hazardous waste does not meet the applicable treatment standards of the land disposal restrictions (LDRs), the unit owner/operator must obtain a no-migration variance before applying any waste to the unit (*see:* the section on land disposal restrictions later in this chapter).

The elements specified in the **permit** include

- Waste analyses to be performed
- Wastes that can be treated
- Design and maintenance of the land treatment unit to maximize treatment
- Soil monitoring
- Hazardous waste constituents that must be degraded, transformed, or immobilized by treatment
- Size and depth of the treatment zone

Prior to the application of waste, a treatment demonstration must be conducted to verify that the hazardous constituents are adequately treated by the unit. Use of the **interim status** or **permitted** LTU for growing food chain crops in a treated area containing hazardous constituents other than cadmium is permitted only after extensive demonstration of no substantial health risk (40 CFR 264 and 265.276). Application of cadmium is restricted to specific annual and cumulative quantities per §§ 264, 265.276(b).

The standards for **interim status** and **permitted** land treatment units include extensive unsaturated zone monitoring requirements. A monitoring program must be established to detect the migration of any hazardous constituents (§§ 264, 265.278. If migration from a **permitted** LTU is detected, a **permit** modification must be submitted outlining changes in operating practices to resolve the problem [§ 264.278(g)]).

Underground Disposal

As noted earlier, underground disposal of hazardous waste in salt dome formations, mine shafts, and injection wells has been and is practiced. Of these, the most widely recognized, and the subject of the Subpart R regulations, is deep-well injection of liquid hazardous wastes (Class I injection wells). A great variety of shallow "injection wells," many of them illegal, continue in operation throughout the nation and territories and are the sources of major groundwater quality problems. An excellent summary of the shallow-well problem has been prepared for the EPA by the Cadmus Group, Inc. (Cadmus 1991). Disposal in salt formation deposits, worked-out salt mines, and other open-pit and shaft mines is generally considered unsatisfactory due to water and groundwater contamination problems and will not be discussed further here.

Deep-well injection has achieved some degree of acceptance in the U.S. and elsewhere. A 1993 EPA database shows some 84 Class I[26] injection wells in operation in the U.S. (EPA, 1993b). Of these, more than half are located along the Texas-Louisiana Gulf Coast, with 31 in Texas and 17 in Louisiana. The area has suitable injection zones and large numbers of hazardous waste generators.

Figure 7.38 is a cross-sectional diagram of a typical injection well. Advocates and practitioners rationalize the practice with the proposition that:

> ... liquid wastes can be injected into, and contained by, confined geologic strata not having other actual or potential uses of a more beneficial nature, thereby providing long-term isolation of the waste material from man's usable environment. The validity of this concept depends on two basic factors: (1) the presence of suitable receptor zones, and (2) the existence of adequate confinement (Walker and Cox 1976, p. 2).

Capacity to accept an injected waste is a function of the amount of void space within the formation (its porosity) and its ability to transmit fluid (its permeability). In the usual case, the void space is already occupied with natural water, either fresh or mineralized to some extent. Thus, injection usually involves compression or displacement of existing fluids. Since the compressibility of water is small, there must be large spaces within the strata to accept the wastes and/or the displaced water (Walker and Cox 1976, p. 2).

Deep-well disposal involves the use of limited formation space, is expensive in construction and operation, and is the subject of ever-tightening regulation. Early criteria and professional conviction was that the method should only be used for

---

[26] Class I Injection wells are completed below the lower-most underground sources of drinking water. Class I wells are further classified for hazardous or nonhazardous wastes. Note that the 1993 Class I well count is down from the 245 count in 1989 (see Table 7.5), a very substantial reduction.

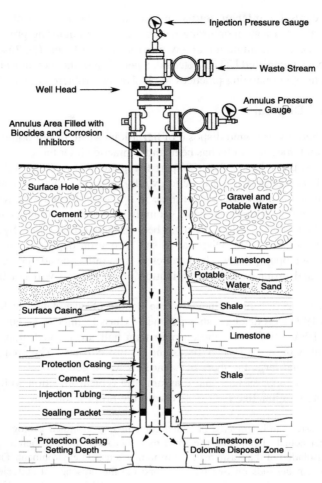

FIGURE 7.38 Cross-section design of a hazardous waste injection well. (Adapted from *Hazardous Waste Management,* © 1989, McGraw-Hill, New York. With permission from Charles A. Wentz.)

those wastes for which there are no other feasible management options.[27] Intensive operational oversight and monitoring is necessary to preclude contamination of nearby aquifers. Sudden changes in operating pressure or annulus pressure are major concerns. The former may indicate hydrofracturing and break-out from the confining formation. The latter usually indicates failure of the well integrity, introducing the possibility of contamination of other formations. Another major concern is the fact that Class I injection wells must penetrate drinking water aquifers to reach deep disposal zones. If the casing is damaged during installation, or later by seismic activity, the penetrated aquifer may become contaminated.

---

[27] As seen in the paragraphs that follow, the EPA has moved to the opposite view, by making deep-well injection of hazardous wastes subject to the treatment standards of the land disposal restrictions.

TABLE 7.5
EPA Classifications of Injection Wells

U.S. EPA Classification	Injection Well Description	1989 EPA Active Inventory
Class I	Wells used to inject liquid hazardous wastes beneath the lowermost USDW	245
		233
	Wells used to inject industrial nonhazardous liquid wastes beneath the lowermost USDW	76
	Wells used to inject municipal waste waters beneath the lowermost USDW	
Class II	Wells used to dispose of fluids associated with the production of oil and natural gas	38,152
		121,086
	Wells used to inject fluids for enhanced oil recovery	918
	Wells used for the storage of liquid hydrocarbons	
Class III	Wells used to inject fluids for the extraction of minerals	21,027[a]
Class IV	Wells used to dispose of hazardous or radioactive wastes into or above a USDW (the EPA has banned the use of these wells)	20
Class V	Wells not included in the other classes used to generally inject nonhazardous fluid into or above a USDW	173,159[b]

Note: A similar later tabulation was not available from the EPA.

[a] Located in 192 facilities.
[b] Inventory from the EPA Class V Report to Congress (U.S. EPA 1984).

Subpart R — Underground Injection

The EPA defines injection wells of Class I through Class V. These are briefly described in Table 7.5. Class I deep wells can be used for injection of hazardous and nonhazardous wastes. Underground injection of hazardous wastes is jointly regulated by RCRA (40 CFR 265, Subpart R, and Part 264, Subpart X) and by the Safe Drinking Water Act (SDWA) (40 CFR Parts 144 to 148). Class I and Class V wells used for injecting hazardous waste must have authorization under both SDWA and RCRA. The wells may be permitted by delegated state agencies or by the EPA. Owners and operators of these facilities must also meet the general standards outlined in Subparts A through E of 40 CFR 265 and the closure and post-closure requirements of SDWA.

Class I injection wells are defined as: "Wells used by generators of hazardous waste or owners or operators of hazardous waste management facilities to inject hazardous waste beneath the lowermost formation containing, within one quarter mile of the well bore, an underground source of drinking water" [40 CFR 144.6 (a)]. The Part 146, Subpart B, "Criteria and Standards Applicable to Class I Wells," paraphrased are

- *Construction Requirements* — New wells must be sited so that they inject into a separate formation from underground sources of drinking water, free of faults, or fractures. Drilling logs and similar tests must be used to ensure that this requirement is met. Both new and existing wells must be cased and cemented to protect sources of drinking water, so that the injection well does not create a significant risk to human health (§ 146.12).

- *Operating, Monitoring, and Reporting Requirements* — The injection pressure of the well must not be so great as to fracture the disposal formation. The owner or operator must monitor the injection well to ensure the integrity of the well bore. Pressure, flow rate, and cumulative volume of the injected material must be periodically monitored and reported to the EPA (§ 146.13).

The EPA has recently banned the underground injection of wastes that do not meet the applicable treatment standards of the land disposal restrictions. These standards are codified at 40 CFR 148. The owner /operator must successfully demonstrate that wastes which do not meet the Part 148 standards will not migrate from the injection zone for as long as the wastes remain hazardous (§ 148.1). On December 7, 1999, the agency published revisions to the Class V injection well standards which eliminate or reduce the injection of wastes through motor vehicle waste disposal wells and large-capacity cesspools. Construction of motor vehicle waste disposal wells is prohibited after April 5, 2000, and existing wells may be required to close or undergo modification. Similarly, large-capacity cesspools existing on April 5, 2000 must be phased out over 5 years. The revisions can be found at 64 FR 68546 and were published as a new Part 144, Subpart G.

OTHER TREATMENT AND DISPOSAL METHODS

Many treatment and some disposal methods, practices, or processes have not been discussed in this chapter. There are too many of them to cover, and they are less significant in terms of numbers, popularity, or utility than those discussed. Three methods that have been frequently employed, but are in declining acceptability, require brief mention.

Ocean Dumping

The U.S., with 46 other nations, is signatory to the Convention on the Prevention of Marine Pollution by Dumping of Wastes and Other Matter, generally known as the London Dumping Convention (LDC). The Convention requires the member nations to establish national systems to control the dumping at sea of wastes and other matter. The Convention was negotiated in 1972 and became effective in 1975. The Marine Protection, Research, and Sanctuary Act (MPRSA) of 1972 was enacted to implement the provisions of the LDC. The MPRSA and its 1988 amendment, the Ocean Dumping Ban Act (ODBA), make ocean dumping of industrial waste and municipal sewage sludge unlawful after December 31, 1991 (EPA 1991, p. 9).

Ocean dumping of all manner of wastes has been practiced by many nations, and the practice continues in some areas of the world. In U.S. waters, and from U.S. ports, dumping was on the decline due to implementation of the MPRSA and was vigorously opposed by environmentalists and the public. The MPRSA permitted some dumping of industrial wastes, and proponents of ocean dumping continued to argue, in the courts and elsewhere, that the practice of dumping in deep ocean strata is less harmful than land disposal. Amendments in 1974 and 1980 tightened dumping restrictions, and dumping of industrial waste was ended in 1988. Dumping of sewage sludge continued until the ODBA implementation date of December 31, 1991. The Act permits the continued dumping of dredged material. District engineers of the U.S. Army Corps of Engineers (COE) issue permits for dumping of dredged materials after an EPA review and approval of the permit application. The COE and EPA share joint responsibility for monitoring to ensure that permit conditions are met and that the marine environment is protected.

Ocean Incineration

As noted earlier, the incineration of hazardous wastes in ocean-going ships was originated by the (former West) Germans and was initially well received. An incinerator vessel was acquired by a U.S. firm and began a series of test burns. The operating permit was never issued by the EPA, and the future of the practice is in doubt.

A number of operating problems and environmental concerns have become problematical. The most serious technical problems appear to be

1. The constraints of ocean-going vessel design do not permit incinerator design that will provide the minimum 2-sec residence time.
2. Similar constraints preclude the necessary stack-gas scrubbers or other emission control devices that are required to capture heavy metals and neutralize the HCl produced in the incineration of chlorinated hydrocarbons.

Concerns for the environmental impacts of a collision or sinking at sea were also a factor, and the enactment of the Ocean Dumping Ban Act in 1988 effectively eliminated further consideration of incineration-at-sea by the EPA (EPA 1991, p. 41).

LAND DISPOSAL RESTRICTIONS

The awakening to the fact that land disposal of hazardous wastes can be the cause of major groundwater contamination problems came to us in the early to mid-1970s. Prior to 1984, efforts to restrict land disposal were focused in regulatory restrictions on land disposal *facilities*. These restrictions were (and are) seen in the increasingly complex 40 CFR 264 and 265 regulations as modified over the years. The 1984 enactment of HSWA mandated stringent new land disposal limitations. In § 3004(m), Congress specified that the EPA should: "… promulgate regulations specifying those levels or methods of treatment, if any, which substantially diminish the toxicity of the waste or substantially reduce the likelihood of migration of hazardous constituents from the waste."

The original and basic purpose of the LDRs was to discourage activities that involve placing untreated wastes in or upon the land when a better treatment or destruction alternative exists. For each hazardous waste, the EPA was required to establish treatment standards that are protective of human health and the environment when the wastes are land disposed. As noted earlier, *land disposal* is specifically declared to include placement in a landfill, surface impoundment, waste pile, injection well, land treatment facility, salt dome or salt bed formation, and underground mine or cave.

The land disposal restrictions (LDRs) are codified at 40 CFR 148 for underground injection (UIC) facilities and in Part 268 for other forms of land disposal. Since enactment of HSWA in 1984, these regulations have progressed through a complex set of progressively restrictive requirements, initially softened by a variety of exceptions, variances, and extended compliance deadlines. The gradual approach was necessitated in part by the large number of waste codes for which standards had to be developed, by the limited national capacity for some forms of treatment or secure disposal, and by the need to allow time for the regulated community to come into compliance with the new regulations.[28] As the deadlines passed, the complexity of the regulations and their implementation have diminished somewhat, but they continue to require the practitioner to be thoroughly familiar with the Parts 148 and 268 requirements.

The general categories of wastes, the completion dates set by Congress in HSWA, and the Federal Register notices were

- First priority — assigned to the solvents (because of the high volumes generated) and dioxins (because of their toxicity); completion date, November 7, 1986; 51 FR 40572.
- The "California List" wastes — a group of liquid hazardous wastes based upon a list established earlier by the California Department of Health Services. The group includes cyanides, acids having pH _ 2.0, polychlorinated biphenyls (PCBs), halogenated organic compounds (HOCs), and metals; completion date, July 8, 1987; 52 FR 25760.
- The "Thirds" — Congress required that the EPA develop a plan to establish treatment standards for all identified or listed hazardous wastes by November 8, 1984. The requirement included ranking the listed wastes from high to low priority based on the wastes' intrinsic hazard and volume generated. The EPA divided the wastes into three groups, scheduled effective dates of the "thirds," and published the standards as follows:
  - First Third — high volume, high hazard; completion date, August 17, 1988; 53 FR 31138
  - Second Third — intermediate volume/hazard; completion date, June 23, 1989; 54 FR 26594
  - Third Third — low volume, lower hazard; completion date, June 1, 1990; 55 FR 22520
- Treatment standards for newly identified or listed wastes — requires that additional wastes listed after November 8, 1984 be evaluated on a case-

---

[28] An excellent step-by-step history of the developmental sequences can be found in McCoy and Associates, Inc., 1994. *The RCRA Land Disposal Restrictions: A Guide to Compliance, 1993.* Elsevier Science Publishing Company, New York.

by-case basis. The EPA must make a determination of whether the waste may be land-disposed within 6 months of the identification or listing.

In recognition of shortages of treatment capacity for some waste categories, the EPA initially provided for "National Treatment Capacity Variances." Exemptions from various aspects of the rules were also allowed in special cases. The variances have diminished as regulatory deadlines have passed; as new facilities began operation; as treatment processes have developed or improved; and as waste minimization efforts have intensified. At this writing, the only capacity variances in effect are those for debris.[29]

The LDRs had, by the turn of the century, evolved to "treatment standards." The EPA chose to base treatment standards on technical practicability rather than upon risk assessment.[30] The agency conducts, contracts, and reviews research on the known, available treatment technologies and selects the one that best minimizes the mobility and/or toxicity of hazardous constituents. This technology is then designated as Best Available Demonstrated Technology (BDAT) for the specific waste. The EPA then establishes a waste code-specific treatment standard based on the performance of BDAT. The standard is then expressed as a concentration level or a required technology (EPA 1999).

Treatment standards expressed as concentration levels, based on the BDAT used as the basis for the standard, do not require that the BDAT treatment technology be used to treat the waste. The regulated facility may use any treatment technology[31] that will achieve the standard. When a treatment standard is a required technology, the regulated facility must use that technology to treat the waste unless the facility can demonstrate that an alternative method can achieve a level of performance equivalent to the required technology (EPA 1997e).

In the standards setting process, many hazardous waste *constituents* were found in both wastewaters and non-wastewaters. The resulting standards can be found in § 268.40, tabulated separately as wastewaters or non-wastewaters. Moreover, many of the constituents are present in more than one coded hazardous waste, each having different indicated BDATs. To ameliorate this problem, the EPA adjusted those coded wastes to establish one constituent concentration limit for wastewaters and one limit for non-wastewaters for each such waste code. These standards are designated Universal Treatment Standards (UTS) and are codified at 40 CFR 268.48. The EPA expects to be able to immediately apply these standards to UTS constituents present in newly identified hazardous wastes.

Dilution of a restricted waste, or the residual from treatment of a restricted waste, as a substitute for appropriate treatment is generally prohibited (§ 268.3). However, in certain circumstances, dilution of wastes that are hazardous only because they exhibit a characteristic is permissible [*see:* § 268.3(b)].

Characteristic wastes are considered to be *de-characterized* once they have been treated to remove the characteristic that caused the waste to become restricted.

---

[29] Consideration of alternatives has generally ended except for debris. Land disposal has become almost entirely contingent upon meeting the LDR standards.

[30] *See:* Chapter 4.

[31] Dilution is impermissible in most situations.

Owners/operators of regulated facilities must examine de-characterized wastes for *underlying characteristics* and treat the waste as prescribed for the applicable waste code(s). Land disposal restrictions are applicable at the "point-of-generation," i.e., the point at which a waste is identified as a RCRA hazardous waste. All Part 268 requirements continue to apply to the waste, even if it is subsequently de-characterized or excluded from the definition of hazardous or solid waste.

The treatment standards for dioxin-containing wastes are based on incineration as BDAT. While any treatment technology (other than dilution) is permissible for achieving the required contaminant levels, only incineration has been able to achieve them (EPA 1997e).

In order to facilitate site cleanup and remediation, the EPA has developed special standards for the management of certain remediation wastes. In the February 16, 1993 *Federal Register* (58 FR 8658), the agency promulgated regulations on the use of corrective action management units (CAMUs) and temporary units (TUs) to manage remediation wastes generated during a site cleanup. CAMUs are discussed in Chapter 11.

Extensive testing, tracking, and record keeping requirements accrue to generators and treatment and disposal facilities handling restricted wastes. These requirements do not lend themselves to useful summarization. The practitioner is advised to examine § 268.7 for these detailed requirements.

TOPICS FOR REVIEW OR DISCUSSION

1. Simple aeration of liquid hazardous wastes, or pumped groundwater, is not considered to be acceptable practice. Why not? What must be done to cause the practice to become acceptable?

2. Carbon adsorption processes are generally most effective for what kinds of hazardous wastes? The effectiveness of activated carbon in removing waste constituents from aqueous streams depends on what characteristic of the carbon?

3. What is the usual objective of the introduction of a hydroxide ion in the neutralization and precipitation of a metal-bearing waste?

4. Air or steam stripping is frequently employed to remove from wastewaters or contaminated groundwater. Early applications of this process were environmentally unsound because they _____

5. Biological treatment of toxic organic components in industrial wastewaters requires considerably more sophisticated controls than are required in similar domestic sewage treatment processes. Why?

6. RCRA permitted hazardous waste surface impoundments must meet three basic requirements. They are: _____

7. The validity of the concept that ... "liquid wastes can be injected into, and contained by, confined geologic strata no having other actual or potential uses of a more beneficial nature, thereby providing long-term isolation of the waste material from man's environment" ... depends upon two basic factors that are: _____ Discuss.

APPENDIX A
MACT Emission Standards for Hazardous Waste Combustion Units

Constituent	189 HW Incinerators 64 FR 52860	33 HW Cement Kilns 64 FR 52875	10 HW Aggregate Kilns 64 FR 52891
Dioxin/furan (ng/dscm TEQ)[a]	0.20 or 0.40 plus off-gas T quench to 400°F[b]	0.20 or 0.40 plus off-gas T quench to 400°F[b]	0.20 or 0.40 plus T quench to 400°F[b]
Particulate matter (mg/dscm)	34	0.15 kg/mg dry feed and opacity less than 20%	57
Mercury (μg/dscm)	130	120	47
Semi-volatile metals (μg/dscm) — cadmium and lead	240	240	250
Low volatility metals (μg/dscm) — antimony arsenic, beryllium, chromium	97	56	110
HCl and Cl_2 (ppmv)	77	130	230
Hydrocarbons	See 64 FR 52860	See 64 FR 52875	See 64 FR 52891
Destruction and removal efficiency (DRE)	See 64 FR 52860	See 64 FR 52875	See 64 FR 52891

Note: MACT, maximum achievable control technology.

[a] TEQ refers to 2,3,7,8-TCDD toxicity equivalence, the method of relating the toxicity of various dioxin/furan cogeners to the toxicity of 2,3,7,8-tetrachlorodibenzo(p)dioxin.
[b] When gas temperatures are quenched to 400°F, dioxin/furan emissions are typically 0.20 ng/dscm.

Source: Federal Register Online via GPO Access (wais.access.gpo.gov) EPA "Final Rule on Hazardous Waste Combustion Emission Standards — Questions and Answers" (www.epa.gov/hwcmact).

8. The heavy metal constituents in a sludge can usually be destroyed in a well-designed and -operated hazardous waste incinerator. True? False? Why?

9. A well-designed and -operated incinerator of chlorinated hydrocarbon wastes will produce at least three residues which are environmentally harmless. Identify the three. One other residue must be managed. What is that product and how may it be managed?

10. Where in the RCRA regulations does the owner/operator of a generator facility find the applicable requirements for contingency planning for his/her facility?

REFERENCES

Alexander, Martin. 1999. *Biodegradation and Bioremediation.* Academic Press, NY.

Barton, Robert G., W. D. Clark, and W. R. Seeker. 1992. "Fate of Metals in Waste Combustion Systems," in *Incineration of Hazardous Waste: Toxic Combustion By-Products,* W. Randall Seeker and Catherine P. Koshland, Eds., Gordon and Breach, Philadelphia.

Brunner, Calvin R. 1988. "Industrial Waste Incineration," *Hazardous Materials Control,* July–August 1988:26ff.

Cadmus Group, Inc. 1991. Drinking Water Contamination By Shallow Injection Wells. (Prepared for U.S. EPA, Office of Drinking Water, Washington, D.C.) Waltham, MA.

Combs, George D. 1989. *Emerging Treatment Technologies for Hazardous Waste,* Section XV. Environmental Systems Company, Little Rock, AR.

Dawson, Gaynor W. and Basil W. Mercer. 1986. *Hazardous Waste Management.* John Wiley & Sons, NY.

DeCamp, Gregory C. 2000. "RCRA Overview: A Generator Perspective," in *Hazardous Materials Management Desk Reference,* Doye B. Cox and Adriane P. Borgias, Eds., McGraw-Hill, NY.

Dempsey, Clyde R. and E. Timothy Oppelt. 1993. "Incineration of Hazardous Waste: A Critical Review Update," *Air and Waste,* January 1993:25ff.

DuPont, Andre. 1988. "Treating Liquid Waste with Lime," *Hazardous Materials Control,* July–August 1988:24ff.

Ehlers, Victor M. and Ernest W. Steel. 1958. *Municipal and Rural Sanitation.* McGraw-Hill, NY.

George, Gazi A. 2000. "Treatment Technologies in Hazardous Waste Management," in *Hazardous Materials Management,* Doye B. Cox and Adriane P. Borgias, Eds., McGraw-Hill, NY.

Gouldin, F. C. and E. M. Fisher. 1997. "Incineration and Thermal Treatment of Chemical Agents and Chemical Weapons," in *Emerging Technologies in Hazardous Waste Management 7,* D. William Tedder and Frederick G. Pohland, Ed., Plenum Press, NY.

Govind, Rakesh, Uma Kumar, Rama Puligadda, Jimmy Antia, and Henry Tabak. 1997. *Emerging Technologies in Hazardous Waste Management 7,* D. William Tedder and Frederick G. Pohland, Eds., Plenum Press, NY.

Gupta, Vinod K., Dinesh Mohan, Saurabh Sharma, and Kuk T. Park. 1999. "Removal of chromium (VI) from electroplating industry wastewater using bagasse fly ash — a sugar industry waste material," in *The Environmentalist 19,* pp. 129–136, Kluwer, Boston.

Haas, Charles N. and Richard J. Vamos. 1995. *Hazardous and Industrial Waste Treatment.* Prentice-Hall, Upper Saddle River, NJ.

Johnson, Nancy P. and Michael G. Cosmos. 1989. "Thermal Treatment Technologies for Haz Waste Remediation," *Pollution Engineering,* October 1989:66ff.

Lewandowski, Gordon A., and Louis J. DeFilippi. 1998. *Biological Treatment of Hazardous Wastes.* John Wiley & Sons, NY.

Manahan, Stanley, E. 1994. *Environmental Chemistry, Sixth Edition.* CRC Press, Boca Raton, FL.

McCoy and Associates, Inc. 1994. *The RCRA Land Disposal Restriction: A Guide to Compliance 1993.* Elsevier Science Publishing Co., NY.

Ozbilgin, Melih M., Jennifer L. Goodell, Joseph P. LeClaire, and Michael C. Kavanaugh. 1992. "The Use of Existing Water Supply Wells to Evaluate the Hydrogeologic and Transport Characteristics of Alluvial Aquifers," in *Hazardous Waste Site Investigations,* Richard B. Gammage and Barry A. Bervin, Eds., Lewis Publishers, Chelsea, MI.

Picardal, Flynn W., Sangoo Kim, Anna Radue, and Debera Backhus. 1997. "Anaerobic Transformations of Carbon Tetrachloride: Combined Bacterial and Abiotic Processes," in *Emerging Technologies in Hazardous Waste Management 7,* D. William Tedder and Frederick G. Pohland, Eds., Plenum Press, NY.

Rappe, Christoffer, Gangadhar Choudhary, and Lawrence Keith. 1986. *Chlorinated Dibenzofurans and Dioxins.* Lewis Publishers, Chelsa, MI.

U.S. Environmental Protection Agency. 1981. Guidance Document for Subpart F Air Emission Monitoring — Land Disposal Toxic Air Emissions Evaluation Guideline. National Technical Information Service, Springfield, VA, PB87-155578.

U.S. Environmental Protection Agency. 1982. RCRA Guidance Document: Landfill Design, Liner Systems, and Final Cover. National Technical Information Service, Springfield, VA, PB87-157657.

U.S. Environmental Protection Agency. 1987a. Background Document on Bottom Liner Performance in Double-Lined Landfills and Surface Impoundments. National Technical Information Service, Springfield, VA, PB87-182291.

U.S. Environmental Protection Agency. 1987b. Background Document on Proposed Liner and Leak Detection Rule. National Technical Information Service, Springfield, VA, PB87-191383.

U.S. Environmental Protection Agency. 1988. Design, Construction, and Evaluation of Clay Liners for Waste Management Facilities. National Technical Information Service, Springfield, VA, PB89-181937.

U.S. Environmental Protection Agency. 1991. Report to Congress on Ocean Dumping 1987 — 1990. Office of Water, Washington, D.C., EPA 503-9-91-009.

U.S. Environmental Protection Agency. 1993a. Report to Congress on Cement Kiln Dust. Solid Waste and Emergency Response, Washington, D.C., EPA 530-S-94-001.

U.S. Environmental Protection Agency. 1993b. RICRIS National Oversight Database.

U.S. Environmental Protection Agency. 1997a. Introduction to Groundwater Monitoring. EPA 530-R-97-055.

U.S. Environmental Protection Agency. 1997b. Introduction to Closure/Post-Closure. EPA 530-R-97-048.

U.S. Environmental Protection Agency. 1997c. Introduction to Miscellaneous and Other Units. EPA 530-R-97-060.

U.S. Environmental Protection Agency. 1997d. Introduction to Land Disposal Units. EPA 530-R-97-059.

U.S. Environmental Protection Agency. 1999. Introduction to Land Disposal Restrictions. EPA 530-R-99-053.

U.S. Environmental Protection Agency. 1998. RCRA Orientation Manual, 1998 Edition. EPA 530-R-98-004.

Voice, Thomas C. 1989. "Activated Carbon Adsorption," in *Standard Handbook of Hazardous Waste Treatment and Disposal,* Harry M. Freeman, Ed., McGraw-Hill, NY.

Walker, William R. and William E. Cox. 1976. *Deep Well Injection of Industrial Wastes: Government Controls and Legal Restraints.* Virginia Resources Research Center, Blacksburg, VA.

Watson, J. S. 1999. *Separation Methods for Waste and Environmental Applications,* Marcel Dekker, New York.

Wentz, Charles A. 1989. *Hazardous Waste Management.* McGraw-Hill, NY.

Wilson, R. D. and C. H. Thompson. 1988. "Activated Carbon Treatment of Groundwater: Results of a Pilot Plant Program," *Hazardous Materials Control,* July–August 1988:17ff.

Woodside, Gayle. 1999. *Hazardous Materials and Hazardous Waste Management, Second Edition.* John Wiley & Sons, NY.

8 Pollution Prevention, Waste Minimization, Reuse, and Recycling

OBJECTIVES

At completion of this chapter, the student should:

- Understand the basic operational approaches to waste minimization, i.e., product changes, source controls, use and reuse, and reclamation.
- Be familiar with the principles, process, and practice of waste reduction assessment.
- Understand the imperatives of waste minimization, reduction, reuse, and recycling.
- Be familiar with the RCRA regulatory mechanisms and program incentives to achieve waste minimization, the national policy aspects, and the local impediments.
- Be similarly familiar with the objectives of the Pollution Prevention Act and the implementing mechanisms.

INTRODUCTION

We now take up the most important issue in the study of hazardous waste management — the elimination or reduction in the quantity of waste generated. Throughout the previous chapters, we have emphasized the fact that much of what has passed for hazardous waste management ultimately came to little more than moving it around, transferring it from one environmental medium to another, changing its form, or hiding it.

Great strides have been made in the sophistication of regulatory programs, treatment and destruction technology, and secure disposal. The thrust of industries and government, until recently, has been toward ever-tightening pollution control rather than pollution prevention. Politicians (and others) are fond of referring to this traditional sanitary engineering approach as the "end-of-the-pipe mentality."

The legislate-regulate-treat-dispose approach has three primary roots:

1. As hazardous wastes became a more serious aspect of industrial management, they were initially handled in a manner similar to the handling of sewage and refuse. There is little that can be done to reduce the amount of sewage generated, so we taught ourselves to treat it to make it less threatening to our health and aesthetic sensibilities and to our environment. Refuse management was based upon similar thought processes, but with somewhat less validity. Sanitary engineers did not advocate adding hazardous wastes to our sewerage systems and had little to do with the dumping of hazardous wastes into whatever refuse management systems were in use. The sewers, atmosphere, and dumping grounds were there, and our use of them was dictated by the politics and the economics of the free enterprise system.

2. During and after the Vietnam War, former President Lyndon Johnson and his Secretary of Defense Robert McNamara, were criticized for their failure to mobilize the nation and vigorously prosecute the war. The policy was referred to as "gradualism," meaning that the resources (men and materials) were added in small increments, to which the enemy was able to accommodate. The parallel with the nation's approach to hazardous waste management is unmistakable. When hazardous wastes began to require our attention, we did not mobilize to deal with them. We did not examine the sources to determine their necessity; or whether there might be alternative processes, raw materials, or end products; or even good operating practices that might *reduce* quantities or strengths of wastes. The feeble impact of the resurrected 1899 Rivers and Harbors Act and the early efforts of environmentalists led us to put in equally feeble "treatment" schemes to transfer pollutants between environmental "media" or to hide our dumping more carefully. As regulatory pressures increased, we added new treatment units, upgraded existing ones, and created the treatment, storage, and disposal industry. With the advent of the Hazardous and Solid Waste Amendments (HSWA), we pushed innovative treatment and destruction and tried to reduce our dependence upon disposal. Only recently have we begun to seriously consider new approaches.

3. The third of these roots is, of course, economics. The economic pressures upon U.S. industry have ranged over the ever-escalating labor-wage demands of the 1960s and 1970s; the profit greed of the 1970s and 1980s; the overseas competition of the late 1980s; and the siren calls of minimal or no environmental controls and cheap labor in "developing" countries. Industry representatives and lobbyists have been highly effective in softening environmental legislation and regulatory issue. Fears of job losses, recessions, stockholder demands, and debt have been the dominant themes. Industrial decision makers tended to opt for the least expensive option of the moment and in hazardous waste management that frequently translated into the purchase of a *treatment* unit or a new contract with a *disposal* facility.

Until recently, there have been few, if any, economic incentives to examine major changes in products, raw materials, materials handling, or process controls to eliminate a waste stream or reduce it in volume or strength. The economic incentives of poor corporate image, Superfund nomination, tort filings, and criminal penalties have much to do with the newly found interest in waste minimization. Ever-diminishing availability of space for disposal services; public resistance to siting of any kind of hazardous waste management facility; and increasingly stringent regulation add further pressures to rethink our traditional approaches. More pointedly, the Pollution Prevention Act of 1990 (PPA) intensifies requirements for reporting of releases and analysis of progress in achieving waste minimization goals. The new Act and the EPA's implementation program have stopped just short of *mandated* reductions of releases. The agency refers to the waste minimization, source reduction, and recycling/reuse program emphasis implemented immediately after enactment of the PPA as "Phase I" of the pollution prevention programs.

"Phase II" of the PPA implementation has been embodied in a flurry of initiatives, strategies, and policy statements that are designed to persuade, coerce, and/or require industries and hazardous waste managers to "virtually eliminate," reduce generation of, or find environmentally safe substitutes for hazardous wastes.[1] These new thrusts are discussed in a later section of this chapter. The new initiatives are well meant and some are, or will be, effective. Nevertheless, many of the traditional and more mundane waste minimization, reuse, and recycling measures and technologies remain valid, useful, and necessary. Accordingly, we will attempt to provide the reader/student/practitioner with a balanced overview of the proven waste minimization, reuse, and recycling techniques and practices, along with the more recent approaches that the EPA is emphasizing.

Hazardous Waste Minimization Techniques

As noted earlier, the statutory authorities for waste minimization programs and for pollution prevention strategies do not include mandatory controls or mechanisms to regulate waste minimization programs. In lieu thereof, the EPA developed a large number of good "how-to" publications that deal with program organization and management, as well as technical approaches. The "Phase I" pollution prevention programs were, and continue to be, focused upon extensive reporting requirements, goal setting, and performance evaluation. The U.S. Congress' Office of Technology Assessment produced an informative critique of the program, entitled "Serious Reduction of Hazardous Waste." We now borrow from these publications, and others, to provide some structure to the topic. Figure 8.1 diagrams an organized way to think about the *waste minimization* techniques. We then follow with examples of each of the diagrammed techniques.

Source Reduction

In the previous chapter, we offered Dr. George Combs' version of the hierarchy of preferable waste management options and priorities. Following enactment of the HSWA in 1984, the EPA waste minimization program offered a similar hierarchy

---

[1] Particularly for priority persistent, bioaccumulative, and toxic (PBT) pollutants, as will be seen later herein.

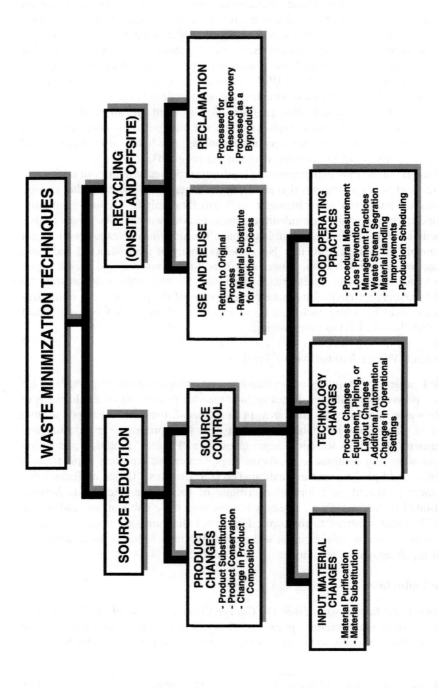

FIGURE 8.1 Waste minimization techniques. (From the U.S. Environmental Protection Agency.)

that may be helpful in thinking about approaches to hazardous waste reduction, or minimization:

1. *Waste Reduction*: Reduce the amount of waste at the source through changes in industrial processes.
2. *Waste Separation and Concentration*: Isolate wastes from mixtures in which they occur.
3. *Waste Exchange*: Transfer wastes through clearinghouses so that they can be recycled in industrial processes.
4. *Energy/Material Recovery*: Reuse and recycle wastes for the original or some other purpose, such as for materials recovery or energy production.
5. *Incineration/Treatment*: Destroy, detoxify, and neutralize wastes into less harmful substances.
6. *Secure Land Disposal*: Deposit wastes on land using volume reduction, encapsulation, leachate containment, monitoring, and controlled air and surface/subsurface water releases.

This hierarchy was the rationale for EPA waste minimization, recycling, and reuse policies and directives from enactment of HSWA in 1984 until the 1990 PPA was passed.

Product Changes

Product Substitution. Changes in the design, composition, or specifications of end-products that allow fundamental changes in the manufacturing process or in the use of raw materials can directly lead to waste reduction. Such changes are also the most difficult approach to waste reduction for several reasons, including

- Concerns on the part of the manufacturer regarding customer acceptance, cost of the conversion, cost of the new product, and quality control
- Concerns on the part of the customer regarding acceptability of the product, quality control, and changes in application made necessary by the substitution, general uncertainty, and fear of the unknown
- Concerns on the part of both manufacturer and customer regarding regulatory and liability impacts

For example, Monsanto (St. Louis, MO) reformulated a specialized industrial adhesive so that hazardous particulates remained in the product, thus eliminating the need to use and dispose of filters and particulates as waste. The company then had to convince its customers that the particulate matter formerly removed by the filters could remain in the product without affecting its adhesive qualities. From the time the idea of reformulating the product was originated, 2 years of effort by Monsanto's Research and Marketing Division was required before the reluctance of the purchaser to accept a different product was overcome and the change could be made (Office of Technology Assessment 1986, p. 83).

Product Conservation. One of the most fruitful areas of waste minimization through product conservation is the effective management of inventory having specific shelf-lives. Holston Army Ammunition Plant reduced waste pesticide disposal from 440 kg to 0 kg in 1 year by better management of stocks (Mills 1988).

Changes in Product Composition. Dow Chemical Company changed the way it packaged a product and achieved waste reduction in doing so. A wettable powder insecticide, widely used in the landscape maintenance and horticulture business, was originally sold in 2-lb metal cans that had to be decontaminated prior to disposal, thereby creating a hazardous waste. Dow now packages the product in 4-oz water-soluble packages which dissolve when the product is mixed with water for use (Office of Technology Assessment 1986, p. 83).

Source Control

Input Material Changes

Material Substitution. A classic issue of material substitution is the question of disposable wipes or reusable towels in thousands of industrial facilities using millions of shop towels daily. The shop towels come in contact with a variety of chemicals, some of which are hazardous materials; thus, disposal of the towels may bring the user under RCRA regulation. The EPA has deferred making decisions on the regulatory status of reusable textile wipes to the EPA regional offices and states. Reusable towels are usually rented from industrial towel services ("... a contractual/closed loop cleaning service"). Most of the state agencies have either exempted or limited the scope of RCRA regulation, where reusable shop towels are contaminated with listed or characteristically hazardous solvents. However, the states and EPA regional offices granting exemptions require that specific reusable shop towel management criteria be followed. The criteria vary from state to state, but most require that (1) the laundry be in compliance with its wastewater discharge permit and (2) the towels not contain any free liquids. These arrangements reduce the customer RCRA liability and the substantially larger volume of hazardous waste created by the use of disposable wipes (Smith, 1998, pp. 36ff).

Material Purification. A U.S. Air Force facility annually generated about 6500 gal of waste 1,1,1-trichloroethane (TCA) from vapor degreasing operations. Chemical laboratory personnel discovered that the TCA was being disposed of because it did not meet an acid acceptance value of 0.10 wt% NaOH. Oil contamination levels were less than 10% at the time of disposal, far less than the expected 30% level. To restore acid acceptance levels, 1,2-butylene oxide was added to the solvent. No adverse reactions or detectable problems were observed when the butylene oxide was added to the vapor degreasers. This example of purification of input material is expected to enable reduction of disposal volumes by 4000 gal (60%) and savings of $30,000/year (EPA 1989, p. 19).

Technology Changes

Process Changes. An example of a classic process change, resulting in reduced waste generation, is staged use of solvent. An electronics firm switched from using three different solvents — mineral spirits for degreasing machine parts; perchloroethylene for computer housings; and a fluorocarbon-methanol blend for printed circuit boards — to a single solvent system. Fresh solvent is used for the printed circuit boards, is then reused to degrease the computer housings, and last is reused to degrease the machine parts. This practice not only reduced solvent consumption and waste, it eliminated potential cross contamination of solvents; generated a single waste stream that can be recycled; simplified safety and operating procedures; and increased purchasing leverage (EPA 1989, p. 17).

Equipment, Piping, or Layout Changes. Equipment changes can be equally beneficial in waste reduction programs. In an electronic circuit manufacturing plant, flexible electronic circuits are made from copper sheeting which must be cleaned before use. Cleaning had been accomplished by spraying with ammonium persulfate, phosphoric acid, and sulfuric acid. This cleaning operation created a hazardous waste stream that required special handling and disposal. Equipment for cleaning by chemical spraying was replaced by a specially designed machine with rotating brushes which scrubbed the copper sheet with pumice. The resulting pumice slurry was not hazardous and could be disposed in a sanitary landfill. Savings of $15,000 in raw material, disposal, and labor costs were achieved in the first year. This process change also eliminated 40,000 lb of hazardous liquid wastes per year (Dupont et al. 2000, p. 357).

Automation. Process automation assists or replaces human employees with automatic devices. Automation can include the monitoring and subsequent adjusting of process parameters by computer or mechanical handling of hazardous substances. Minimizing the probability of employee error (which can lead to spills or "off-spec" products) and increasing product yields through the optimum use of raw materials can reduce waste. Bar-coded labels (Figure 8.2) can link containers and materials to a computer through all stages of a container's life. This improves the accuracy of material tracking and inventory accounting. Bar codes allow material monitoring during use and can prevent materials from being lost or becoming outdated.

Good Operating Practices

Procedural Measures. The Occupational Safety and Health Administration (OSHA) requires businesses to maintain files of Material Safety Data Sheets (MSDS) for all hazardous materials. The sheets contain the manufacturer's information regarding:

- Identity of the chemical and the Chemical Abstracts Service (CAS) number
- Physical characteristics
- Physical and health hazards
- Primary routes of entry
- Exposure limits
- Precautions
- Controls
- Emergency and first aid procedures
- Name of the manufacturer or importer

A major industrial facility uses MSDS to screen all material coming into their plant. Before the material is requisitioned, medical and hazardous materials experts must approve it. This approval ensures that a substance has been researched and evaluated

FIGURE 8.2 Bar-coding as a process tracking tool. (From ROMIC Chemical Corporation, 2081 Bay Road, Palo Alto, CA 94303.)

for its hazardous characteristics prior to its use. This potentially reduces generation of hazardous wastes by eliminating their use.[2]

Material Loss Prevention. Loss prevention programs are designed to reduce the chances of spilling a product. The key point is that a hazardous *material* becomes a RCRA hazardous *waste* when it is spilled, and all cleanup material and cleaned-up material must be managed as hazardous waste. A long-term, slow-release spill is often difficult to find and when found may have caused the creation of a large amount of hazardous waste. A material loss prevention program may include the following procedures:

- Use properly designed tanks and vessels only for their intended purpose.
- Pressure-test underground piping.
- Install overflow alarms for all tanks and vessels.
- Reduce dragout from process/cleaning baths.
- Maintain physical integrity of all tanks and vessels.
- Set up written procedures for all loading, unloading, and transfer operations.
- Install sufficient secondary containment areas.
- Forbid operators to bypass interlocks, alarms, or significantly alter set-points without authorization.

---

[2] This practice may also apply to the category of product substitution.

- Install electrolysis (anode and cathode) to recover metallic components in wastewater.
- Isolate equipment or process lines that leak or are not in service.
- Have interlock devices to stop flow to leaking sections.
- Use seal-less pumps.
- Use bellows-seal valves and a good valve layout.
- Pressure-test valves and fittings.
- Document all spillage.
- Perform overall material balances and estimate the quantity and dollar value of all losses.
- Install leak detection systems for underground storage tanks according to RCRA Subtitle I.
- Use floating-roof tanks for VOC control.
- Use conservation vents on fixed-roof tanks.
- Use vapor recovery systems.

Management Practices. Good operating practice involving management is exemplified by a large consumer product company which adopted a corporate policy to minimize the generation of hazardous waste. The company mobilized quality circles made up of employees representing areas within the plant that generated hazardous waste. The company experienced a 75% reduction in the amount of wastes generated by instituting proper maintenance procedures suggested by the quality circle teams. Since the team members were also line supervisors and operators, they made sure the procedures were followed (EPA 1988, p. 16).

Segregating Waste Streams. Hazardous waste sent off-site to be disposed of often includes a mixture of two or more different wastes. Segregating materials and wastes can decrease the amount of wastes to be disposed. Good operating practices for successful waste segregation include the following program ingredients:

- Prevent mixing of hazardous wastes with nonhazardous wastes.
- Isolate hazardous wastes by contaminant.
- Isolate liquid wastes from solid waste.

These measures can result in lower volumes of waste haulage and easier disposal of the hazardous waste. Recyclers and waste exchanges are more receptive to wastes not contaminated with other substances. One company altered dust collection equipment to collect waste streams from different processes separately. Each collection can now be recycled back to the process from which it originates. The firm has eliminated over $9000/year in disposal costs and recovered useable material worth $2000/year.

Material Handling Improvement. A major national company has reduced organics in wastewater by 93% through 4 separate changes in its handling of phenol and urea resins, as follows:

1. The company altered its method of cleaning the filters which remove large particles of resinous material as the resin product is loaded into tank cars.

They began collecting the rinse water instead of sending it down the floor drains and into the company's on-site wastewater treatment plant. This rinse water can be reused as an input in the next batch of phenolic resin.

2. When loading urea resin, they began reversing the loading pump at the end of each load so that resin on the filters would be sucked back into the storage tank and would not be rinsed out as waste.

3. The company revised rinsing procedures for reactor vessels between batches. Previously, 11,000- to 15,000-gal chambers had been cleaned by filling them with water, heating and stirring the water to remove resin residues, and then draining the rinse water into the plant's wastewater. The plant now has a two-step process. A small, first rinse of 100 gal of water removes most of the residue from the containers. Then a second, full-volume rinse is used to complete cleaning. The first 100 gal of rinse water is reused as input material for a later batch of resin. Water from the second rinse is discharged as wastewater, but has a lower phenol concentration than the previous volume of wastewater.

4. Procedures for transferring phenol from tank cars to storage tanks have been altered. Formerly, when the hose used to transfer the phenol from car to tank was disconnected, a small amount of phenol dripped down the drain — enough to cause problems given the strict regulatory limitation of phenol. Now, the hose is flushed with a few gallons of water to rinse the last bit of phenol into the storage tank.

In addition to greatly reducing wastewater volumes, these fairly simple changes have eliminated most of the hazardous solid wastes generated by the resin manufacturing processes because the company was able to discontinue use of the on-site evaporation pond to treat these wastewaters (Office of Technology Assessment 1986, p. 81).

Production Scheduling. Management should, wherever possible, devise and incorporate good operating practices to improve production scheduling and planning. Improved production techniques may include maximizing batch size, dedicating equipment to a single product, or altering batch sequencing to reduce cleaning frequency. Production runs of a given formulation should be scheduled together to reduce the need for equipment cleaning between batches. Careful examination of workload distribution may reveal opportunities for waste reduction. Dense loading may result in localized instability of the process solution. In other situations, maximizing batch size may minimize waste generated. Optimizing production schedules can greatly reduce waste in a production facility. Such options may offer easy implementation and immediate evidence of results.

Hazardous Waste Recycling

In hazardous waste management practice and in the RCRA regulations, "recycling" refers to the effective use or reuse of a waste as a substitute for a commercial product or use of a waste as an ingredient or feedstock in an industrial process. It also refers to reclaiming useful constituent fractions within a waste material or removing contaminants from a waste to allow it to be reused. The traditional EPA definition of

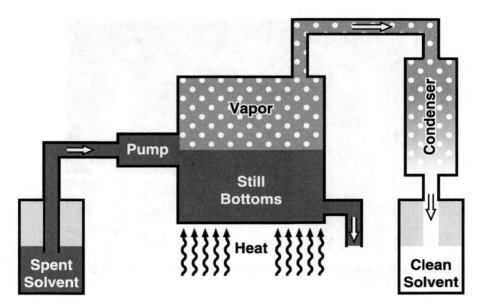

FIGURE 8.3 The distillation process.

recycling implies *use, reuse,* or *reclamation* of a waste, either on-site or off-site,[3] after it is generated by a particular process [40 CFR 261.1(c)].

One of the most basic and frequent applications of hazardous waste recycling is the distillation of spent solvents. Large numbers of companies are engaged in the solvent reclamation business and much of the solvent in use has been reclaimed. Figure 8.3 diagrams the process. Figure 8.4 is of typical distillation columns. Figure 8.5 illustrates "before and after" appearance of spent and reclaimed solvent (*see also:* Allen and Rosselot 1997, Chapter 6).

Use and Reuse

Return of a Waste to the Original Process. A printer of newspaper advertising purchased an ink recycling unit to produce black newspaper ink from its various waste inks. The unit blends the different colors of waste ink together with fresh black ink and black toner to create the black ink. This mixture is then filtered to remove flakes of dried ink and is used in lieu of fresh black ink. The need for shipment of waste ink to off-site disposal is eliminated. The price of the recycling unit was recovered in 9 months, based upon savings in fresh ink purchases and costs of disposal of the waste ink (EPA 1988, p. 17).

In the microelectronics industry, the high purity requirements for wafer fabrication make recycling and reuse of the solvents difficult. However, waste solvents can be recycled and used for the steps in which ultrahigh purity is not required. Examples

---

[3] As will be seen shortly hereafter, the Pollution Prevention Act and the EPA implementing initiatives now exclude "out-of-process" recycling as a form of pollution prevention.

FIGURE 8.4 Distillation columns. (From ROMIC Chemical Corporation, 2081 Bay Road, Palo Alto, CA 94303.)

FIGURE 8.5 Before and after — spent solvent and reclaimed solvent. (From ROMIC Chemical Corporation, 2081 Bay Road, Palo Alto, CA 94303.)

of those steps are the wafer-washing step before wafer lapping and the washing step after etching and before polishing (Dupont et al. 2000, p. 356).

Substitution for Raw Material in Another Process. A U.S. Air Force solvent reclaiming operation is successfully reclaiming polyurethane paint thinners. The original material contains 40% cellusolve acetate, 12% toluene, 30% methyl ethyl ketone (MEK), and 10% n-butyl acetate. The distillate, which contains only toluene and MEK, is used for wipedown and cleanup of painting equipment (Harris 1988).

(For discussion and references on solvent recovery and reuse in laboratories, *see:* Reinhardt et al. 1996).

Reclamation

Processing Hazardous Waste for Product Recovery. Sand used in the casting process at foundries contains residues of heavy metals such as copper, lead, and zinc. If these concentrations exceed Toxicity Characteristic Leaching Procedure (TCLP) standards, the sand is a hazardous waste and must be managed as such. Researchers are investigating various techniques for reclaiming the metal values from the sand. Recent experiments demonstrated that 95% of the copper could be precipitated and recovered in minutes (McCoy and Associates 1989, pp. 1–23). Sand may also be processed in smelters to recover metal values.

A printed wiring board (PWB) operation uses ammoniacal etchants to etch patterns on PWBs. The spent etchant is sent back to the chemical supplier, where the copper is extracted with an organic solvent to create a copper-rich organic layer and copper-lean aqueous solution. The aqueous phase is regenerated by the addition of ammonia and other additives to create fresh etchant. The organic layer is treated with sulfuric acid to remove the copper from the organic solvent. Regenerated solvent is fed back into the process, and the copper in the aqueous stream is recovered as copper sulfate pentahydrate via crystallization or as copper metal via electrowinning. Copper sulfate recovered by this process can be used to manufacture other copper-based chemicals or used directly in applications such as wood preservatives or algicide. A simplified schematic is shown in Figure 8.6 (Milliman and Luyten 1999, pp. 32–35).

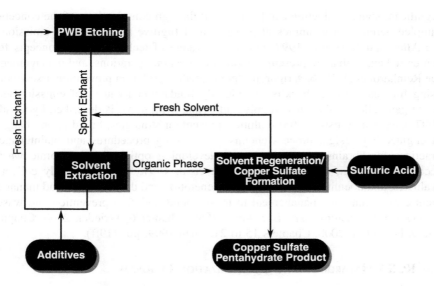

FIGURE 8.6 Alkaline ammonia etchant cycle. (Copyright ©1999, by Kevin E. Milliman and Henry C. Luyten, "Waste Not Want Not," published in *Environmental Protection,* May 1999. All rights reserved by Stevens Publishing Corporation. Reprinted with permission.)

Processing Hazardous Waste As a By-Product. Some classic uses of hazardous waste as a by-product, raw material, or feedstock have been mentioned. The most common include wastewaters used for irrigation and oil field pressurization; sludges used as fertilizers or soil matrix; lime generated by the carbide process for acetylene production used and routinely marketed for many purposes; and sulfuric acid from smelters used for a variety of purposes. These uses have stimulated entrepreneurs to the development of the "waste exchange."

The waste exchange serves as a clearinghouse for data on available wastes and raw materials needed. When the exchange identifies a match between an available waste commodity and a need, the parties are notified and allowed to consummate an arrangement suitable to both. Waste exchanges became numerous in the U.S. and Canada during the early and mid-1980s.

In practice, the exchange has enjoyed only limited acceptance. The potential participants tend toward secrecy, fearing compromise of trade secrets by their competitors. The liability implications of transferring control to other than a permitted treatment, storage, and disposal facility are also significant impediments to acceptance of exchanges. The EPA initially supported several exchanges, through grant programs, but some of these operations had not gained sufficient momentum and failed when the grant support was discontinued (Hild 1988) (*see also:* Higgins 1989, Chapters 1 to 8; Breen and Dellarco 1992; Alexander 1992; U.S. EPA 1997); Dupont et al. 2000, pp. 306–308).

MULTI-CONCEPTUAL APPROACHES

Significant source reduction can be achieved through combinations of the concepts sampled herein. For examples of prevention of fugitive and secondary emissions, see Allen and Rossellot, 1997. For a wide range of techniques and concepts for source reduction that are present in laboratory design, operations, and management, see Reinhardt et al. 1996. A rigorous "multi-media" pollution prevention assessment using four categorical checklists to identify fundamental causes of emission and waste generation, rather than simply addressing symptoms, is described by Chada 1997. Chada uses lists of 100 pollution prevention strategies, based upon changes in engineering design, process chemistry, operating procedures and maintenance practices for "brainstorming ideas and developing options." A corporate waste accounting system, described by Nizolek et al. 1997, "... consistently collects, evaluates, and documents essential waste generation and disposal data, and management costs," enabling management to make sound pollution prevention and waste management decisions (*see also:* Wentz 1989, Chapter 6; Wrieden 2000, Chapter 13; DuPont et al. 2000, Chapters 15 to 21; Shen 1999, pp. 219ff).

THE RCRA HAZARDOUS WASTE MINIMIZATION PROGRAM

As noted, RCRA/HSWA do not put forth a mandatory hazardous waste minimization program. Early RCRA implementation reflected Congressional sentiment that imposition of specific hazardous waste reduction requirements would amount to an unacceptable intervention in business and industrial practice. Nevertheless, in Section 1003 of RCRA, the Congress stated succinctly:

The Congress hereby declares it to be the national policy of the U.S. that, wherever feasible, the generation of hazardous waste is to be reduced or eliminated as expeditiously as possible. Waste that is nevertheless generated should be treated, stored, or disposed of so as to minimize the present and future threat to human health and the environment.

This national policy was, and continues to be, implemented through three specific activities which were mandated by the 1984 RCRA amendments (HSWA). These specific requirements apply to generators who manage their wastes on-site or those who ship wastes off-site and to permitted facilities:

- *Reporting Procedures:* Generators subject to reporting requirements were to include in their annual or biennial reports "... efforts undertaken during the year to reduce the volume and toxicity of waste generated; and ... the changes in volume and toxicity of waste actually achieved during the year ... in comparison to previous years ...."
- *Manifest System:* A section on waste minimization was added to require a generator's certification on the manifest for all regulated off-site shipments to state that "the generator of the hazardous waste has a program in place to reduce the volume or quantity and toxicity of such waste to the degree determined by the generator to be economically practicable; and ... the proposed method of treatment, storage, or disposal is that practicable method currently available to the generator which minimizes the present and future threat to human health and the environment."
- *Permits:* Effective September 1, 1985, any permit issued for the treatment, storage, or disposal of hazardous waste on the premises where the waste was generated required that the permittee certify no less often than annually that "the generator of the hazardous waste has a program in place to reduce the volume or quantity and toxicity of such waste to the degree determined by the generator to be economically practicable; and ... the proposed method of treatment, storage, or disposal is that practicable method currently available to the generator, which minimizes the present and future threat to human health and the environment."

The biennial reports of generators who ship their wastes off-site, and of permitted and interim status TSD facilities, are the only data collection mechanisms by which the effectiveness of the waste minimization program can be judged. Until recently, efforts to aggregate this information did not produce good results, due to absence of consistent definitions, procedures, and measurements, among states. The quality of the reporting and aggregation has improved since 1989, and the later data reflects new rigor in the process.

The EPA thus continued the RCRA *hazardous waste minimization* program in the nonmandatory format. Nevertheless, the agency intensified its efforts toward greater achievement in waste minimization by a series of pronouncements and initiatives:

- In May 1993, the Administrator announced the Draft Hazardous Waste Minimization and Combustion Strategy and placed a "temporary capacity freeze" on new incinerators and other combustion units[4] as discussed in the previous chapter.
- In November 1993, the Administrator sent letters to 22,000 large quantity generators that were required to certify that they had a waste minimization program in place in 1991. Letters were also sent to 12,000 chief executives of the parent corporations of those generators. The letters referenced current requirements for waste minimization programs and encouraged the companies to make response information available to the public (EPA 1994b).
- In May 1994, the EPA released a Draft RCRA Hazardous Waste Minimization National Plan which generally incorporated the thrust of the Hazardous Waste Minimization and Combustion Strategy (EPA 1994d). The final "Waste Minimization National Plan" was released by the EPA in November 1994. The goals of the plan are summarized in a later section of this chapter.

The Combustion Strategy is translated, in the final plan, to the setting of ... "initial national priorities for metals contained in hazardous wastes treated by combustion facilities and metals in releases from combustion facilities" (EPA 1994a, p. ES-1). The Biennial RCRA Hazardous Waste Report (EPA 1994c) shows that "thermal treatment" accounted for 1.1% of the 306 million tons of hazardous waste generated in 1991. As noted earlier, the rationale for this high profile focus on 1.1% of the hazardous waste generated (in 1991) was apparently never made clear.

RCRA REGULATION OF RECYCLING

As emphasized in Chapter 1 and elsewhere in this text, the Congress and the EPA have justifiably focused upon the accumulation of hazardous waste as an activity rich with potential for mismanagement. At the date of this writing, nearly three decades after enactment of RCRA, unscrupulous operators continue to attempt to convince the EPA and state inspectors that the stack of drums or the pile of waste

---

[4] The EPA, the regional offices, and the states thereupon began an intensive inspection and enforcement campaign directed toward combustion facilities, requiring permit applicants to perform full risk assessments, including assessment of the risk of indirect exposure to emissions through the food chain, as part of all new combustion facility permits. As this scenario unfolded, 27 incinerator and 22 boiler/industrial furnace facilities withdrew permit applications, abandoned interim status, or otherwise capitulated regarding their plans or efforts to incinerate hazardous wastes (EPA 1994a,b).

on the back lot or in the shed are destined for recycling. Sham recycling of "hazardous waste fuel" remains a problem. The EPA continues the quest for the regulatory definitions and formulae which will finally end deceptive activity and questionable practice, without imposing still heavier record keeping and reporting burdens. Significant progress has been made in this regard, but the regulatory scheme is complex, contorted, and ambiguous. Representatives of regulated industry chafe at the ambiguity of the regulations and the interpretations thereof. These regulations continue to occupy target lists of industry focus groups, trade associations, and political office holders and candidates. The recycling regulations also incur the opposition of environmental activists, some of whom look upon the combustion of hazardous waste as a major threat to public health.

Part 266 of the 40 CFR originally provided much of EPA's regulatory program for recycling of hazardous wastes. Perusal of the table of contents for Part 266 will reveal that Subparts A, B, D, and E[5] are now "reserved" and that Subparts C, F, and G deal with relatively simple issues. Subpart H houses the very complex and controversial regulations for "Hazardous Waste Burned in Boilers and Industrial Furnaces." Over time, the EPA has vacated much of the language of Part 266 and has increasingly relied on the definitions of solid and hazardous waste, as set forth in 40 CFR 261, and on the Land Disposal Restrictions of Part 268 to regulate the recycling of wastes.

The Subpart H standards were published on February 21, 1991 (56 FR 7208). In Subpart H, the EPA attempted to deal with the issue of hazardous wastes being burned in boilers and industrial furnaces (BIFs) for fuel content vs. destruction of hazardous waste constituents. Although HSWA requires the EPA to develop technical standards for burning of hazardous wastes in BIFs for heat recovery, the agency had not, prior to this date, promulgated the regulation. "Sham recycling" had been commonplace and Subpart H was intended to meet the HSWA requirement and gain control of the sham recycling problem. The regulation establishes standards for controlling emissions of organic compounds, metals, and HCL from BIFs that burn hazardous waste irrespective of the purpose of the burning, but a significant grouping of burners remain exempt from the standards.

The complexities of the Subpart H standards, *including 12 appendices,* greatly exceed the scope of this text, and it must be left to the student or practitioner to examine the details of the standards according to his/her needs.

EPA Implementation of The Pollution Prevention Act (PPA) of 1990

The most notable feature of the PPA is a Congressional restatement of the national goals regarding waste minimization:

---

[5] The Used Oil Management Standards, formerly found at 40 CFR 266, Subpart E, were published as 40 CFR 279 on September 10, 1992 (57 FR 41612).

> The Congress hereby declares it to be the national policy of the U.S. that pollution should be prevented or reduced at the source whenever feasible; pollution that cannot be prevented should be recycled in an environmentally safe manner, whenever feasible; pollution that cannot be prevented or recycled should be treated in an environmentally safe manner whenever feasible; and disposal or other release into the environment should be employed only as a last resort and should be conducted in an environmentally safe manner [PPA § 13101(b)].

Congress also redefined the term *source reduction* to exclude "out-of-process" recycling and "any practice which alters the physical, chemical, or biological characteristics or the volume of a *hazardous substance, pollutant, or contaminant* through a process or activity which itself is not integral to and necessary for the production of a product or the providing of a service" [PPA § 13102(5)(B)]. In other words, a waste that has been released cannot be *prevented* — a highly controversial redefinition. Stated another way, "in-process" or "closed-loop" recycling, the direct reintroduction of a waste into the same process, apparently qualifies as source reduction or pollution prevention. The text of the Act may be accessed at <http://www.epa.gov/opptintr/p2home/uscode.htm> (*see:* Phipps 1995, p. 2).

In "Phase II" of the EPA implementation of the PPA, the agency has left standing most of the "Phase I" policy instruments and has brought forth and/or embraced a new generation of initiatives, policy statements, and programs. These instruments have a common thread, the deemphasis of the earlier waste minimization/recycling/reuse rubric, and nearly total emphasis on Pollution Prevention (P2). The more significant initiatives, in summary form, are

- *The Waste Minimization National Plan:* The Plan is described by the EPA as a long-term national effort to reduce the quantity and toxicity of hazardous wastes. The goals of the plan are
  1. To reduce, as a nation, the presence of the most persistent, bioaccumulative, and toxic constituents by 25% by the year 2000 and by 50% by the year 2005
  2. To avoid transferring these constituents across environmental media
  3. To ensure that these constituents are reduced at their source whenever possible or, when not possible, that they are recycled in an environmentally sound manner (EPA 1994a; a summary outline of the Plan may be accessed at <http://www.epa.gov/epaoswer/hazwaste/minimize/waste.txt>)
- *Toxics Release Inventory:* The TRI originated with the implementation of the Emergency Planning and Community Right-to-Know Act (EPCRA) in 1986. Congress specifically required manufacturing facilities having Standard Industrial Codes 20 through 39 to report annually the quantities of toxic chemicals they release into the environment. The EPA acted upon

the increased reporting requirements of the PPA by lowering selected TRI reporting thresholds and requiring greater detail in reporting on recycling and progress on source reduction. In 1999, the EPA published a final rule (40 CFR 372) under EPCRA § 313 lowering the reporting thresholds for persistent, bioaccumulative, toxic (PBT)[6] chemicals then subject to the TRI reporting requirements and added other PBT chemicals to the TRI (EPA 1999a; 64 FR 58665). The TRI can be accessed at a variety of sites and in a variety of formats. The user may access the EPA homesite and use the search or publications tab.

• *Common Sense Initiative:* The CSI is an attempt by the EPA to reach consensus with representatives ("stakeholders") of industrial categories ("sectors") on ... "opportunities to change complicated and inconsistent environmental policies into comprehensive sector environmental strategies for the future" (EPA 1999b). The objective is to eventually revamp environmental regulation from a pollutant or environmental media focus to an industry sector focus, in the belief that this strategy would lead to less costly, less adversarial way of regulating industry. Stakeholders from six industry sectors[7] have been organized into sector teams and work groups which meet frequently to work on projects, policy considerations, and other issues. Output of these groups, in the form of reports, decisions, issues, and data are forwarded to the CSI Council, which includes high-level decision makers from the stakeholder groups and industries. In addition to sector-specific innovations, the teams explore solutions to common issues, including alternative flexible regulatory systems, pollution prevention as a standard business practice, improved reporting for public consumption, and enhanced compliance and public participation in the permitting process. A concise explanation of the initiative can be accessed at <http://www.epa.commonsense/bckgrd.htm>.

• *Multimedia Strategy for Priority Persistent, Bioaccumulative, and Toxic Pollutants:* The PBT strategy is intended to ... "overcome the remaining challenges in addressing priority PBT pollutants .... Due to a number of adverse health and ecological effects linked to PBT pollutants — especially mercury, PCBs, and dioxins — it is key for EPA to aim for further reductions in PBT risks" (EPA 1999, ES). This strategy reinforces and builds on existing EPA commitments to priority PBTs, i.e., the 1997 Canada-U.S. Binational Toxics Strategy (BNS). The EPA approach to the PBT Strategy is summarized as follows:

1. *Develop and Implement National Action Plans for Priority PBT Pollutants.* The EPA is initially focusing on 12 BNS Level 1 substances: aldrin/dieldrin, benzo(a)pyrene, chlordane, DDT, hexachlorobenzene, alkyl-lead, mercury and compounds, mirex, octachlorostyrene, PCBs, dioxins and furans, and toxaphene. The agency plans to focus initially

---

[6] *See:* Multimedia Strategy for PBTs below.
[7] Automobile manufacturing, computers and electronics, iron and steel, metal finishing, petroleum refining, and printing.

on achieving reductions in mercury releases, as an element of a pre-viously initiated mercury reduction plan. Elements in progress or planned include

  a. Conduct process-specific and (P2) projects under the mercury action plan, including regulatory actions and voluntary reductions.
  b. Focus enforcement and compliance assistance activities on PBTs.
  c. Develop or revise water quality criteria for mercury and other priority PBTs and revise methodology for mercury water quality criteria.
  d. Conduct research and analyses on PBTs, especially on mercury emission controls for coal-fired utility boilers, and on the transport, fate, and risk management of mercury.
  e. Continue active participation in international efforts beyond the BNS to reduce PBT risks.

2. *Screen and Select Additional Priority PBT Pollutants for Action.*
3. *Prevent Introduction of New PBTs.* The agency is acting to prevent new PBT chemicals from entering commerce by a variety of testing and restriction proposals, rulemaking to control reintroduction of out-of-use PBTs, incentives for development of lower-risk substitutes, and monitoring screening criteria for new and re-registering pesticides.
4. *Measure Progress.* The EPA is defining measurable objectives and instruments, such as human health and environmental indicators; sur-veys and studies of chemical residues in fish; chemical release, waste generation and use indicators; and program activity measures such as compliance and enforcement data (EPA 1998, ES). The EPA Strategy can be accessed at <http://www.epa/opptintr/pbt/execsum.htm> (and /pbt/pbtstrat.htm).

- *The 33/50 Program:* In 1991, in concert with the PPA, the Clean Air Act Amendments of 1990, and the RCRA waste minimization program, the EPA established the "33/50 Program." The program targeted 17 priority chemicals and set a goal of 33% reduction in releases and transfers of these chemicals by 1992 and a 50% reduction by 1995, measured against a 1988 baseline. The intent was to demonstrate that voluntary partnerships with industry could augment the agency's traditional command-and-control approach by bringing about targeted reductions more quickly than would regulations alone. The EPA reported that toxic releases were reduced by about 40% by 1992 and more than 55% by 1995 (EPA 1999c).[8] The 33/50 program can be accessed at <http://www.epa.gov/opprintr/3350/33fin01.htm> (*see also:* Bolstridge 1992, Chapter 12).
- *The Virtual Elimination Project:* Programs to "virtually eliminate" priority persistent, bioaccumulative, and toxic (PBT) and other "selected" pollut-ants are another EPA approach to reduction of risks to human health and

---

[8] The Government Accounting Office (GAO), in two September 1994 reports to Congress, is critical of the fact that the program embodied no means for the EPA to verify the reported quantities or even that the reductions are attributable to the 33/50 Program. The reader wishing further detail may obtain the reports by contacting GAO. The reports are GAO/RCED-94-93 and GAO/RCED-94-207.

the environment. This initiative seeks to prevent any new releases into the environment, from all pathways (land, air, and water), and to eliminate the use of these target compounds wherever possible, thereby minimizing future releases. The EPA states in the source document; "This approach create (sic) opportunities for immediate reductions, *without the need for additional research or regulatory action.*"[9] The concept of virtual elimination has been endorsed by the governments of Canada and the U.S. The 1994 Biennial Report of the International Joint Commission (IJC) identified three stages of virtual elimination: (1) controlling releases, (2) preventing use or generation, and (3) developing sustainable industry and product/material use (EPA 1998b). The EPA can be expected to seek P2 measures in permits, TRI and Biennial reporting by generators, enforcement decrees and settlements, interagency agreements, etc.

- *Other P2 Linkages:* Federal and state governments, departments and agencies, corporations, non-governmental organizations, interest groups, and environmental advocacy groups have formed an array of inter- and intra-agency programs, partnerships, and other linkages which have as their focus the promotion and implementation of P2 concepts, plans, and programs. Their numbers and operating mechanisms are too numerous and diverse to attempt an organized presentation. Most are accessible on the internet, and most are eager to provide information and recruit participants and/or support. These resources are listed in the EPA Office of Solid Waste publication "Waste Minimization/Pollution Prevention Resource Directory" (EPA 1999c), which can be accessed at <http://www.epa.gov/epaoswer/haz-waste/minimize/p2.htm>.

Other sources of P2 information, concepts, and linkages are provided in Appendix B (*see also*: Phipps 1995; Shen 1999; Dupont et al. 2000).

TOPICS FOR REVIEW OR DISCUSSION

1. The text refers to a classic process change — the use of a single solvent for several purposes, reusing the solvent in succeeding processes which require decreasing purity. Name at least three advantages of such a modification.
2. There are apparently several reasons why many hazardous waste exchanges have not prospered. What are three of the reasons?
3. What format could be used to make a regulatory distinction between burning of hazardous waste for the fuel value and incinerating to destroy the hazardous constituents?
4. The EPA waste minimization program embodies a hierarchy of preferable options for hazardous waste management. Proceeding from most desirable to least desirable, list those options.

---

[9] Thus, the EPA apparently interprets the language of the PPA to provide authority to impose new exposure criteria without awaiting the findings of exposure research and to add new permitting discretion to that previously provided by the omnibus authorities of RCRA § 3005(c)(3).

APPENDIX A
State Agency P2 Linkages and Resources

AL Department of Environmental Management	205-250-2779	www.@adem.state.al.us
AK Department of Environmental Conservation, P2 Office	907-269-7582	www.state.ak.us
AZ Department of Environmental Quality, P2 Unit	602-207-4235	www.adeq.state.az.us
AR Department of P2 and Ecology, HazWaste Division	501-570-0018	www.adeq.state.ar.us
CA EPA, Department of Toxic Substance Control, OP2	916-322-3670	www.dtsc.ca.gov/txpollpr
CO Department of Public Health and Environment, P2 Unit	303-692-3003	www.sni.net/light/p3/
CT Department of Environmental Protection, Bureau of Waste Management	203-566-5217	dep.state.ct.us/deao/ca/assist
DE Department of Natural Resources and Environmental Controls, P2 Program	302-739-5071	www.dnrec.state.de.us
FL Department of Environmental Protection, P2 Program	904-488-0300	www.dep.state.fl.us/waste/programs/p2
GA Department of Natural Resources, P2 Division	404-651-5120	www.ganet.org/dnr/p2ad/
HI Department of Health, Environmental Management Division, Office of SW Management	808-586-8143	www.state.hi.us/doh/eh
ID Department of Environmental Quality, P2 Program	208-373-0502	www.state.id.us/deq/ptwo.htm
IL Environmental Protection Agency, Office of P2	217-782-8700	www.epa.state.il.us/p2/index.html
IN Department of Environmental Management, Office of P2 and Technical Assistance	317-232-8172	www.ai.org/idem/oppta/
IA Department of Natural Resources, Waste Management Assistance Division	515-281-8927	www.iwrc.org/programs.html
KS Department of Health and Environment, Division of Environment, P2 Program	785-296-0669	www.ink.org/public/kdhe
KY Division of Waste Management, P2 Program	502-564-6716	www.state.ky.us/agencies/nrepc/programs/p2
LA Department of Environmental Quality, Technical Program Support	225-765-0720	www.deq.state.la.us/osec/latap.htm
ME Department of Environmental Protection, P2 Office	207-287-3811	www.state.me.us/dep/p2home.htm
MD Department of the Environment	410-631-4119	www.mde.state.md.us.permit/p2prog.html
MA Department of Environmental Protection, Bureau of Waste Prevention	508-767-2775	www.state.ma.us/dep/bwp/dhm

APPENDIX A   *(Continued)*
State Agency P2 Linkages and Resources

MI Department of Comm. & Natural Resources, Office of Waste Red Services	517-373-1871	www.deq.state.mi.us
MN Pollution Control Agency, Environmental Assessment Office	612-296-8643	www.pca.state.mn.us/programs/p2_p
MO Department of Natural Resources, Division of Environmental Quality	573-526-6627	www.dnr.state.mo.us/deq/tap
MS Department of Environmental Quality, Waste Min Unit	601-961-5321	www.deq.state.ms.us/domino/erowb
MT Department of Environmental Quality, P2 Bureau	888-678-6822	www.montana.edu/wwwated
NE Department of Environmental Control, Hazardous Waste Section	402-471-4217	
NV Division of Environmental Protection, Bureau of Waste Management	702-667-4870	www.scs.unr.edu/nsbdc/bep.htm
NH Department of Environmental Service, Waste Management Division, P2 Program	603-271-2902	www.state.nh.us/des/nhppp/
NJ Department of Environmental Protection, P2 Office	609-292-1122	E: mdower@dep.state.nj.us
NM Environmental Department, P2 Program	505-827-0197	
NY State Department of Environmental Conservation, P2 Unit	518-457-7267	www.dec.state.ny.us/website/ppu/
NC Division of Pollution Prevention and Environmental Assisstance	919-715-6500	www.p2pays.org
ND Department of Health, Environmental Health Section, P2 Program	701-328-5153	E: jburgess@state.nd.us
OH Environmental Protection Agency, Office of P2	614-644-3469	www.epa.ohio.gov/opp/oppmain.html
OK Department of Environmental Quality, P2 Technical Assisstance	405-271 1400	www.deq.state.ok.us/p2intro.htm
OR Department of Environmental Quality, P2 Coordinator	503-229-5458	www.deq.state.or.us/hub/p2.htm
PA Department of Environmental Resources, Office of Air & Waste Management	717-783-0540	www.dep.state.pa.us/dep/
RI Department of Environmental Management, P2 Supervision	401-222-6822	www.state.ri.us/dem/org/otca.htm
SC Department of Health & Environmental Control, Center for Waste Management	803-734-4715	www.state.sc.us/dhec/
SD Department of Environmental & Natural Resources, P2 Coordination	605-773-4216	

APPENDIX A   *(Continued)*
State Agency P2 Linkages and Resources

TN Department of Environment & Conservation, P2 Division	605-741-3657	www.state.tn.us/environment/p2.htm
TX Water Commission, Office of P2 & Conservation	512-239-3166	www.tnrcc.state.tx.us/exec/oppr/index
UT Department of Environmental Quality	801-536-4480	www.eq.state.us/eqoas/p2/p2_home.htm
VT Department of Environmental Conservation, P2 Division	802-241-3629	
VA Department of Environmental Quality, Office of P2	804-371-3712	www.deq.state.va.us/opp.html
WA Deptartment of Ecology, P2 Services	360-407-6702	www.wa.gov/ecology/pie/98overvu/98aohwtr
WI Department of Natural Resources, Haz P2 Audit	608-267-3125	www.dnr.state.wi.us/org/caer/cea
WY Department of Environmental Quality, SW Management Program	307-777-7752	www.deq.state.wy.us/outreach 1.htm

APPENDIX B
P2 Information, Concepts, and Linkages

Pollution Prevention Information Clearinghouse (PPIC)	http://www.epa.gov/opptintr/library/libppic.htm
Design for the Environment (DfE)	http://www.epa.gov/dfe
Incorporation of Pollution Prevention Principles into Chemical Science Education	http://www.umich.edu/-nppcpub/resources/chemabstract.html
EPA/OSW Waste Minimization Products and Documents	http://www.epa.gov/epaoswer/hazwaste/minimize/docs.htm
EPA Waste Minimization National Plan	http://www.epa.gov/epaoswer/hazwaste/minimize/waste.txt
EPA Meeting the Challenge: A Summary of Federal Agency Pollution Prevention Strategies	http://es.epa.gov/oeca/fedfac/initiati/airfed/95fed-p2.html
Strategic Environmental Management List	http://www.umich.edu/-nppcpub/resources/ResLists/SEM.html
Voluntary Standards Network (including ISO 14000 series)	http://www.epa.gov/opptintr/p2home/vns
Life Cycle Analysis/Life Cycle Assessment (LCA)	http://www.epa.gov/ordntrt/ORD/NRMRL/std/SAB/lca_brief.htm
ISO 12001: A Discussion of Implications for Pollution Prevention	http://www.p2.org/inforesources/iso.html

5. RCRA requires hazardous waste generators who treat on-site or transport off-site to certify that they have waste minimization programs in place. How is this certification accomplished?
6. How would you expect the EPA or a state environmental regulatory authority to state a regulatory requirement that RCRA facilities reduce the quantities of hazardous waste generated?

REFERENCES

Alexander, Henry, P. E. 1992. "Source Reduction and Waste Minimization for Hazardous Wastes," in *Environmental Management.* May/June, 1992:37ff.

Allen, David T. and Kirsten Sinclair Rosselot. 1997. *Pollution Prevention for Chemical Processes.* John Wiley & Sons, NY.

Bolstridge, June C. 1992. *EPCRA Data on Chemical Releases, Inventories, and Emergency Planning.* Van Nostrand Reinhold, NY.

Breen, Joseph J. and Michael J. Dellarco. 1992. *Pollution Prevention in Industrial Processes — The Role of Process Analytical Chemistry.* American Chemical Society, Washington, D.C.

Chada, Nick. 1997. "Develop Multimedia Pollution Prevention Strategies," in *Environmental Management and Pollution Prevention,* Gail F. Nalven, Ed., American Institute of Chemical Engineers, NY.

Dupont, R. Ryan, Louis Theodore, and Kumar Ganesan. 2000. *Pollution Prevention The Waste Management Approach for the 21st Century.* CRC Press, Boca Raton, FL.

Harris, Margaret. 1988. "In-House Solvent Reclamation Efforts in Air Force Maintenance Operations," in Hazardous Waste Minimization within the Department of Defense, Joseph A. Kaminski, Ed., Office of the Deputy Assistant Secretary of Defense (Environment), Washington, D.C.

Higgins, Thomas E. 1989. *Hazardous Waste Minimization Handbook.* Lewis Publishers, Chelsea, MI.

Hild, Nicholas R. 1988. Professor, Information and Management Technology, Arizona State University East, Personal Communication.

Milliman, Kevin E. and Henry C. Luyten. 1999. "Waste Not, Want Not," in *Environmental Protection.* May, 1999:32ff. Stevens Publishing, Waco, TX.

Mills, Michael B. 1988. "Hazardous Waste Minimization in the Manufacture of Explosives," in Hazardous Waste Minimization within the Department of Defense, Joseph A. Kaminski, Ed., Office of the Deputy Assistant Secretary of Defense (Environment), Washington, D.C.

McCoy and Associates. 1989. "Detoxifying Foundry Waste," in *The Hazardous Waste Consultant* March/April, 1989, McCoy and Associates, Inc., Lakewood, CO.

Nizolek, Donald C., W. Corey Trench, and Mary E. McLearn. 1997. "Set Up a Waste Accounting System to Track Pollution Prevention," in *Environmental Management and Pollution Prevention,* Gail F. Nalven, Ed., American Institute of Chemical Engineers, NY.

Phipps, Erica. 1995. *Pollution Prevention Concepts and Principles.* National Pollution Prevention Center for Higher Education, University of Michigan, Ann Arbor.

Reinhardt, Peter A., K. Leigh Leonard, and Peter C. Ashbrook. 1996. *Pollution Prevention in Laboratories.* CRC Press, Boca Raton, FL.

Shen, Thomas T. 1999. *Industrial Pollution Prevention.* Springer-Verlag, Berlin.

Smith, J. D. 1998. "Once Is Not Enough," in *Environmental Protection.* October, 1998:36ff.

U.S. Congress, Office of Technology Assessment. 1986. *Serious Reduction of Hazardous Waste,* Superintendent and Documents, U.S. Government Printing Office, Washington, D.C.

U.S. Environmental Protection Agency. 1988. Waste Minimization Opportunity Assessment Manual, Hazardous Waste Engineering Laboratory, Cincinnati, OH, EPA 625/7-88/003.

U.S. Environmental Protection Agency. 1989. Waste Minimization in Metal Parts Cleaning, Office of Solid Waste and Emergency Response, Washington, D.C., EPA 530-SW-89-049.

U.S. Environmental Protection Agency. 1994a. The Waste Minimization National Plan, Office of Solid Waste and Emergency Response, Washington, D.C., EPA 530-R-94-045.

U.S. Environmental Protection Agency. 1994b. Strategy for Hazardous Waste Minimization and Combustion, Washington, D.C., EPA 530-R-94-044.

U.S. Environmental Protection Agency. 1994c. The Biennial RCRA Hazardous Waste Report (Based on 1991 Data) Executive Summary, Office of the Solid Waste and Emergency Response, Washington, D.C., EPA 530-S-94-039.

U.S. Environmental Protection Agency. 1994d. Draft RCRA Waste Minimization National Plan Summary. Office of Solid Waste and Emergency Response. Washington, D.C., EPA 530-S-94-002.

U.S. Environmental Protection Agency. 1997. Waste Minimization National Plan — Reducing Toxics in our Nation's Waste. Office of Solid Waste and Emergency Response. Washington, D.C., EPA 530-F-97-010.

U.S. Environmental Protection Agency. 1998a. EPA's Multimedia Strategy for Priority Persistent, Bioaccumulative, and Toxic (PBT) Pollutants, Executive Summary. Office of Pollution Prevention and Toxics, Washington, D.C.

U.S. Environmental Protection Agency, 1998b. The Virtual Elimination Project. Region 5 — Air and Radiation Division, Chicago, IL.

U.S. Environmental Protection Agency. 1999a. Toxics Release Inventory 1998 Data Release Washington, D.C.

U.S. Environmental Protection Agency. 1999b. The Common Sense Initiative: A New Generation of Environmental Protection, Washington, D.C.

U.S. Environmental Protection Agency. 1999c. 33/50 Program: The Final Record. Washington, D.C., EPA 745-4-99-004.

U.S. Environmental Protection Agency. 1999 ES. A Multimedia Strategy for Priority Persistent, Bioaccumulative, and Toxic (PBT) Pollutants. Office of Pollution Prevention and Toxics, Washington, D.C.

U.S. Government Accounting Office. 1994. Toxic Substances EPA Needs More Reliable Source Reduction Data and Progress Measures. Washington, D.C., GAO/RCED-94-93.

U.S. Government Accounting Office. 1994. Toxic Substances Status of EPA's Efforts to Reduce Toxic Releases. Washington, D.C., GAO/RCED-94-207.

Wentz, Charles A. 1989. *Hazardous Waste Management.* McGraw-Hill, NY.

Wrieden, Mary. 2000. "Source Reduction," in *Pollution Prevention The Waste Management Approach for the 21st Century.* R. Ryan Dupont, Louis Theodore, and Kumar Ganesan, Eds., CRC Press, Boca Raton, FL.

9 RCRA Permits, Compliance, and Enforcement

OBJECTIVES

At completion of this chapter, the student should:

- Understand the basic outline of the RCRA permitting process.
- Be familiar with the four steps of the RCRA corrective action process and the application of each of the steps.
- Understand the goals of the RCRA Enforcement Program and the actions which may be taken to achieve these goals.
- Be familiar with the administrative, civil, and criminal enforcement provisions of RCRA.

INTRODUCTION

In the previous chapters, we have attempted to first present materials on the generally accepted practice pertaining to the hazardous waste management subject at hand. We have followed the general (or "generic") material with an overview of the regulatory requirements of the Resource Conservation and Recovery Act (RCRA) and other pertinent statutes, as they apply to the subject. This chapter deals with three related aspects of RCRA which have no generic counterpart.

Similarly, we have attempted to present the highly complex subject of hazardous waste management, and the respective components of RCRA, in an orderly flow of compartmentalized subjects. We now find it necessary to present an important set of materials that do not "fit together" as nicely. Certainly, compliance is required of permit holders, and enforcement actions are taken against those not in compliance. However, compliance with RCRA (and other) statutes and regulations is required of all who handle hazardous wastes (not just permit holders or applicants), and enforcement actions may be taken against those who do not comply with the regulations. We will attempt to keep these aspects clear, but the reader should approach the subject with care.

The requirement to apply for, and obtain, an operating permit to treat, store, and dispose of hazardous waste is the subject of § 3005 of RCRA. The implementing regulations for § 3005 are codified at 40 CFR 270. The related authority of EPA and/or state inspectors to enter upon the premises of any "… person who generates, stores, treats, transports, disposes of, or otherwise handles or has handled hazardous waste …" to inspect, obtain samples, and copy records is contained in RCRA § 3007. Section 3008 provides authorities for enforcement of RCRA provisions. Section 3013 provides the EPA with authorities to require owners or operators of Treatment, Storage, and Disposal Facilities (TSDFs) to conduct monitoring, testing, analysis, and reporting and to take enforcement action against any person who fails or refuses to comply with an order issued under this section. The operating requirements for TSDFs are found in Parts 264, 265, and 266. Facilities that have not received a permit and are operating under interim status must comply with the Part 265 standards. The administrative procedures that apply to the permitting process, including procedures for issuing, modifying, revoking, reissuing, or terminating permits, are provided in 40 CFR 124 (EPA 2000; DeCamp 2000).

PERMITS TO TREAT, STORE, OR DISPOSE OF HAZARDOUS WASTE

Permits identify the administrative and technical standards that must be met by TSDFs. Permits are issued by the EPA or by a state agency that has been authorized by the EPA to administer the program. The permit specifies the operating requirements for the facility based upon the general and technical standards of Part 264, as well as requirements for corrective actions.

Facilities Permitted

RCRA requires every owner or operator of a TSDF to obtain an operating permit. Congress and the EPA recognized that several years would be required to issue permits to all TSDFs and made provision for granting "interim status" to TSDFs that were in operation on November 19, 1980, and had "notified" the EPA prior to that date. Other TSDFs that are in operation "… on the effective date of statutory or regulatory amendments, under the Act, that render the facility subject to the requirement to have a RCRA permit shall have interim status and shall be treated as having been issued a permit …" provided they have "notified" the EPA of hazardous waste activity and comply with applicable operating standards (40 CFR 270.70).

Interim status facilities are allowed to operate in that status until a final permit is issued or denied. New facilities or existing facilities that failed to qualify for interim status are ineligible for interim status and must obtain a permit before commencing operations. Only in a very limited number of circumstances can a person treat, store, or dispose of hazardous waste without interim status or a permit. Such circumstances include

- Generators (LQGs and SQGs) storing waste on-site for time periods shorter than prescribed by § 262.34
- Farmers disposing of their own pesticide wastes on site as provided by § 262.70
- Owners or operators of totally enclosed treatment facilities, wastewater treatment units, and elementary neutralization units as defined by § 260.10

- Transporters storing manifested hazardous waste in containers meeting the requirements of § 262.30 at a transfer facility for a period of 10 days or less
- Owners and operators performing containment activities during an immediate response to an emergency per § 264.1
- Persons adding absorbent material to hazardous waste in a container and persons adding hazardous waste to absorbent material in a container[1] (EPA 1998d, Chapter 8)

Permits are also issued for research, development, and demonstration projects; post-closure of land disposal facilities; emergency situations involving imminent and substantial endangerment to human health or the environment; temporary permits for incinerators to conduct trial burns; and for land treatment facilities to demonstrate acceptable performance.[2] Ocean disposal vessels and barges regulated by the Marine Protection Research and Sanctuaries Act (MPRSA), UIC wells regulated by the Safe Drinking Water Act (SDWA), and Publicly owned Treatment Works (POTWs) regulated by the Clean Water Act (CWA) are considered to have "permits-by-rule." RCRA provides for these facilities' non-RCRA permits to serve in place of a RCRA permit, provided that the facilities are in compliance with their issued permits and the basic RCRA administrative requirements (adapted from EPA 1998c; EPA 1998d, Chapter 8; and EPA 2000).

The Permitting Process

Owners and operators of TSDFs must submit a comprehensive permit application consisting of two parts (Figure 9.1). Part A of the application is a short form (EPA Form 8700-23) that calls for basic information about the facility, such as name, location, nature of business conducted, regulated activities, and topographic map of the site. The Part B, in narrative form, is much more extensive than Part A and requires the submission of substantially more detailed technical information. General requirements are provided by 40 CFR 270.14. Specific information requirements for containers, tanks, surface impoundments, incinerators, land treatment facilities, landfills, and miscellaneous units are provided by §§ 270.15 through 270.23. The applicant must become familiar with the requirements of Parts 264 and 270 in order to determine the nature of the data required for the particular facility.

Technically, Part A may be submitted initially, to be followed by Part B when "called in" by the EPA. This procedure grew from the necessity for then-existing facilities to apply prior to November 19, 1980 in order to be granted interim status.[3]

---

[1] Provided that the container meets the definition of § 260.10 and that these actions occur at the time the waste is first placed in the container and §§ 264.17(b), 264.171, and 264.172 are complied with.

[2] Although there are many hundreds more storage facilities nationwide than either treatment or disposal facilities, it is easy to lapse into an association of permitting with the "T" and "D" of the TSDF acronym. Generator (large and small) personnel should keep in mind the ease with which status can shift to that of a storage facility and the permitting requirements that apply.

[3] In recent years, the EPA has granted interim status to some applicant facilities burning hazardous wastes in boilers and industrial furnaces.

```
┌─────────────────────────────────────────────────────┐
│                        Part A                        │
│  • Activities conducted that require a permit         │
│  • Facility name, mailing address, and location       │
│  • Facility standard industrial classification (SIC) codes │
│  • Treatment, storage, and disposal processes         │
│  • Design capacity of waste management units          │
│  • List of wastes to be managed at facility           │
│  • Permits received or applied for under other regulatory │
│    programs                                           │
│  • Topographic map                                    │
└─────────────────────────────────────────────────────┘
```

```
┌─────────────────────────────────────────────────────┐
│                        Part B                        │
│  • General facility description                       │
│  • Analyses of wastes to be managed                   │
│  • Facility security procedures                       │
│  • Inspection schedule                                │
│  • Contingency plan                                   │
│  • Procedures and precautions to prevent release of   │
│    waste into environment                             │
│  • Procedures and precautions to prevent accidental   │
│    ignition or reaction of waste                      │
│  • Facility location information                      │
│  • Closure plans                                      │
└─────────────────────────────────────────────────────┘
```

FIGURE 9.1 Example Parts A and B, RCRA permit application requirements (EPA 1998d, Chapter 8).

Some authorized states no longer want Part A submissions, requiring Part B as the initial submission. Others require that Parts A and B be submitted simultaneously. Owners and operators of new facilities must submit Parts A and B simultaneously, at least 180 days prior to the expected date of start of construction. Construction may not begin until the application is reviewed and a final permit is issued. Except in the case of very simple or primitive facilities, the assigned permit writer will require even further submissions after reviewing the Part B. The supplemental data are requested through a "notice of deficiency" (NOD) letter to the applicant. As noted in earlier chapters, the omnibus authorities of RCRA § 3005(c)(3) provide the permit writer with considerable discretion to require the applicant to develop data, to conduct risk assessments, or to prescribe conditions not specifically addressed by the RCRA regulations. Figure 9.2 is an example of Parts A and B permit application requirements.

The applicant should carefully coordinate preparation of the required submissions with the EPA or state program manager and/or assigned permit writer. The process is time-consuming, detailed, and exacting. For large and/or complex facilities, the process may require several years to complete. Public participation has become integral to each step of the permitting process and increasingly involves contentious and/or protracted issues.

After a complete RCRA permit application is filed, the 40 CFR 124 regulations establish the procedure for processing the application and issuing the permit. The process includes

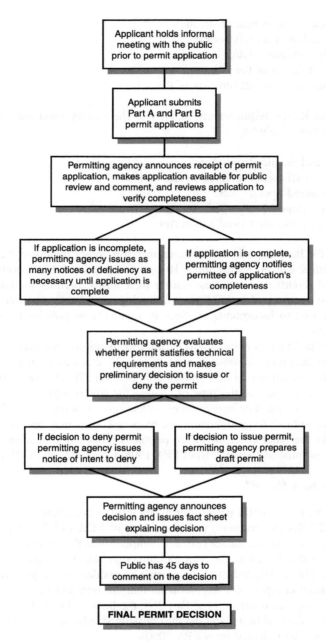

FIGURE 9.2 The permitting process (EPA 1998d, Chapter 8).

- Review of the permit application
- Preparation of a draft permit
- Public comment and/or hearing
- Issue or denial of the permit
- Maintenance and termination of the permit

In addition to RCRA requirements, activity at the facility must not conflict with other federal laws, including

- Wild and Scenic Rivers Act
- National Historic Preservation Act of 1966
- Endangered Species Act
- Coastal Zone Management Act
- Fish and Wildlife Coordination Act

Permits for land disposal facilities, incinerators, other treatment facilities, and storage facilities can be issued for a 10-year fixed term. While permits may be reviewed and modified at any time during their terms, permits for land disposal facilities must be reviewed within 5 years following issue. When reviewed, the permit may be modified to incorporate changes in standards or policies regarding land disposal facilities.

The permitted facility may request permit modifications for a variety of reasons, ranging from relatively inconsequential "Class 1" changes to major "Class 3" changes such as creation of a new landfill unit. Section 270.42, Appendix I, assigns classifications according to the type of change for which approval is sought. The EPA or the authorized state agency may initiate modification, in which case only the conditions subject to modification are reopened, or it may revoke and reissue the permit. In the latter case, the entire permit is reopened and, when reissued, a new term is established (adapted from EPA 1998, Chapter 8, and EPA 2000).

The "Permit As a Shield"

Compliance with an RCRA permit during its term is considered compliance (for purposes of RCRA enforcement) with Subtitle C of RCRA [§ 270.4(a)]. This provision means that an owner and/or operator complies with the requirements specified in the permit, rather than with the corresponding regulations as promulgated in Parts 264 and 266. This is referred to as the **permit-as-a-shield** provision. Nevertheless, a permittee must comply with requirements that are imposed by the statute itself, e.g., the land disposal restrictions of Part 268, the liner and leak detection requirements for land disposal units, and the air emission standards of Subparts AA, BB, and CC of Part 265 (adapted from EPA 2000).

Closure Plans and Post-Closure Permits

Owners and/or operators of TSDFs submitting Part B **permit** applications must include closure **plans** in accord with §§ 264, 265.112. The plan must explain in detail how the performance standard of §§ 264, 265.111 is to be achieved. The approved closure plan then becomes an enforceable component of the issued permit. **Interim status** facilities must have a written closure plan on the premises within 6 months after the facility becomes subject to § 265.112.

If all hazardous waste and contaminants, including contaminated soils and equipment, can be removed from a site or unit at closure, the site or unit can be "clean-closed," meaning that post-closure care is not required (*see:* Chapter 11 or Glossary). Land disposal/treatment facilities with contaminated equipment, structures, and soils that cannot be clean-closed must obtain closure permits, thereby ensuring that appropriate monitoring and maintenance requirements will be met. Owners and/or operators must submit a post-closure plan for the site or unit to be closed as part of the post closure permit application. The plan must include

- A description of planned groundwater activities
- A description of planned maintenance activities
- The name, address, and telephone numbers of the person or office to contact during the post-closure plan (40 CFR 264 and 265.117-118)

Post-closure care consists primarily of groundwater monitoring and maintaining waste containment systems. The post-closure period is normally 30 years, but may be extended or shortened by the regulatory agency (adapted from EPA 1997; EPA 1998d, Chapter 8; and EPA 2000; *see also:* EPA 1998a: EPA 1998e).

RCRA Permits As a "Virtual Elimination" Tool

As noted in the previous chapter, the EPA has made clear the intent to use the RCRA permit program (among others) to eliminate the use of target compounds[4] wherever possible, thereby minimizing future releases. The agency has indicated that intent in the document "The Virtual Elimination Project" (EPA 1998b), unambiguously stating that reductions in releases can be achieved without the need for additional research or regulatory action. The EPA has also published a "handbook" entitled "Pollution Prevention Solutions During Permitting, Inspections and Enforcement," which provides examples of P2 inclusions in permits and enforcement settlements in air, water, and hazardous waste actions by the EPA and state agencies (EPA 1998a).

THE CORRECTIVE ACTION PROCESS

Note: As this text was being prepared, the EPA was responding to Congressional, public, and stakeholder pressures to increase the pace of RCRA cleanups with a set of administrative reforms known as the RCRA Cleanup Reforms. The reforms are designed to achieve faster, more efficient cleanups at RCRA Treatment, Storage, and Disposal sites that have potential environmental contamination. The goals of the reforms focus on 1712 RCRA facilities identified by the EPA and states because of the potential for unacceptable exposure to pollutants and/or for groundwater

---

[4] The current emphasis being on elimination of persistent, bioaccumulative toxic (PBT) chemicals.

contamination. This group is referred to as the "RCRA Cleanup Baseline." The goals are that by 2005, the states and the EPA will verify and document that 95% of these 1712 RCRA facilities will have "current human exposures under control"[5] and 70% will have "migration of contaminated groundwater under control."[5] A number of guidance and related documents are being propagated and are accessible on the Office of Solid Waste Web site <www.epa.gov/epaoswer/osw/cleanup.htm> (EPA 1999b). Thus, much of the following material is subject to change. Additional notes relative to the reforms are provided in the following paragraphs.

The 1984 Hazardous and Solid Waste Amendments (HSWA) expanded the authorities of the EPA and the authorized states to address releases of hazardous waste through corrective actions beyond those then contained in 40 CFR 264, Subpart F. Corrective action requirements are imposed upon **permitted** or **nonpermitted** facilities through a permit, an enforcement order, or lawsuit. RCRA facilities generally are brought into the RCRA corrective action process when there is an identified release of hazardous waste or hazardous constituents or when the regulatory agency is considering the permit application submitted by the facility. The agency can incorporate corrective action requirements in an **existing permit** or a **newly issued permit**. At a minimum, permits including corrective action requirements will include schedules for compliance and provisions for financial assurance to cover the cost of implementation of the required cleanup.

In addition to the Part 264, Subpart F, requirements for groundwater monitoring and correction of any releases from land disposal units, HSWA added statutory provisions for addressing corrective action in permits as follows:

- Section 3004(u) provides authority to the EPA to require corrective action for releases of hazardous waste or hazardous constituents from solid waste management units (SWMUs) on or within the facility.
- Section 3004(v) authorizes the agency to impose corrective action requirements for releases that have migrated beyond the facility boundary.
- The omnibus permitting authority of § 3005(c)(3) authorizes the EPA or the state agency to modify a permit as necessary to require corrective action for any potential threat to human health or the environment.

The EPA has additional statutory authorities to order corrective actions that are not contingent upon a facility permit:

- RCRA § 3008(h) authorizes the EPA to require corrective action or other necessary corrective measures in the form of an administrative enforcement order or to seek a court order, in case of a release of hazardous waste or constituents from an **interim status** facility.
- Section 7003 provides the EPA broad enforcement authority, upon finding evidence of past or present handling of solid or hazardous waste, to require any action necessary to abate potential imminent and substantial hazards caused by releases from any source.

---

[5] Two guidance documents termed "Environmental Indicators" (EIs).

The decision regarding these alternatives is made by the EPA on a case-by-case basis, taking into account the nature and magnitude of the release.

Corrective actions presently proceed through one or more of six steps. The procedures are detailed and tailored to the situation at the facility in question. The steps, briefly, involve:

- RCRA Facility Assessment (RFA) — a review of existing information on contaminant releases, including information on actual or potential releases (The RFA may include sampling if needed.)
- A follow-up investigation referred to as a release investigation or Phase I RCRA Facility Investigation (Phase I RFI) — may be useful before full-scale characterization for a variety of reasons or purposes, such as confirming dated information
- RCRA Facility Investigation (RFI) — wherein the owner or operator of a facility may be required to conduct further investigations to verify and/or characterize a release or releases or to conduct a full-scale site characterization
- Interim Measures — short-term actions to control ongoing risks while site characterization is underway or before a final remedy is selected
- Corrective Measure Study (CMS) — in which the owner or operator is required to identify, evaluate, and recommend specific corrective measures that will remediate the site
- Corrective Measures Implementation (CMI) — may include design, construction, maintenance, and monitoring of the selected corrective measures

Interim corrective measures may be required, at any point in the process, where the EPA or the authorized state agency believes that expedited action should be taken to protect human health or the environment (EPA 1998d).

Remediation Waste Management Units[6]

Until recently, the EPA had implemented the corrective action program primarily through direct use of statutory authorities and by the issue of guidance and policies developed pursuant to those authorities. In 1993, the Agency codified rules pertaining to corrective action management units (CAMUs) and temporary units (TUs) at 40 CFR 264, Subpart S (58 FR 8683). The subpart was revised and § 264.554 (staging piles) was added in 1998 (63 FR 65939).[7]

---

[6] "Remediation waste" means all solid and hazardous wastes and all media (including groundwater, surface water, soils, and sediments) and debris that contain listed hazardous wastes or that themselves exhibit a hazardous characteristic and are managed for implementing cleanup (40 CFR 260.10).

[7] The EPA initially proposed to replace CAMUs with "remediation piles" (61 FR 18779; April 29, 1996). The term was replaced with "staging piles" when § 264.554 was added (63 FR 65939). The Agency did not follow through with the replacement proposal and has announced that it will not take final action on the proposed Subpart S; however, the portions of Subpart S that have been finalized (CAMUs) will remain in effect (64 FR 54604; *see also:* Porter 1999).

A CAMU[8] is a physical, geographical area designated by the EPA or an authorized state agency for managing remediation wastes during corrective action. One or more CAMUs may be designated at a facility. The CAMU enables the facility to manage the remediation waste in a unit without having to comply with LDR treatment standards or the minimum technical requirements for land-based treatment, storage, or disposal units (40 CFR 264.552).

TUs are containers or tanks that are designed to manage remediation wastes during corrective action at **permitted** or **interim** status facilities. The TU regulations for non-land-based units were promulgated at the same time as the CAMU regulations for land-based units. TUs may operate for 1 year, with opportunity for a 1-year extension (40 CFR 264.553).

A staging pile is an accumulation of solid, non-flowing remediation waste that is not a containment building and is used only during remedial operations for temporary storage within the contiguous property under the control of the owner/operator, where the wastes originated. The staging pile must be operated according to the design criteria designated by the regulatory agency. The staging pile must not operate for more than 2 years, unless an operating term extension is granted by the regulatory agency (40 CFR 264.554; *see also:* EPA 1998c; EPA 1998d; EPA 1997a).

COMPLIANCE REQUIREMENTS OF RCRA

The goals of the RCRA enforcement program are to ensure that the regulatory and statutory provisions of RCRA are met and to compel corrective action, where necessary. Facility inspections by the EPA and/or state agency officials are the primary tool by which compliance is monitored; however, self-monitoring and reporting activity are important elements of the program.

EPA Regional Administrators and officials of authorized state agencies have some discretion in reaction to findings of noncompliance by a facility that is subject to RCRA regulations. When noncompliance is detected, the range of enforcement options include the use of administrative orders, civil lawsuits, or criminal indictments, depending upon the nature and severity of the offense.

Federal and state administrators must have reliable compliance data in order to make fair and equitable decisions regarding enforcement options and to assess the overall effectiveness of the RCRA program. Competent monitoring acts as a deterrent, by determining the extent to which a facility is in or out of compliance; by

---

[8] On February 11, 2000, the EPA, the Environmental Defense Fund (EDF), the Natural Resources Defense Council (NRDC), and the Environmental Technology Council (ETC) reached a settlement agreement on the pending litigation over the CAMU regulations for remediation waste. The settlement calls for the Agency to amend the 1993 rule to establish CAMU-specific treatment and design standards and minimum liner and cap standards for CAMUs (EPA 2000a).

identifying potential and actual problems; and by generating credible data for use as leverage in negotiated settlements and as evidence in judicial proceedings.

Self-Monitoring

Many regulatory agencies have neither the field nor laboratory resources to conduct definitive compliance monitoring of each and every potential and actual source of release of pollutants to the environment. Most environmental laws and regulations require the regulated entity to perform self-monitoring and to report or maintain the data in files. RCRA is no exception to that generality.

With some limited exceptions, owners and operators of permitted surface impoundments, waste piles, land treatment units, and landfills must comply with the groundwater monitoring requirements of 40 CFR 264.91 through 264.100. The requirements are prescribed by the permitting authority and are tailored to the type and configuration of the facility, the type(s) of wastes to be managed, the geological and hydrogeological conditions of the site, and other variables. The general groundwater monitoring requirements of § 264.97, paraphrased, are

 a. The groundwater monitoring system must consist of a sufficient number of wells, installed at appropriate locations and depths to yield groundwater samples from the uppermost aquifer that:
  1. Represent the quality of backgroundwater that has not been affected by leakage from a regulated unit
  2. Represent the quality of groundwater passing the point of compliance[9]
  3. Allow for the detection of contamination when hazardous waste or hazardous constituents have migrated from the waste management area to the uppermost aquifer (EPA 1997c)

Figure 9.3 illustrates placement of clusters of monitoring wells to provide data on backgroundwater and on downgradient water quality.

In general, the groundwater monitoring requirements are met by three categories of monitoring:

 1. *Detection Monitoring Program:* The owner or operator must monitor for indicator parameters, waste constituents, or reaction products that provide a reliable indication of the presence of hazardous constituents in groundwater (§ 264.98).
 2. *Compliance Monitoring Program:* If the Detection Monitoring Program indicates contamination of the uppermost aquifer, a permit modification establishing a *compliance* monitoring program must be initiated. The owner or operator must determine whether there is statistically significant evidence of increased contamination by any chemical parameter or hazardous constituent specified in the permit (§ 264.99).

---

[9] The "point of compliance" is a "vertical surface located at the hydraulically downgradient limit of the waste management area that extends down into the uppermost aquifer underlying the regulated units" (40 CFR 264.95).

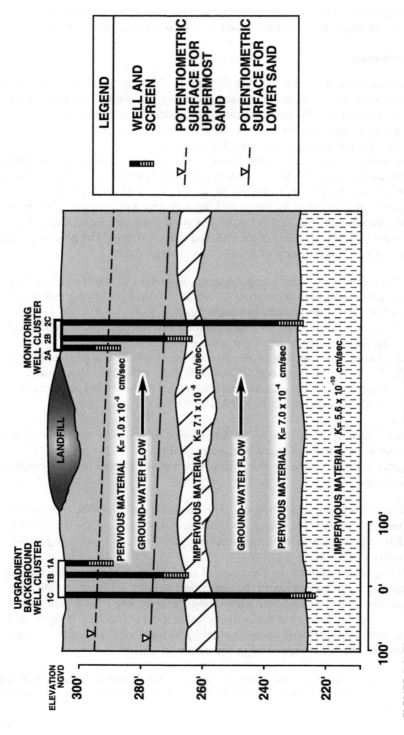

FIGURE 9.3 Placement of background and monitoring wells (EPA 1986).

3. *Corrective Action Program:* If the Compliance Monitoring Program verifies that any concentration limit specified in the permit is being exceeded, the owner or operator must notify the EPA or authorized state agency administrator within 7 days and initiate a permit modification to establish a corrective action program. The program, when approved, will require the owner or operator to take action to remove the hazardous waste constituents or treat them in place (§ 264.100).

In each of the above steps, the permit or permit modification will detail the specific corrective measures that are to be implemented. Interim status land treatment and disposal facilities are subject to somewhat less stringent self-monitoring requirements. Regional administrators, authorized state agency officials, and permit writers have wide discretion regarding the type, extent, and frequency of monitoring that may be required. The EPA has published a number of guidance documents that inject a degree of consistency into the monitoring programs, but the system is designed to enable the regulatory authority to tailor the monitoring requirements to the individual facility (*see also:* EPA 1986; EPA 1997c; EPA 1998, Chapter 10; NRC 1997, ES-7, Chapter 1; Corbitt 1989, pp. 9.78ff; Sara 1993, Chapter 10).

Inspections

The preferred method of obtaining compliance data is by the conduct of an inspection of the regulated facility. RCRA § 3007 provides authorities for representatives of the EPA or authorized state agencies to enter any premises where hazardous waste is handled to observe operations, examine records, and take samples of the wastes. In instances where criminal activity is suspect, investigators from the EPA National Enforcement Investigations Center (NEIC), Office of Criminal Investigation at EPA Headquarters, and state attorney general's staff, or combinations thereof, may become involved. Department of Transportation (DOT) investigators may participate in inspections involving transportation of hazardous wastes, and Customs Officers will play a major role in cross-border cases. Contractor personnel may be tasked to perform inspections; however, complications may arise in the use of contractor-obtained evidence, particularly where criminal penalties are sought.

HSWA requires that all federal- or state-operated facilities be inspected annually. All TSDFs must be inspected at least once every 2 years. Facilities may also be inspected at any time the EPA or the state agency has reason to suspect that a violation has occurred (adapted from EPA 1998d, Chapter 10).

Types of Inspections. Several types of inspections have been developed to meet RCRA requirements. The formats and descriptors for RCRA inspections change frequently due to changes in the statutes, regulations, and policies. Court decisions also play a major role in the processes of evidence gathering. The currently used formats include

- *Compliance Evaluation Inspection (CEI):* The CEI is a routine inspection to evaluate compliance with RCRA. These inspections usually encompass a file review prior to the site visit; an on-site examination of generation, treatment, storage, or disposal areas; a review of records; and an evaluation of the facility's compliance with RCRA.

- *Case Development Inspection (CDI):* The CDI is conducted when significant RCRA violations are known, suspected, or revealed and is structured to gather data in support of a specific enforcement action.
- *Comprehensive Groundwater Monitoring Evaluation (CME):* The CME is a detailed evaluation of the adequacy of the design and operation of groundwater monitoring systems at RCRA facilities.
- *Compliance Sampling Inspection (CSI):* Samples are collected for laboratory analysis. A sampling inspection may be conducted in conjunction with or in support of other inspection formats.
- *Operation and Maintenance Inspection (O&M):* The O&M inspection is structured to determine whether or not groundwater monitoring and other systems are functioning properly after closure of a land-disposal facility.
- *Laboratory Audits:* Laboratory audits are inspections of laboratories performing groundwater analyses. The audit is intended to ensure that the laboratory is using proper sample handling and analysis protocols.
- *RCRA Facility Assessment (RFA):* As noted earlier in this chapter, the RFA is performed at a TSDF to identify releases or potential releases of hazardous constituents from solid waste management units that may require corrective action. RFAs are usually conducted as part of the permitting process (EPA 1997b; EPA 1998d).

ENFORCEMENT OF RCRA REGULATIONS

RCRA provides a variety of enforcement options to the EPA and authorized state agencies. The goals of these provisions are, quite simply, to compel:

- Proper handling of hazardous waste
- Compliance with RCRA record keeping and reporting requirements
- Necessary corrective action

The enforcement program is carried out through evaluation of compliance monitoring data and the various levels of inspection activity discussed earlier. The enforcement options include **administrative actions, civil actions, and criminal actions**. Administrative actions may be informal or formal. The decision to pursue one of these options is based upon the nature and severity of the problem.

Informal Administrative Actions

The agencies initiate informal administrative actions by notifying owners or operators of waste handling facilities of a problem with their compliance status. Such actions may involve no more than a telephone call or a face-to-face conversation. They include issuance of a "notice of violation" or "notice of deficiency." This type of action is appropriate where the violation is of a minor nature, such as a record keeping error or omission.

If the owner or operator does not take steps to comply within a reasonable or specified time period, a "warning letter" may be sent. The warning letter also sets forth the enforcement actions that will follow if the recipient fails to take the necessary steps to bring the operation into compliance.

Formal Administrative Orders

More severe violations or failure to respond to an informal action can be the basis for the agency to issue a formal administrative order. Such an order, issued under RCRA authority, imposes enforceable legal duties. Orders can be used to force a facility to comply with specific regulations; to take corrective action; to perform monitoring, testing, and analysis; or to address a threat of harm to human health and the environment. Four types of orders can be issued under RCRA:

Compliance Orders. A Section 3008(a) Order may be issued to any person who is in noncompliance with a RCRA requirement. The order may require immediate compliance or may set a timetable to be followed in achieving compliance. The order may specify penalties as great as $27,500/day for each day of noncompliance and can suspend or revoke the permit or interim status of the facility.

Corrective Action Orders. A Section 3008(h) Order may be issued requiring corrective action at an interim status facility when there is evidence of a release of a hazardous waste to the environment. Such orders can be issued to require corrective action ranging from investigative activity to repairing liners or pumping and treating a plume of contaminated groundwater. The order may utilize "reach back" authorities to require cleanup of previously caused problems. These orders can also impose penalties as great as $27,500/day for each day of noncompliance.

Section 3013 Orders. Section 3013 provides authorities for the EPA to issue an Administrative Order to correct a "substantial hazard to human health and the environment." The order requires that the nature and extent of the problem be evaluated through monitoring, analysis, and testing. The order may be issued to the current owner of the facility or to a past owner or operator, as appropriate.

Section 7003 Orders. The 7003 Order is used to order cleanup of an "imminent and substantial endangerment to health or the environment" that is or has been caused by the handling of nonhazardous or hazardous waste. The order may be issued to any contributing party, including past or present generators, transporters, or owners or operators of the site. Violation of a Section 7003 Order can result in penalties of as much as $5500/day.

Civil Actions

Formal lawsuits may be brought in civil jurisdictions to seek court-ordered compliance with RCRA provisions, cleanup following a release, or to obtain court orders to persons whom have failed to comply with Administrative Orders issued under Sections 3008, 3013, or 7003. Civil actions are generally employed in situations that present repeated or significant violations or where there are serious environmental concerns.

Criminal Action[10]

Criminal actions resulting in fines or imprisonment may be taken in seven specific instances. These are *knowingly:*

- Transporting hazardous waste to a non-permitted facility
- Treating, storing, or disposing of waste without a permit or in violation of a material condition of a permit or an interim status standard
- Omitting important information from, or making a false statement in, a label, manifest, report, permit, or compliance document
- Generating, storing, treating, or disposing of waste without complying with the RCRA record keeping and reporting requirements
- Transporting waste without a manifest
- Exporting waste without the consent of the receiving country
- Treating, disposing of, or exporting any hazardous waste in such a way that another person is placed in imminent danger of death or serious bodily injury

---

In the 30-year history of the EPA, enforcement policies and strategies of the Agency have undergone periods of increasing and diminishing intensity and aggressiveness. Factors responsible for these fluctuations include the ideological bent of the Administration in office, public opinion and pressures, numbers and seriousness of headline environmental outrages, and a matrix of budget variables, new investigative technologies, and investigative and prosecutorial capabilities, to name a few. Generally, throughout the push and pull of these periods, the Agency and the Department of Justice have reserved criminal prosecution for only the most egregious violations. In the mid-1990s, however, both agencies began focusing on criminal activity by policy statements, increases in numbers of criminal investigators, and numbers of criminal prosecutions.

During the same period, the EPA announced a variety of "incentive" policies and programs designed to achieve voluntary compliance with environmental laws

---

[10] The key word in the designation of the seven offenses that are subject to criminal prosecution, under RCRA, is *knowingly.* A *knowing* violation of RCRA does not necessarily coincide with our common, everyday understanding of the term. In *United States v. Hoflin,* 880 F.2d 1033, 1038-1040 (9th Cir. 1989), the defendant was convicted of disposing of hazardous waste without a permit even though he neither knew the material was an RCRA hazardous waste nor that the party actually doing the disposal lacked a RCRA permit. The court found *knowing* disposal of a hazardous waste when, even though the defendant did not know the paint was an RCRA hazardous waste, he did know the paint was not "an innocuous substance like water." (From Wasson, Eugene R. 2000. *Hazardous Materials Management Desk Reference,* Doye B. Cox, Editor-in-Chief, Adriane P. Borgias, Technical Editor, McGraw-Hill, New York, Chapter 2. With permission.)

while promoting the public's right-to-know.[11] The Compliance Assistance program claims that the nine compliance centers on-line received an average of 750 "hits" per day during FY 99. Other compliance assistance efforts reached approximately 330,000 entities through on-site visits, hotlines, workshops, training, distribution of checklists, and guides (EPA 1999; EPA 2000a).

The annual announcement of enforcement statistics for FY 99 indicated that the Agency was pursuing an aggressive, intensive enforcement policy and program. The numbers included $3.6 billion for environmental cleanup, pollution control equipment, and improved monitoring, an 80% increase over 1998; $166 million in civil penalties, 60% higher than 1998; and 3935 civil judicial and administrative actions, the highest in 3 years. Criminal defendants were sentenced to a record 208 years of prison time for committing environmental crimes. The Agency claimed an impressive tally of environmental improvements as a result of these enforcement actions and the compliance assistance activities (EPA 2000b).

TOPICS FOR REVIEW OR DISCUSSION

1. What was the rationale on the part of the EPA for grouping treatment, storage, and disposal facilities as the third element in the "cradle-to-grave" system of hazardous waste control?
2. Why would the EPA or a state regulatory agency insist that groundwater monitoring continue for 30 years after closure of a land disposal facility?
3. Why was Congress concerned about conflicts with the National Historic Preservation Act of 1966 at an RCRA facility?
4. What rationale does the EPA have for entrusting facilities to conduct self-monitoring of critical compliance parameters?
5. How do permit inspections under the Clean Water Act compare with those of RCRA? Explain some of the differences.

REFERENCES

Corbitt, Robert A. 1989. *Standard Handbook of Environmental Engineering.* McGraw-Hill, NY.

DeCamp, Gregory C. 2000. "RCRA Overview: A Generator Perspective," Chapter 35, in *Hazardous Materials Management Desk Reference,* Doye B. Cox, Editor-in-Chief, Adrianne P. Borgias, Technical Editor, McGraw-Hill, NY.

National Research Council. 1997. *Innovations in Ground Water and Soil Cleanup.* National Academy Press, Washington, D.C.

Porter, Amy. 1999. "EPA Suspends Consideration of RCRA Bill While Settling Corrective Action Suit," in *Environment Reporter* February 19, 1999, pp. 2074–2075. Bureau of National Affairs, Washington, D.C.

Sara, Martin N. 1993. *Standard Handbook for Solid and Hazardous Waste Facility Assessments.* Lewis Publishers, Ann Arbor, MI.

---

[11] The name of the EPA Office of Enforcement was changed to Office of Enforcement and Compliance Assurance.

U.S. Environmental Protection Agency. 1986. RCRA Ground-Water Monitoring Technical Enforcement Guidance Document, OSWER Directive Number 9950.1. Office of Waste Programs Enforcement, Washington, D.C.

U.S. Environmental Protection Agency. 1997. Introduction to Closure/Post-Closure, Office of Solid Waste and Emergency Response, Washington, D.C., EPA 530-R-97-048.

U.S. Environmental Protection Agency. 1997a. Introduction to RCRA Corrective Action, Office of Solid Waste and Emergency Response, Washington, D.C., EPA 530-R-97-065.

U.S. Environmental Protection Agency. 1997b. Introduction to RCRA Enforcement and Compliance, Office of Solid Waste and Emergency Response, Washington, D.C., EPA 530-R-97-066.

U.S. Environmental Protection Agency. 1997c. Introduction to Groundwater Monitoring, Office of Solid Waste and Emergency Response, Washington, D.C., EPA 530R-97-055.

U.S. Environmental Protection Agency. 1998. A Multimedia Strategy for Priority Persistent, Bioaccumulative, and Toxic (PBT) Pollutants, Office of Pollution Prevention and Toxics, Washington, D.C.

U.S. Environmental Protection Agency. 1998a. Pollution Prevention Solutions during Permitting, Inspections and Enforcement, Office of Solid Waste and Emergency Response, Washington, D.C., EPA 530-R-98-015.

U.S. Environmental Protection Agency. 1998b. The Virtual Elimination Project, Region 5 — Air and Radiation, Chicago, IL.

U.S. Environmental Protection Agency. 1998c. Management of Remediation Waste under RCRA, Office of Solid Waste and Emergency Response, Washington, D.C., EPA 530-F-98-026.

U.S. Environmental Protection Agency. 1998d. RCRA Orientation Manual, Office of Solid Waste, Washington, D.C., EPA 530-R-98-004.

U.S. Environmental Protection Agency. 1998e. Elizabeth Cotsworth, Acting Director, Office of Solid Waste. Memorandum TO: Senior Policy Advisors Regions I–X, SUBJECT: Risk-Based Clean Closure, March 16, 1998, Washington, D.C.

U.S. Environmental Protection Agency. 1999. RCRA Enforcement and Compliance, Office of Solid Waste and Emergency Response, Washington, D.C., EPA 530-R-99-060.

U.S. Environmental Protection Agency. 2000. Introduction to Permits and Interim Status, Office of Solid Waste and Emergency Response, Washington, D.C., EPA 530-R-99-057.

U.S. Environmental Protection Agency. 2000a. "EPA Reaches Settlement on 'CAMU' Rule," Note to Correspondents, February 11, 2000. Communications, Education, and Media Relations, Washington, D.C.

U.S. Environmental Protection Agency. 2000b. "EPA Sets Enforcement Records in 1999," Environmental News, January 19, 2000. Communications, Education, and Media Relations, Washington, D.C.

Wasson, Eugene R. 2000. "Overview of the Law in an Environmental Context," Chapter 2, in *Hazardous Materials Management Desk Reference,* Doye B. Cox, Editor-in-Chief, Adriane P. Borgias, Technical Editor, McGraw-Hill, NY.

10 Assessment Techniques for Site Remediation

OBJECTIVES

At completion of this chapter, the student should:

- Be cognizant of the necessity for an appropriate form of site environmental assessment where individuals or organizations have "care, custody, and control" of real property or contemplate assumption of same.
- Be familiar with the general format for site assessments for property transactions and for remediation by regulatory agencies.
- Be familiar with the kinds of background information that are needed for establishing compliance history, assessing need for additional or new data, designing new information/data gathering activity, ensuring safety of the investigators and public, and protecting the rights of the responsible parties.
- Be familiar with site assessment factors such as information-gathering activity appropriate to the problem site, behavior of site owner/manager, severity of the health/environmental threat, health and welfare of the public, safety of workers on the site, and the applicable laws and regulations.
- Understand the importance of record keeping, documentation, and chain-of-custody procedures, irrespective of the nature of the corrective action contemplated.

INTRODUCTION

Our focus, in previous chapters, has been upon the management of hazardous waste as it is generated, transported, stored, treated, destroyed, or disposed. The primary objective of hazardous waste management is the handling of the waste in a manner that prevents harm to the public health and the environment. Whatever the degree or numbers of our successes in attaining this goal may be, the fact remains that large numbers of sites have been contaminated with hazardous waste(s). Contaminated sites must be remediated, whether preparatory to transfer of ownership or as a result of regulatory requirements. Similarly, prospective landowners must have reliable mechanisms for evaluating the extent or absence of contamination of potential acquisitions. Individuals and organizations having responsibility for remediating contaminated sites must have generally recognized and accepted procedures for assessment of site clean-up needs. In this and the following chapter, we will overview techniques and regulatory procedures for accomplishing these tasks.

Two sets of considerations bring about the need for definitive evaluation of site contamination:

1. The Comprehensive Environmental Response, Compensation, and Liability Act (CERCLA)[1] and several court findings impose strict, joint and several liability[2] upon owners or operators of hazardous waste sites, i.e., sites where a release of hazardous substance(s) has occurred. These liabilities can be so severe that avoidance of such liability has become an imperative that transcends most others in commercial property transactions. The acquisition of property, particularly property previously used for industrial or commercial activity, is now made contingent upon a "clean bill of health" determination by a competent environmental investigator. Such determination, variously labeled a due diligence evaluation, an environmental audit, environmental site assessment, or property transfer site assessment,[3] has become a major activity of consultants and attorneys. Significant differences exist between and within the assessment procedures, and these will be discussed briefly. Regardless of the assessment procedure used, the stakes are exceedingly high, and the need for exactness in all aspects of the work is of the highest order. "Super lien" laws in many states (e.g., New Jersey, Massachusetts) allow the state to attach a priority lien to any property to pay for the cost of remediation should environmental contamination be discovered (Hopper 1989).

2. Sites suspected or known to be contaminated are subject to cleanup under state and federal laws. Such cleanup may be carried out voluntarily by the "responsible party(ies);"[4] as a result of a negotiated agreement; in response to an administrative order; under a court-ordered settlement; or by a regulatory agency implementing a funded cleanup (i.e., Superfund) type provision of a statutory authority. Moreover, there are seemingly limitless variations upon each of these mechanisms.

As in the first instance, exacting standards of investigation and analysis are in order. The cost of cleanup of a contaminated site is nearly always measured in the

---

[1] The acronym CERCLA is traditionally used to identify the statute. The nickname "Superfund" usually refers to the program that implements CERCLA. The Superfund site remediation process will be overviewed in the next chapter.

[2] "Strict" liability means that no showing of actual fault is required in order to assign liability; joint and several liability means that multiple contributors of hazardous waste to a site may all be held equally liable unless one or more can demonstrate that its wastes can be separately identified or could not possibly have contributed to the harm (*see:* CERCLA § 107).

[3] Or, increasingly, a "risk-based site assessment" (*see:* discussions of risk-based site assessment in Chapter 4 and in paragraphs below).

[4] Persons having caused, permitted, or contributed to the contamination of a site that is caught up in the Superfund process are referred to as "potentially responsible parties" (PRPs). As the Superfund process continues to the stage that responsibility(ies) has (have) been established, the term becomes "responsible parties" (RPs).

millions of dollars. By 1995, the average Superfund site cleanup cost was more than $10 million (Priznar 1995, p. 168). That cost inevitably falls upon property owners, "responsible parties," or the taxpayers or is passed along to consumers or users of products and services in the form of higher prices. The costs (and environmental impacts) of a misdirected, inadequate, or overdone cleanup, resulting from erroneous or incomplete assessment data, can be unacceptably high.

Formal risk assessment techniques can be used to help estimate the risk posed by a contaminated site if remediation cannot be accomplished or must be delayed. They can be used to focus on the exposure pathways, media, and chemicals that pose the greatest risk; establish cleanup or treatment goals; compare alternatives for most cost-effectively achieving the cleanup goals and other remedial action objectives; and develop postremediation monitoring plans. (Washburn and Edelmann 1999; *see:* discussion of risk assessment in the standards-setting process, Chapter 4).

It is not possible within the scope of this text to provide either a detailed study of site assessment procedures or of the many risk assessment techniques now finding their way into each of the steps of the procedures. The student should become conversant in the general concepts and actions involved and understand their importance.[5] In the next section, we provide the generic approach to identification of problem sites and some approaches to obtaining necessary background information for the conduct of site assessments. In the following section, we discuss the procedural organization of a site assessment as conducted in the private sector. In subsequent sections, we outline the corollary procedure for site assessments leading to remediation under the Resource Conservation and Recovery Act (RCRA) or CERCLA and outline the regulatory site evaluation process established by the Environmental Protection Agency (EPA) to carry out CERCLA mandates.

IDENTIFYING PROBLEM SITES AND OBTAINING BACKGROUND INFORMATION

Purpose

Given either of the two most likely situations — (1) a site is under consideration for acquisition or (2) there is concern or doubt regarding regulatory compliance — a review of background information is needed. The background report for the acquisition site will contain similarities to that done for the suspect site, but the objectives and follow-on activity may be greatly divergent. An authority and practitioner provides the following explanation:

- "In an audit, an auditor is seeking to verify expectations. More specifically, the auditor is seeking to confirm or deny a specific condition.

---

[5] A well-written, easily understood "Technical Information Package" entitled "Risk Assessment" has been published by the EPA Information Office. It is oriented to human health and ecological risk assessment, includes a comparison of risk assessment and risk management, and provides references. The eight-page document can be accessed at <http://www.epa.gov/oiamount/tips.htm>. A useful two-page summary, with references, of risk assessment techniques applicable to each step of the Superfund process can be accessed at <http://www.epa.gov/oerrpage/superfund.htm>.

Typically, in an environmental context, the auditor seeks to understand whether regulatory or policy requirements are being met (compliance audit). The answers to an auditor's questions are limited to 'yes,' 'no,' and 'do not know'."

- "Greater judgement is involved in an assessment, which is similar to an appraisal. An assessor seeks to estimate, or judge, environmentally important factors that affect value or character .... An assessment does not measure expectations of conditions documented in standards against actual conditions, whereas audits should ..." (Priznar 1995, p. 160).

In either event, a background data collection process is necessary. Background information is that which is available or can be obtained from existing records. Such records may contain:

- Cultural history related to man's activities as differentiated from technical data
- Technical details including environmental data, natural phenomena, well logs, etc.
- Regulatory history

Cultural History

Records or histories of man's activities that may be significant with respect to hazardous waste contamination of a site include

- Land-use patterns, e.g., former agricultural use with pesticide residues in fields, container disposal, or heavily contaminated mixing areas
- Site use, e.g., type(s) of industrial or commercial activity
- Records of catastrophic events
- Interviews with former or present residents, employees, owners, labor unions, local officials, or historical societies, regarding past activities on the site

Technical Information

Technical information may be nonexistent, primitive, or otherwise questionable. The investigator should seek out corroborating or coincident data to strengthen existing technical data if possible. Useful data may include

- Geological studies, soil tests, groundwater pump tests, ground or surface water quality data
- Ground and/or surface water hydrological data
- Irrigation history
- Utility and right-of-way maps

Regulatory History

Files of regulatory agencies may contain highly pertinent data, including

- Zoning and ordinance changes
- Tax assessments and business licenses
- Building permits
- Fire code, sanitation, or health violations

In some areas, the U.S. Department of Agriculture has extensive aerial photographic coverage. It is possible (again, in some areas) to view 5- or 10-year sequences of the property under consideration. Other governmental agencies also maintain remote sensing and aerial imagery files which can be very helpful in constructing historical use patterns.

If past or present hazardous waste activity is known or acknowledged, the state or local regulatory agencies should have records indicating:

- Types and volumes of wastes on the site
- Sources and processes associated with the waste
- Chronology and location of waste disposal activity
- Records of violations, spills and other releases, cleanup activity, monitoring activity, and disposition of cleanup or treatment residues

The historical review of an acknowledged hazardous waste site should include records of inspections, site investigations, and regulatory activity on adjacent and nearby sites.

Background Report

All data must be carefully reviewed, conflicts in data must be reconciled, and questionable data must be set aside and so noted. The background report is the basis for decisions regarding follow-on activity and should focus on the findings, the additional data needed, and recommended strategies for dealing with the site.

Consulting firms with extensive historical involvement in the area of the site may be able to confidently advise their client based upon a simple background report. At the opposite extreme, a background report for a site being brought into the federal Superfund process is highly structured and voluminous. In the Superfund lexicon, the background report is called a Preliminary Assessment (or PA) and is but the first step in a protracted process that can lead to remedial action and cost recovery. Although the sophistication of the background report may vary according to the needs of the client and the history of the site, the importance of thoroughness and accuracy cannot be overemphasized.

SITE ASSESSMENT PROCEDURES IN THE PRIVATE SECTOR

Site assessments for property transactions are usually less exhaustive than those for regulatory purposes. As will be seen, a site investigation leading to a remedial action under CERCLA or RCRA may consist of many more steps or "phases" than is normally the case in a Property Transaction Site Assessment (PTSA). Moreover, risk assessments have become essential (required) elements of each step of site assessments in regulatory programs and are becoming synonymous with "due diligence" in the conduct of PTSAs. Most practitioners of the PTSA structure the activity in at most three phases. The following, adapted from Burby (1989), is an example. References for other examples are provided. These structures are for guidance and are flexed to meet the client's needs.

Pre-Phase I

Variously called a "scoping step," the Pre-Phase I is intended to provide a preliminary environmental survey. This activity might involve a 1-day site visit to make a preliminary assessment of the hazardous materials situation. The site visit includes observing general physical conditions, collecting readily available materials such as copies of permits and records, and interviewing present and past employees, managers, and/or owners. The Pre-Phase I may culminate in a letter report to the client that summarizes site observations and findings and includes a scope of recommended additional work.

Phase I

Phase I is a thorough qualitative review of the site, based on field observations and reasonably available existing information. A typical Phase I investigation includes

- Review of appropriate files (e.g., title search, property records, regulatory permits, and databases) to investigate past or current activities at the site or adjacent properties with respect to wastewater, site drainage, air emissions, and toxic substance and hazardous material/waste handling, storage, treatment, disposal, and spills
- Review of reasonably available historic aerial photographs of the site and adjacent properties to identify the timing of past activities in the area and associated significant topographic changes
- Reconnaissance visit(s) to the site and adjacent properties, including the interiors of any on-site buildings, to inspect the general condition of the property and surrounding area for evidence or suspicion of contaminant releases to the soil, surface, and/or groundwater from spills, dumping, or burial of hazardous materials
- Interviews of available personnel and past or present site owners and operators
- Risk assessments as appropriate for the data obtained

The Phase I assessment concludes with a written report that summarizes the observations and findings made and includes recommendations for sampling or other investigative work if needed.

Phase II

Phase II may consist of air, soil, surface, and/or groundwater sampling on and near the property and analyses, as needed to characterize the site. A sampling plan is developed which includes quality assurance and quality control (QA/QC) criteria.

Upon completion of the sampling and chemical characterization and the related data, a Phase II report is prepared detailing the procedures and protocols followed and the findings. Depending upon the findings, the report may include recommendations for additional investigative activity, a scope of recommended remedial actions (Phase III), or other recommendations (Burby 1989) (*see also:* EPA 1998, VI-2; Williams 1998; Washburn and Edelmann 1999; Woodside 1999, Chapter 16; Liner 2000).

Standardized Environmental Site Assessments

Various entities, public and private, have sought to achieve some standardization of environmental site assessment (ESA) procedure. The motive(s) for standardization generally center upon assurances of acceptance, by any/all parties and courts of review, of the findings produced by an ESA. A thorough and structured format has been developed by The American Society for Testing and Materials (ASTM) for the conduct of an *environmental site assessment.*[6] The E 1527-93 Standard Practice for Environmental Site Assessments: Phase I, Environmental Site Assessment Process has filled the described need and has become widely accepted as the standard for Phase I assessments. The ASTM standard cannot be reproduced herein, but the overall structure can be seen from the Recommended Table of Contents and Report Format, which is provided as Appendix A to this chapter. The full text of the standard may be obtained by writing:

The American Society for Testing and Materials
1916 Race Street
Philadelphia, PA 19103

(*See also:* Consulting Engineers Council 1989; Von Oppenfeld 1990; Turim 1991; Sara 1994; Priznar 1995; Petts et al.,1997, Chapters 2 to 4; Douben 1998; Uliano 2000.)

Environmental Audits

The concept of the environmental audit has, at this writing, at least a 20-year history characterized by uncertainty. The uncertainty is seen in the current ASTM definition of an environmental audit:

> ... the investigative process to determine if the operations of an existing facility are in compliance with applicable environmental laws and regulations. This term should not be used to describe Practice E 1528 or this practice, although an environmental audit

---

[6] The italics are used by the ASTM to indicate any terms that are specifically defined in the standard. The ESA is defined by the ASTM as "the process by which a person or entity seeks to determine if a particular parcel of real *property* (including improvements) is subject to *recognized environmental conditions.* At the option of the user, an environmental site assessment may include more inquiry than that constituting *appropriate inquiry* or, if the user is not concerned about qualifying for the *innocent landowner defense,* less inquiry than that constituting *appropriate inquiry* .... An environmental site assessment is both different from and less rigorous than an *environmental audit.*"

may include an environmental site assessment or, if prior audits are available, may be part of an environmental site assessment (ASTM E-1527-93, paragraph 3.3.10).

The concept, whether carried out in a management format or in a regulatory/enforcement context, is regarded by regulatory and enforcement program directors as a means of extending limited staff resources and by corporate officials and industry managers with suspicion and hesitance. The EPA began proposing environmental audits as elements of enforcement case settlements in the mid-1970s, and such settlements have now become routine (EPA 1994; EPA 1998a).

Early discussions and proposals contained language similar to that now seen in descriptions of site assessments. A current definition reads:

> A process that seeks to verify documented expectations, typically regulations and policies, by conducting interviews, reviewing records, and making first-hand observations ... (Priznar 1995, p. 160)

The EPA now defines environmental audits as:

> ... a systematic, documented, periodic and objective review by regulated entities of facility operations and practices related to meeting environmental requirements. Audits can be designed to accomplish any or all of the following: verify compliance with environmental requirements; evaluate the effectiveness of environmental management systems already in place; or assess risks from regulated and unregulated materials and practices (59 FR 38455).

Thoughtful industry managers and executives turned to internal audits as state and federal enforcement programs became active. The practice was seen as a "heads-up" management technique which would detect noncompliance, enable timely correction, and generally avoid problems. Priznar states the reasoning succinctly:

> ... typically, requestors want assurance that their organization will not be surprised by fines, negative publicity, and related distractions if they are caught in noncompliance by regulatory agencies. Audits also serve to demonstrate to internal staff and external entities the organization's good faith with regard to environmental management (Priznar 1995, p. 160).

With time, however, executives and managers have become wary of the process. Concerns center upon revelation of sensitive information; compromise of business confidential/trade secret information; discovery by regulatory agencies; and personal risks of executives and managers. The EPA continues to advocate and propose auditing and has attempted to assuage doubts, but a recent restatement of policy does little to achieve that end:

> Corporate culpability may be indicated when a company performs an environmental compliance or management audit, and then knowingly fails to promptly remedy the non-compliance and correct any harm done. On the other hand, EPA policy strongly encourages self-monitoring, self-disclosure, and self-correction. When self-auditing

has been conducted (followed up by prompt remediation of the non-compliance and any resulting harm) and full, complete disclosure has occurred, the company's constructive activities should be considered as mitigating factors in EPA's exercise of investigative discretion. Therefore a violation that is voluntarily revealed and fully and promptly remediated as part of a corporations systematic and comprehensive self-evaluation program generally will not be a candidate for the expenditure of scarce criminal resources (59 FR 38455).

Practices employed by practitioners to offset some or all of the concerns include use of internal auditors to improve protection from "leaks," use of attorneys to perform the audits thus enabling the protection of the attorney-client privileges, and elimination of written audit reports (Priznar 1995, pp. 166ff; *see also:* Lipscomb and McKeeman 2000).

Environmental Management Systems

Similar to environmental auditing, environmental management systems (EMS) have found favor among some industrial facilities and consultants. Systems such as the International Organization for Standardization (ISO) 14000 series are chosen by companies to achieve internal management system efficiencies, waste reduction, and proactive regulatory compliance (Cascio 1996). The ISO 14001 standard involves extensive commitments by a company to pollution prevention, compliance with environmental regulation and legislation, and continuing improvement in environmental management. The system brings in active participation by employees at all levels, evaluation in the form of monitoring and corrective procedures, management review, and certification to the standard. The proactive nature of such systems is intended to obviate the necessity for compliance assessments by regulatory agencies, but may be negotiated as settlement items or reduced penalties in enforcement actions (*see also:* Kemp et al. 2000).

COMPLIANCE INSPECTIONS/INVESTIGATIONS BY REGULATORY AGENCIES

Purpose

CERCLA § 104(e) contains authorities enabling the EPA to conduct inspections at sites and facilities where hazardous substances are or may have been generated, stored, treated, disposed, or transported to or from. As overviewed in previous chapters, RCRA §§ 3007 and 3008 authorize entry, inspections, data collection, sampling, file review, etc. These authorities are the basis for compliance inspections and/or investigations conducted in the implementation of the CERCLA and RCRA regulations.

In practice, the CERCLA authorities are used to request information, inspect, obtain samples, investigate, monitor, survey, test, and study actual or suspected releases on or from a site. These actions are taken pursuant to a finding in a PA that a release may have occurred and are performed as a Site Inspection (or SI). The SI provides portions of the input data to a hazard-ranking procedure which is overviewed later in this chapter.

The RCRA authorities are generally directed toward more immediate enforcement actions such as the assessment of an administrative penalty, commencement of a permit action, or institution of civil or criminal proceedings. As noted in the previous chapter, they may also be used in a closure or post-closure investigation.

Although each investigation must be tailored to specific objectives, many of the technical procedures are similar, whether conducted under CERCLA or RCRA (see Figure 10.12 for a comparison of RCRA and CERCLA remedial processes).

The Inspection Plan

Site inspections may be carried out by consultants having appropriate expertise or by federal, state, or local regulatory agency personnel. The inspector is trained and drilled in the necessity (1) for a detailed plan for the conduct of the inspection and (2) for careful adherence to the plan. The inspector must always anticipate that the plan may be subjected to intense review in a court of law and that seemingly minor deviance from the plan may sharply affect the credibility of evidence developed in the inspection. Courts and defense attorneys take a dim view of vaguely planned and/or executed inspections that appear to be "fishing expeditions."

Items that the plan should address include

- Scope of the inspection that depends upon the purpose and objectives: The scope is expressed in terms of issues to be addressed, areas to be inspected, depth of detail required, time allocated to conduct the inspection, etc. For example, inspections performed in response to information received concerning alleged violations will generally be comprehensive in scope and entail a detailed evaluation of all RCRA regulated activities at the site.
- Coordination required with other offices, agencies, or services
- Procedure regarding prior notification, denial of entry, denial of access to records, areas, units, etc.
- Entering the facility: Is there to be an opening conference? What items should be covered? Should the inspector proceed with a visual inspection immediately to preclude hasty adjustments or concealment?
- Summarized findings from the background report
- Applicable regulations, policies, and guidance documents
- Procedure regarding records review, i.e., review on-site, copy for later review, or other arrangement
- Personnel assigned and duties
- Sampling plan, including list of equipment
- Protective clothing and safety equipment requirements
- Site safety plan — as appropriate
- Contingency plan for emergencies that may arise during the inspection
- Checklists (if any) to be used
- Data Quality Objectives[7]

---

[7] *See:* Data Quality Objectives Process for Hazardous Waste Site Investigations, EPA 600-R-00-007.

In general, the plan should lay out the activities and sequences to be followed, the resources required, and it should highlight any particular issues that may pertain to the site.

Conduct of the Inspection/Investigation

In some cases, the inspector will have limited information on the facility or may be inspecting an uncontrolled site. The inspector should be prepared to encounter the worst conditions in such cases. *Inspectors should never proceed with inspections involving site conditions for which they are not prepared or do not have the proper safety equipment* (adapted from EPA 1988, pp. 2:17–2:20).

Entry. The RCRA regulations have been interpreted to allow either announced or unannounced inspections. If an inspection is to be announced, the facility is contacted and advised of the forthcoming inspection. Time and date of arrival is usually provided, but is not required. In an unannounced inspection, no notice is provided before the inspector's arrival on-site.

The regulations require that the inspector:

- Enter the premises at a reasonable time and complete the inspection as promptly as is possible.
- Issue receipts for samples collected.
- Provide duplicate samples.
- Furnish the owner, operator, or agent a copy of any sample analysis conducted.

Upon arrival the inspector should:

- Locate the owner, operator, or agent as soon as possible and determine that this official has the proper authority to speak and act for the facility.
- Present identification to the owner, operator, or agent, even if it is not requested.
- Document entry activity in a logbook or field notebook, noting date, time, and the names and titles of facility personnel encountered.

Inspectors may be requested to sign a log or passbook and may do so. Such documents are useful in the event of fire or other emergency. The EPA instructs inspectors *not* to sign waivers or other legal documents that limit the facility's liabilities in the event of an accident. Inspectors are also instructed *not* to sign documents that may limit the inspector's rights or the owner's responsibilities.

The owner or agent in charge at the time of the inspection either gives or denies consent to inspect the premises. Consent may be withdrawn at some point during the inspection. Such action is considered denial of access. Other actions that amount to denial include not allowing the inspector to bring in necessary equipment (e.g., a camera) or not allowing the inspector access to documents.

When an inspector is denied access, the EPA specifies that step-by-step procedures be followed. The procedures begin with a request for the reason for denial and may culminate in a return to the facility with a search warrant, or company officials may be issued a subpoena (EPA 1992, pp. 29–30).

Opening discussions with the owner, operator, or agent are usually held to:

- Outline the objectives of the inspection or investigation.
- Brief facility management on the applicable elements of RCRA or other regulations, as appropriate.
- Establish the sequence of the operations to be inspected.
- Establish schedules for meetings or other events.
- Arrange for facility personnel to accompany the inspector.
- Arrange to provide duplicate samples and sample receipts.
- Determine whether the owner or operator intends to make confidential business information claims.

Owners, operators, or agents should expect the inspector to be highly inquisitive. The EPA instructs inspectors to "… question, question, and question some more …. Inconsistencies must be pursued until they are resolved" (adapted from EPA 1988, Chapter 4).

Operations, Waste Handling, and Records Review. Early in the inspection/investigation process, the inspector will ask the facility representative to describe operations and waste management practices in detail. The purpose of this discussion will be to:

- Gain a detailed understanding of the operations.
- Answer any questions the inspector may have regarding waste generation, waste flow, and waste management activities.
- Identify changes in operating and/or waste management practices from those indicated in the permit and/or facility files.
- Identify and reconcile any discrepancies between the operations described by the facility representative and those described in the files.

The EPA does not attempt to prescribe a format for the records review; however, Appendix H of the Multi-Media Investigation Manual provides a suggested record/documents request (EPA 1992). The general thrust of inquiry can be expected to follow the record keeping requirements of 40 CFR 262, 263, 265 (or 264), and 270. They include personnel and training records, agreements with local authorities, contingency plans, manifests, biennial reports, exception reports, waste analysis plans, waste analyses and test results, inspection schedules and results, operating records, groundwater monitoring plans, groundwater monitoring records, closure plan, post-closure plans, annual assessment for tanks, certification of major repairs, contingent post-closure plan, land treatment operating record and closure plan, landfill operating record, and contents and organization of land disposal cells.

The records review frequently leads to large numbers of what facility managers disparage as "paperwork" violations. EPA officials consider the records review to be the primary option available to ensure that the "cradle-to-grave" management system is being implemented and have successfully defended that view before Congressional and other inquiries.

Visual Inspection. EPA inspectors attempt to organize the visual inspection in such a way that the flow of waste materials and the related processes can be understood and that the compliance status of each process or unit can be determined. The EPA provides the example of inspection of a plant that generates hazardous waste, stores waste for off-site disposal, and treats some waste on-site. The visual inspection could proceed as follows, in brief:

- Inspect points of waste generation and accumulation. Determine if the owner/operator has identified all hazardous wastes based on generating operations, and determine if accumulation points meet satellite storage area requirements.
- Evaluate in-plant waste transport from generation and accumulation points to storage and treatment units. Determine if there is potential for mislabeling, misplacing, or mishandling wastes and if wastes are adequately tracked to enable proper identification at storage and treatment units.
- Evaluate storage and treatment units for compliance with applicable standards. Determine if wastes in units correspond to those whose points of generation have been inspected, and identify where any other wastes in the units originate. Determine if any hazardous wastes are generated in the unit (e.g., treatment sludge) and evaluate the management of such waste for compliance.

The sequence of this procedure enables the inspector to understand the movement and control of waste within the facility and thereby identify:

- Hazardous wastes that may not currently be considered hazardous by the owner/operator
- Noncomplying procedures or management practices that are part of the facility's routine operations
- Steps in the management process during which wastes may be mishandled or misidentified and in which there are opportunities for spills or releases
- Unusual situations that may be encountered during the inspection, that vary from the facility's stated normal operating procedures, and may indicate potential violations (EPA 1988, Chapter 4)

The EPA has developed a number of general and industry-specific checklists which can help the inspector approach the inspection in an organized way. The RCRA Inspection Manual [OSWER Directive 9938.02(b)] can be accessed at <http://www.epa.gov/oeca/ore/rcra/cmp/110098a.pdf>. Appendix IV to the manual is a detailed set of general site inspection checklists (EPA 1998b). In addition, a large number of guidance documents pertaining to Subtitle C requirements are available to the inspector and to the regulated community. The availability can be ascertained by contacting an EPA regional office, or by calling the RCRA Hotline (800-424-9346) to request a catalog of the documents (EPA 200-B-99-001).

Sampling and Monitoring. If the site is permitted or if sampling has been accomplished in accord with interim status requirements, sufficient data may be available in the files to meet the needs of a Superfund SI or a RCRA compliance investigation. EPA guidance de-emphasizes drum sampling at these stages if the data on hand is believed sufficient for the immediate activity. This policy reflects the fact that (1) drum sampling is resource-intensive and (2) if the investigation proceeds to a removal action, it may be necessary to sample all drums on the site. This requires "staging" of the drums and a highly efficient sampling and analysis scheme.

Similarly, if groundwater monitoring has been performed and the data are available, the inspection or investigation may require only minimal confirming sampling and analysis. If groundwater data are needed, and monitoring wells are not in place, various options may suffice:

- Seeps and springs obviously reflect the near-surface groundwater quality and may reflect the effects of local contamination.
- Nearby water supply wells may similarly reflect local contamination.
- In the absence of other options, and where depth to groundwater permits, driven well points may provide sampling access to leachate plumes or near-surface groundwater.

Comprehensive groundwater monitoring networks are rarely installed specifically for site inspections. They may be installed to support a large-scale RCRA compliance investigation or may come into the Superfund process at the Remedial Investigation/Feasibility Study sequence of activity. The procedures, techniques, and standards for groundwater monitoring network systems are extensive and dependent upon many factors. The technologies are the subject of entire training courses, manuals, and textbooks and cannot be covered here. References are provided at the end of this chapter.

Surface water, including streams and impoundments, may receive contaminated groundwater flow or run-off. Surface water sampling can supplement groundwater monitoring or, in the absence of other monitoring points, be the most practical way of identifying off-site pollutant movement. Downgradient surface water suspected of receiving groundwater inflow should be sampled.

Ambient concentrations of pollutants may be very low, yet their presence in any measurable concentration may be significant. Procedures used or materials contacting the sample should not cause pollutants to be gained or lost. Sampling equipment and sample containers must be fabricated from inert materials and must be thoroughly cleaned before use.

Sampling Equipment and Procedures. Sampling in support of field investigations of hazardous waste sites has advanced in planning, technique, equipment, execution, and analytical support to the extent that an explication of the topic is not possible in a text such as this. Some basics that every practitioner should have well in mind before venturing any on-site activity are presented hereafter. This material is followed by a number of references that will be useful to the uninitiated as well

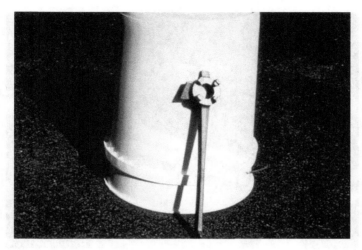

FIGURE 10.1 Spark-proof brass bung wrench.

as the experienced practitioner. It cannot be overemphasized that in-depth training under experienced investigators is essential to beginning the practice and is required by RCRA, OSHA, and state statutes and regulations (*see:* Chapter 15).

Samples must be secured in a manner that will ensure the safety of the sampler, all others working in the area, and the surroundings. Stored, abandoned, or suspect waste will often be containerized in drums or tanks. Such containers pose special safety problems. Care must be exercised in opening drums or tanks to prevent sudden releases of pressurized materials, fire, explosion, or spillage.

Drums should be opened using a spark-proof brass bung wrench such as that shown in Figure 10.1. Drums with bulged heads are particularly dangerous (Figure 10.2). The bulge indicates that the contents are or have been under extreme pressure.

FIGURE 10.2 Bulged drum. (From the Arizona Department of Environmental Quality.)

FIGURE 10.3 Remotely controlled pneumatic wrench and positioning device for bung removal. (From the U.S. Environmental Protection Agency.)

If it is deemed necessary to sample a bulged drum, a remotely operated drum-opening device such as that shown in Figure 10.3 enables the sampler to open suspect drums from a safe distance (Figure 10.4). This device can be fabricated using an ordinary pneumatic impact wrench and a brass bung attachment (Figure 10.5). Such operations should only be carried out by fully trained technicians in full protective gear.

Liquid waste tanks must be sampled in a manner which assures that the sample is representative of the contents of the tank. The EPA continues to specify that the sampling be done with the "Colawassa" sampler. The Colawassa consists of a long tube with a stopper at the bottom end. The stopper is opened and closed by use of the handle at the top of the tube. The device is intended to enable the sampler to retrieve representative material from throughout the depth of the tank. The Colawassa has many shortcomings, the most troublesome of which is the need to completely clean it and remove all residues between each sampling. This is not only difficult and time consuming to accomplish, but also creates another batch of hazardous waste that must then be managed. The Colawassa shown in use in Figure 10.6 is a single-use item that can be disposed of as hazardous waste after use.

A glass Colawassa (Figure 10.7), which eliminates the possibility of sample contamination by metals and stopper materials, is available through technical and scientific supply houses. In situations involving depths of no more than 36 in., ordinary glass tubing (Figure 10.8) can be used and can be discarded after use.

"Bomb samplers" (Figure 10.9), which can be lowered into a liquid waste container and then opened at a selected depth, are useful in special situations.

Long-handled dippers can be used to sample ponds, impoundments, large open tanks, or sumps. An obvious shortcoming is the fact that the device cannot cope with stratified materials. Makeshift devices using tape or other porous or organic material (Figure 10.10) introduce the likelihood of contamination of the sample by the extraneous material.

FIGURE 10.4 Remote operation of the bung removal apparatus. (From the U.S. Environmental Protection Agency.)

FIGURE 10.5 Brass bung fitting used with pneumatic wrench. (From the U.S. Environmental Protection Agency.)

FIGURE 10.6 Single-use "Colawassa" sampler in use. (Courtesy of Safety-Kleen Corporation.)

FIGURE 10.7 Glass "Colawassa" sampler.

FIGURE 10.8 Sampling liquid waste with standard laboratory glass tubing. (Courtesy of Safety-Kleen Corporation.)

FIGURE 10.9 "Bomb" sampler for sampling liquids at a specific depth. (From the Arizona Department of Environmental Quality.)

FIGURE 10.10 Makeshift dipper using improper materials. (From the Arizona Department of Environmental Quality.)

Dry solid samples may be obtained using a thief or trier (Figure 10.11) or an auger or dipper. Sampling of process units, liquid discharges, and atmospheric emissions all require specialized equipment and training that greatly exceed the scope of this text.

The EPA has published many guidance documents and manuals that deal with the various aspects of hazardous waste, soil, surface and groundwater, and waste stream sampling. A few of the early publications continue to be referenced in later EPA publications and in the commercial texts (EPA 1981, 1985, 1986). The analyt-

FIGURE 10.11 Two types of "Thief" samplers for solid waste sampling. (From the U.S. Environmental Protection Agency.)

ical methods manual "Test Methods for Evaluating Solid Waste, Physical/Chemical Methods" (SW-846), originally published in 1983, has undergone many revisions, gained massive proportions (3500 pages), and continues as an authoritative source of guidance for analytical methods, sampling, and quality control (EPA 1983–1998;[8] later pertinent EPA publications include EPA 1993; EPA 1993a,b; EPA 1997; EPA 1998b; EPA 2000).

A recent comprehensive document, covering sampling, sampling and test equipment, field measurements, instrumentation, and selection of analytical methods is the product of a U.S. Navy/EPA collaborative effort. The publication "Field Sampling and Analysis Technologies Matrix and Reference Guide" (USN/EPA 1998) can be obtained by placing an order with the EPA National Service Center for Environmental Publications, (800) 490-9198 or online at <http://www.epa.gov/ncepi-hom/index.html> (see also: De Vera et al. 1980; Evans and Schweitzer 1984; Corbitt 1990, Table 9.7; Gammage and Berven 1992; Manahan 1994, Chapter 24; Breckenridge, R.P. et al., 1996; Lewis et al., 1996; Pohlmann and Alduino 1996; NRC 1997; HMTRI 1997; Butterfield 2000).

SITE EVALUATION

CERCLA Section 105 requires the EPA to issue a revised National Oil and Hazardous Substances Pollution Contingency Plan [commonly referred to as the National Contingency Plan (NCP)] to provide guidance to on-scene coordinators (OSC) and others responsible for removal and remedial action following a release of hazardous substances. The NCP (40 CFR 300) is structured in eight subparts, but we will overview only Subpart F, entitled "Hazardous Substances Response."

National Priorities List

Section 105(a)(8) requires the EPA to establish a National Priorities List (NPL),[9] prioritizing sites for possible removal and remedial action. "Removal" as used in CERCLA means the physical removal of a hazardous substance, pollutant, or contaminant from the location of release. Remedial action means those additional measures that may be necessary to protect the public health, or to protect or prevent harm to the environment.

As this is written, the NPL lists 1246 sites. Two recent General Accounting Office (GAO) reports indicate that 3036 contaminated sites are being tracked by CERCLIS. Of those on the CERCLIS list, 1789 sites await decisions regarding listing on the NPL, and another 1234 sites are considered unlikely candidates for listing. A large number of the potentially eligible sites have contaminated nearby drinking water or drinking water sources. A recurring theme in the reports was the

---

[8] Official printed copies of SW 846 may be obtained from the U.S. Government Printing Office (GPO), Superintendent of Documents, Washington, D.C. 20402, Publication Number 955-001-00000-1, or from the National Technical Information Service (NTIS), U.S. Department of Commerce, 5285 Port Royal Road, Springfield, VA 22161. NTIS also offers SW 846 on CD-ROM. Additional information on SW 846 may be accessed at <http://www.epa.gov/epaoswer/hazwaste/test/sw846.htm>.

[9] The NPL is found at 40 CFR 300, Appendix B. The NPL is revised frequently.

inadequacy of state resources to finance or enforce Superfund cleanups at the potential sites (GAO/RCED-99-8, 22).

Hazard Ranking System

The EPA implemented this requirement by the development of a "Hazard Ranking System" (HRS).[10] This mathematical model enables the EPA and the state agencies to assign "scores" to sites, based upon basic environmental data. The scores, in turn, are ranked to ostensibly assign highest priority to the sites representing the greatest threat to human health and the environment.

The HRS is initiated with a minimal screening[11] to determine whether the CERCLA site assessment process is appropriate for the site or if another option is more appropriate. Once a hazardous waste site is identified as appropriate for site assessment, the EPA and/or a state agency enters information about the site in the CERCLIS[12] database. The EPA uses preliminary investigations such as the PA and the SI, outlined above, and data from generators and transporters and/or others to estimate quantities and kinds of wastes that may have been placed upon the site. During the PA, the Agency collects and reviews readily available information (i.e., site history, drinking water sources, surrounding populations) regarding the site to determine whether a threat or potential threat exists and to determine whether further investigation is needed. The SI is performed to further evaluate the extent to which a site represents a threat to human health or the environment by performing field investigations to determine the presence and/or movement of hazardous substances to the surrounding environment. The site is then scored, using the HRS model, and if the site is scored above a cutoff point, it is nominated for listing on the NPL. A site that receives a HRS score below the cutoff receives a "No Further Remedial Action Planned" (NFRAP) designation (adapted from EPA ICR #1488.04) .

Remedial Investigation/Feasibility Study

Once listed, this initial investigation is followed up by a more formal Remedial Investigation/Feasibility Study (RI/FS). These separate studies, usually performed by contractors, involve major technical investigations, including ground and surface water monitoring, atmospheric emissions, hazards of human contact, or explosion threat. The RI is the mechanism for collecting data to:

- Characterize site conditions
- Determine the nature of the waste
- Assess risk to human health and the environment
- Conduct treatability testing to evaluate the potential performance and cost of the treatment technologies that are being considered

---

[10] The HRS is found at 40 CFR 300, Appendix A.
[11] CERCLIS prescreening.
[12] Comprehensive Environmental Response, Compensation, and Liability Information System.

The FS is the mechanism for the development, screening, and detailed evaluation of alternative remedial actions. The RI and the FS are conducted concurrently, with each study providing information to the other, enabling development of remedial alternatives and identifying additional data collection requirements. The RI/FS process includes five phases:

- Scoping
- Site Characterization
- Development and Screening of Alternatives
- Treatability Investigations
- Detailed Analysis

Human health and environmental risk assessments are integral to the RI/FS process.

Record of Decision

Following listing and evaluation of the site, the EPA makes a determination of the extent of the threatened or actual damage to human health or the environment and of the most feasible and cost-effective remedial action to mitigate the threat or damage. This determination is formally issued as a Record of Decision (ROD). The ROD, a public document, identifies the remedy that the EPA will require and establishes the framework for remedial negotiations between the EPA and the PRPs.

Negotiations, Enforcement

In the course of the negotiations, many complex issues must be settled, for example:

- Release from liability
- Resolution of disputes
- Timing of cleanup
- Stipulated penalties
- *De minimis* settlements
- Liabilities of municipalities

Factors that may affect settlement outcomes include

- Volume and nature of wastes contributed
- Strength of the evidence tracing the wastes
- Ability to pay
- Litigative risks of proceeding to trial
- Public interest considerations
- Precedential value
- Present value of money

Any of the criteria listed can affect the timing of the settlement, the value of the remedy achieved, and the amount of costs recovered.

Where negotiations fail, administrative actions that can be taken pursuant to CERCLA Section 106 include administrative orders requiring PRPs to undertake specific cleanup measures. The EPA may collect daily penalties for noncompliance. Noncompliance also sets the stage for seeking treble damages if the case proceeds to cost recovery.

The EPA initiates judicial action where negotiations are unsuccessful or where an Administrative Order is disregarded. If these actions are unsuccessful or the responsible party(ies) is(are) bankrupt, or if responsible parties cannot be identified, funded cleanup proceeds. All CERCLA cleanup actions are carried out under EPA and/or state agency oversight. RCRA Corrective Action and CERCLA Superfund Processes are compared in Figure 10.12.

TOPICS FOR REVIEW OR DISCUSSION

1. Identify two sets of circumstances that might bring about the necessity for a background report on a former site.
2. A "Phase I" review of a site would not normally include extensive ground-water sampling. True? False?
3. Review the meanings of "strict" liability and "joint and several" liability and their implications for present or potential property owners, "responsible parties," contractors performing site clean-ups, others.
4. Why would facility owners be reluctant to have environmental audits preformed on their facility — even if performed by a private contractor/consultant?
5. How might a facility owner maintain confidentiality of the contractor's findings?
6. Why are inspection policies, sampling procedures, and records review procedures of regulatory agencies of concern to the facility owner/operator?
7. Since you may frequently need to refer to the ASTM standard for Phase I Site Assessments, how would you go about obtaining a copy? Did you try to do so?

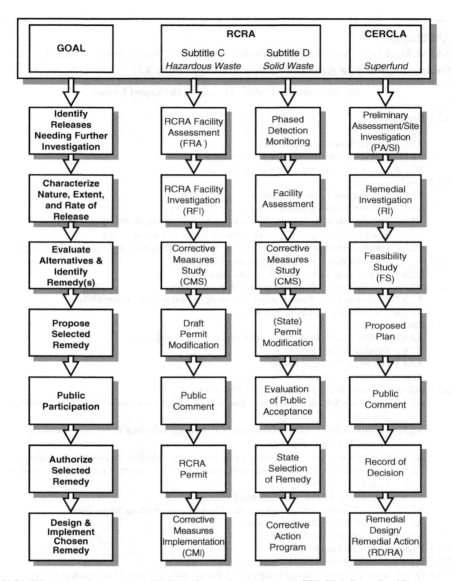

FIGURE 10.12 Comparison of RCRA Corrective Action and CERCLA Superfund Processes.

APPENDIX A
ASTM Standard Practice for Environmental Site Assessments: Phase I Environmental Site Assessment Process (Appendix X2)

X2. Recommended Table of Contents and Report Format

X2.1 Summary

X2.2 Introduction

    X2.2.1 Purpose

    X2.2.2 Special Terms and Conditions

    X2.2.3 Limitations and Exceptions of Assessment

    X2.2.4 Limiting Conditions and Methodology Used

X2.3 Site Description

    X2.3.1 Location and Legal Description

    X2.3.2 Site and Vicinity Characteristics

    X2.3.3 Descriptions of Structures, Roads, Other Improvements on the Site (including heating/cooling system, sewage disposal, source of potable water)

    X2.3.4 Information (if any) Reported by User Regarding Environmental Liens or Specialized Knowledge or Experience (pursuant to Section 5)

    X2.3.5 Current Uses of the Property (to the extent identified)

    X2.3.6 Past Uses of the Property (to the extent identified)

    X2.3.7 Current and Past Uses of Adjoining Properties (to the extent identified)

    X2.3.8 Site Rendering, Map, or Site Plan

X2.4 Records Review

    X2.4.1 Standard Environmental Record Sources, Federal and State

    X2.4.2 Physical Setting Source(s)

    X2.4.3 Historical Use Information

    X2.4.4 Additional Record Sources (if any)

X2.5 Information from Site Reconnaissance and Interviews

    X2.5.1 Hazardous Substances in Connection with Identified Uses (including storage, handling, disposal)

    X2.5.2 Hazardous Substance Containers and Unidentified Substance Containers (including storage, handling, disposal)

    X2.5.3 Storage Tanks (including contents and assessment of leakage or potential for leakage)

    X2.5.4 Indications of PCBs (including how contained and assessment of leakage or potential for leakage)

    X2.5.5 Indications of Solid Waste Disposal

    X2.5.6 Physical Setting Analysis (if migrating hazardous substances are an issue)

    X2.5.7 Any Other Conditions of Concern

    X2.5.8 Site Plan (if available)

X2.6 Findings and Conclusions

X2.7 Signatures of Environmental Professionals

X2.8 Qualifications of Environmental Professionals Participating in Phase I Environmental Site Assessment

X2.9 Optional Appendices (for example)

    X2.9.1 Other Maps, Figures, and Photographs

    X2.9.2 Ownership/Historical Documentation

    X2.9.3 Regulatory Documentation

    X2.9.4 Interview Documentation

    X2.9.5 Contract between User and Environmental Professional

Source: The American Society for Testing and Materials (ASTM), 1916 Race Street, Philadelphia, PA, 19103.

REFERENCES

American Society for Testing and Materials (ASTM). 1993. *Standard Practice for Environmental Site Assessments: Phase I Environmental Site Assessment Process.* E 1527-93, Philadelphia, PA.

Breckenridge, R. P., J. R. Williams, and J. F. Keck. 1996. "Characterizing Soils for Hazardous Waste Site Assessments," Chapter 14, in *EPA Environmental Assessment Sourcebook.* J. Russell Boulding, Ed., Ann Arbor Press, Ann Arbor, MI.

Burby, Brian G. 1989. "An Overview of Methods for Conducting Property Transaction Site Assessments," in *Environmental Reporter* December 29, Bureau of National Affairs, Washington, D.C.

Butterfield, W. Scott. 2000. "Multimedia Sampling," Chapter 41, in *Hazardous Materials Management Desk Reference.* Doye B. Cox, Editor-in-Chief, Adriane P. Borgias, Technical Editor, McGraw-Hill, NY.

Cascio, Joseph. 1996. *The ISO Handbook.* CEEM Information Services, Fairfax, VA.

Consulting Engineers Council of Metropolitan Washington. 1989. *Guidelines for Environmental Site Assessments.* Washington, D.C.

Corbitt, Robert A. 1990. *Standard Handbook of Environmental Engineering.* McGraw-Hill, NY.

De Vera, E. R., B. P. Simmons, R. D. Stephens, and D. L. Storm. 1980. Samplers and Sampling Procedures for Hazardous Waste Streams. Environmental Protection Agency, Cincinnati, OH, EPA 600-2-80-018.

Douben, Peter E. T. 1998. "Perspectives on Pollution Risk," in *Pollution Risk Assessment and Management.* Peter E. T. Douben, Ed., John Wiley & Sons, Chichester, West Sussex, England.

Evans, Roy B. and Glenn E. Schweitzer. 1984. "Assessing Hazardous Waste Problems," in *Environmental Science and Technology,* 18(11).

Gammage, Richard B. and Barry A. Berven, Eds. 1992. *Hazardous Waste Site Investigations.* Lewis Publishers, Ann Arbor, MI.

Hazardous Materials Training and Research Institute (HMTRI). 1997. *Site Characterization Sampling and Analysis.* Van Nostrand Reinhold, NY.

Hopper, David R. 1989. "Cleaning Up Contaminated Waste Sites," *Chemical Engineering* August, p. 95ff.

Kemp, Harry S., Steve Rowley, and Ed Pinero. 2000. "Environmental Management Systems — ISO 14001," Chapter 4, in *Hazardous Materials Management Desk Reference.* Doye B. Cox, Editor-in-Chief, Adriane P. Borgias, Technical Editor, McGraw-Hill, NY.

Lewis, T. E., A. B. Crockett, R. L. Siegrist, and K. Zarrabi. 1996. "Soil Sampling and Analysis for Volatile Organic Compounds," Chapter 15, in *EPA Environmental Assessment Sourcebook.* J. Russell Boulding, Ed., Ann Arbor Press, Ann Arbor, MI.

Liner, Keith. 2000. "Property Assessments," Chapter 18, in *Hazardous Materials Management Desk Reference.* Doye B. Cox, Editor-in-Chief, Adriane P. Borgias, Technical Editor, McGraw Hill, NY.

Lipscomb, Robert and Chris McKeeman. 2000. "Environmental Compliance Audits," Chapter 5, in *Hazardous Materials Management Desk Reference.* Doye B. Cox, Editor-in-Chief, Adriane P. Borgias, Technical Editor, McGraw-Hill, NY.

Manahan, Stanley E. 1994. *Environmental Chemistry, Sixth Edition.* CRC Press, Boca Raton, FL.

National Research Council (NRC). 1997. *Innovations in Groundwater and Soil Cleanup.* National Academy Press, Washington, D.C.

Petts, Judith, Tom Cairney, and Mike Smith. 1997. *Risk-Based Contaminated Land Investigation and Assessment.* John Wiley & Sons, Chichester, West Sussex, England.

Pohlman, K. F. and A. J. Alduino. 1996. "Potential Sources of Error in Ground-Water Sampling at Hazardous Waste Sites," Chapter 16, in *EPA Environmental Assessment Sourcebook.* J. Russell Boulding, Ed. Ann Arbor Press, Ann Arbor, MI.

Priznar, Frank J. 1995. "Environmental Audits and Site Assessments," in *Accident Prevention Manual for Business and Industry Environmental Management.* Gary R. Krieger, Ed., National Safety Council, Itasca, IL.

Sara, Martin N. 1994. *Standard Handbook for Solid and Hazardous Waste Facility Assessments.* Lewis Publishers, Chelsea, MI.

Turim, Jay. 1991. "Environmental Assessments for Property Transfers," in *Environmental Claims Journal,* 3(2):Winter 1990/91.

Uliano, Jr., Tony. 2000. "Environmental Health and Safety Risk Analysis," in *Hazardous Materials Management Desk Reference,* Doye B. Cox, Editor-in-Chief, Adriane P. Borgias, Technical Editor, McGraw-Hill, NY.

U.S. Environmental Protection Agency. 1981. *NEIC Manual for Groundwater/Subsurface Investigations at Hazardous Waste Sites.* National Enforcement Investigations Center, Denver, CO.

U.S. Environmental Protection Agency. 1983–1998. Test Methods for Evaluating Solid Waste. Physical/Chemical Methods, Updates I, II, IIA, IIB, III, and IIIA and Drafts IVA and IVB, Office of Solid Waste, Superintendent of Documents, U.S. Government Printing Office, Washington, D.C., SW 846.

U.S. Environmental Protection Agency. 1985. Characterization of Hazardous Waste Sites — A Methods Manual, Volume II, Available Sampling Methods. Washington, D.C., EPA 600-4-84-075.

U.S. Environmental Protection Agency. 1986. Ground-Water Monitoring Technical Enforcement Guidance Document, OSWER Directive Number 9950.1. Office of Waste Programs Enforcement, Washington, D.C.

U.S. Environmental Protection Agency. 1988. RCRA Inspection Manual. OSWER Directive Number 9938.2A. Office of Waste Programs Enforcement, Washington, D.C.

U.S. Environmental Protection Agency. 1992. Multi-Media Investigation Manual, National Enforcement Investigations Center, Denver, CO, EPA 330-9-89-003-R.

U.S. Environmental Protection Agency. 1993. Draft Field Methods Compendium. Office of Emergency and Remedial Response 9285.2-11. Washington, D.C.

U.S. Environmental Protection Agency. 1993a. Subsurface Characterization and Monitoring Techniques: Desk Reference Guide, Volume 1: Solids and Ground Water Appendices A and B, Office of Solid Waste and Emergency Response, Washington D.C., EPA 625-R-93-003a.

U.S. Environmental Protection Agency. 1993b. Subsurface Characterization and Monitoring Techniques: Desk Reference Guide, Volume 2: Vadose Zone, Field Screening and Analytical Methods, Appendices C and D, Office of Solid Waste and Emergency Response, Washington D.C., EPA 625-R-93-003b.

U.S. Environmental Protection Agency. 1994. Catalog of Hazardous and Solid Waste Publications, Solid Waste and Emergency Response, Washington, D.C., EPA 530-B-93-002.

U.S. Environmental Protection Agency. 1997. Expedited Site Assessment Tools for Underground Storage Tank Sites: A Guide for Regulators, Office of Solid Waste and Emergency Response, Washington, D.C., EPA 510-B-97-001.

U.S. Environmental Protection Agency. 1998. RCRA Orientation Manual, Office of Solid Waste and Emergency Response, Washington, D.C., EPA 530-R-98-004.

U.S. Environmental Protection Agency. 1998a. Pollution Prevention Solutions during Permitting, Inspections and Enforcement, Office of Solid Waste and Emergency Response, Washington, D.C., EPA 530-R-98-015.

U.S. Environmental Protection Agency. 1998b. Revised RCRA Inspection Manual (November 1998 Revision), OSWER Directive No. 9938.02(b), Office of Regulatory Enforcement, Washington, D.C.

U.S. Environmental Protection Agency. 1999. EPA National Publications Catalog, Fifth Edition, Office of Administration and Resources Management, Cincinnati, OH, EPA 200-B-001.

U.S. Environmental Protection Agency. 2000. Data Quality Objectives Process for Hazardous Waste Site Investigations, Office of Environmental Information, Washington, D.C., EPA 600-R-00-007.

U.S. General Accounting Office. 1999. Hazardous Waste: Unaddressed Risks at Many Potential Superfund Sites, Washington, D.C., GAO/RCED-99-8.

U.S. General Accounting Office. 1999. Hazardous Waste: Information on Potential Superfund Sites, Washington, D.C., GAO/RCED-99-22.

U.S. Navy and Environmental Protection Agency. 1998. Field Sampling and Analysis Technologies Matrix and Reference Guide, First Edition, Naval Facilities Engineering Command (NAVFAC) and Environmental Protection Agency (EPA) Technical Innovation Office (TIO), Washington, D.C., EPA 542B-98-002.

Von Oppenfeld, Rolf. 1990. "Environmental Due-Diligence: Risk Assessment and Management," *Arizona Attorney*, Arizona Bar Association, Phoenix, AZ.

Washburn, Stephen T. and Kristen G. Edelmann. 1999. "Development of Risk-Based Remediation Strategies," in *Practice Periodical of Hazardous, Toxic, and Radioactive Waste Management*, pp. 77–82. American Society of Civil Engineers, NY.

Williams, Michael A. 1998. "Phase I and II Environmental Site Assessments," Chapter 11, in *Waste Management Concepts*, Neal K. Ostler and John T. Nielsen, Eds., Prentice-Hall, Upper Saddle River, NJ.

Woodside, Gayle. 1999. *Hazardous Materials and Hazardous Waste Management: Second Edition*. John Wiley & Sons, NY.

11 Site Remedial Technologies, Practices, and Regulations

OBJECTIVES

At completion of this chapter, the student should:

- Be familiar with the technologies that may be employed in site remediation, e.g., on-site containment, solidification/stabilization, chemical treatment, bioremediation and destruction; "pump-and-treat" regimes; natural attenuation; extraction; off-site treatment and disposal; and related RCRA[1] and CERCLA[2] requirements and policies.
- Understand the respective roles of RCRA and CERCLA in site remediation.
- Be familiar with "How Clean is Clean" issues, the basis for them, some resolutions thereof, and the roles assigned to risk assessment in the remediation processes.
- Be familiar with the National Contingency Plan, the "blueprint" role of the NCP in site remediation, how to find the NCP and how to maintain or ensure currency with it.
- Understand the linkages between hazardous waste site remediation, the Brownfields Initiative, and environmental justice issues.

INTRODUCTION

In Chapter 10 we introduced and briefly overviewed the technologies and processes involved in the evaluation of contaminated or suspect sites. The generic, RCRA Corrective Action, and CERCLA (Superfund) approaches to site evaluation were introduced as the necessary precursors to site cleanup. We now continue with the overview of site cleanup procedures. To the extent possible, we will continue the pattern of introduction of technologies and processes in the "generic" or established

---

[1] Resource Conservation and Recovery Act of 1976.
[2] Comprehensive Environmental Response, Compensation, and Liability Act of 1980 (and Superfund Amendment and Reauthorization Act of 1986).

practice format. We will then overview the options and/or requirements as applied to RCRA and Superfund site remediation.

The technologies for site remediation have been developed over a relatively short period of time. Some of the technologies were introduced in the 1970s or earlier and some sites were remediated in the latter part of that decade. However, it is arguable that actual cleanup of Superfund sites did not begin making significant progress until the mid-1980s. With obvious exceptions, the corporate and public cultures that eventually gave impetus to private sector cleanups were similarly timed. Thus, some of the technologies continue to evolve, while some have become proven and standardized. New treatment or cleanup technologies are in plentiful supply and new management philosophies are being put to the test. A few of the more promising new approaches to site remediation, as well as the time-tested ones, will also be overviewed in this chapter. References to those introduced and others will be provided.

Development of treatment technologies has been given support by the EPA Superfund Innovative Technology Evaluation (SITE) program. The Superfund Amendments and Reauthorization Act (SARA) of 1986 authorized $20 million per year, through 1991, to support development of new treatment technologies and to provide sound engineering and cost data on selected technologies. Approximately ten new project awards were made each year to test and/or demonstrate innovative or improved hazardous waste management technologies in laboratory and full-scale operations. The program was extended with the Superfund reauthorization in 1991, but SITE reauthorization died with Superfund reauthorization in 1994. In following years, separate appropriations have enabled continuation of the SITE program (*see also:* EPA 1989; Payne 1998, pp. 17–19).

The national programs for cleanup of uncontrolled hazardous waste sites (e.g., RCRA corrective actions, Superfund removal, and/or remedial actions) have been the focus of great controversy. Both programs were fought tenaciously by lobbyists, in the courts, and by policy makers of the Reagan Administration. To many in Congress and elsewhere, the Superfund program has progressed too slowly and at excessive costs. To others it has been overly aggressive, unyielding, burdened with process, and utopian in cleanup objectives. It has been bedeviled by the "how-clean-is-clean" issue; by charges that it is "anti-business" and/or merely moves the contaminants and creates future Superfund sites; and by the ponderousness of the Superfund process. In 1999, House[3] and Senate[3] Superfund reauthorization bills failed for variations of the above issues and others. At the time of this writing in 2000, neither body had produced a reauthorization bill acceptable to all parties (*see also:* RAND 1989; GAO 1993, 1994a,b, 1999).

Nevertheless, the program is making significant progress and is having some notable successes. Superfund, imperfections notwithstanding, is here to stay and will be a major factor in the nation's hazardous waste cleanup. The National Priorities List (NPL) now includes approximately 1289 sites (65 FR 30482-8), and sites are added to the list several times each year. These sites must be cleaned up, and no preferable program format has been suggested, although the 1994 reauthorization bill contained significant changes to the earlier statute. Moreover, the failures of the

---

[3] House Bill 1300; Senate Bill 1090.

1994 and subsequent annual Superfund bills continue to point up deep unresolved divisions in political, public, professional, and activist notions of the form that a reauthorized Superfund program should take.

REMEDIAL OBJECTIVES

Programmatic Objectives

In the most general sense, hazardous waste site remedial activity is pursued to correct the results of mismanagement and accidental releases. Remedies usually involve removal of contaminated materials and safe disposition thereof; treatment, destruction, and/or containment in-place; or some variation(s) of these.

Remedial actions may be taken by individuals or corporations without the involvement of federal and/or state regulatory agencies. Indeed, privately funded and/or executed cleanup activity preceded the advent of RCRA and Superfund, and both statutes are structured to encourage (leverage) private cleanups.

RCRA corrective actions are an essential element of the national policy objective, i.e., the minimization "... of the present and future threat to human health and the environment." These authorities enable the EPA to address releases to the groundwater and other environmental media at RCRA-regulated sites. The RCRA authorities do not extend to abandoned sites or those for which responsible parties cannot be identified.

Superfund was originally intended to enable timely response to emergency cleanup needs and to provide resources and authorities for cleanup of abandoned sites and those for which responsible parties (1) cannot be identified or (2) refuse or are unable to conduct the necessary cleanup. Over time, government owned and/or operated facilities have been made subject to the law, and the "innocent landowner" provision has been added in an effort to limit the reach of the strict joint and several liability provisions. Provision has been made for *de minimis* settlements for small contributors to Superfund sites (King and Amidaneau 1995, pp. 68–69; *see also:* U.S. GAO 1993, 1994a; EPA 1998).

Technical Objectives

Whether privately funded and/or executed or carried out under statutory mandates, remedial actions must have the protection of human health and the environment as their overall objective.[4] The more applicable objectives are the prevention of further migration of releases that have occurred, amelioration of exposures and impacts caused by those releases, and prevention of further releases. These objectives are pursued by one of two basic operations:

1. **On-site** treatment, destruction, or containment
2. **Off-site** management of hazardous wastes and contaminated materials, followed by treatment, destruction, or safe disposal

---

[4] Studies have shown that higher than expected cancer rates may be associated with proximity to Superfund sites, e.g., the Baird and McGuire site (*Environment Reporter,* November 16, 1990, p. 1359).

While there are many variations and combinations of these two basic techniques, it is useful to categorize remedial actions as "**on-site**" or "**off-site**" operations.

RCRA and CERCLA require use of risk assessment techniques based upon site-specific data and the circumstances of the site. The technical objectives must be stated in terms of the degree of cleanup to be achieved in order to protect human health and the environment (i.e., how much residual contamination at the site is acceptable?). This question is the crux of the "how-clean-is-clean" issue. The answer to the immediate question and the eventual resolution of the issue have far-reaching implications for managers of public health risks and for responsible parties.

There is no single safe level of hazardous chemical concentrations applicable to all chemicals and all sites that, if achieved, would justify a declaration of "clean." Epidemiologists, risk managers, and policy makers initially found it necessary to rely to a great extent upon exposure criteria, such as drinking water and air quality standards, which were never intended for use as hazardous waste site cleanup standards. With time, rationalization of exposure criteria for some carcinogenic and noncarcinogenic substances has been achieved. Where pathways and exposure data exist to support a risk-assessment process, EPA policy is that the level of *total* individual carcinogen risk from exposures attributable to a Superfund site may be in the range of one excess occurrence in 10,000 (10^{-4}) to 1 in 10 million (10^{-7}). The most frequently proposed criteria is 10^{-6}.

Nevertheless, these standards (with a few exceptions) deal with individual inorganic and organic pollutants, whereas the hazardous waste site cleanup criteria must consider a wide variety of inorganic and complex organic compounds and mixtures. Thus, the rigor of the risk assessment processes continue to be limited by the necessity to incorporate a variety of assumptions for critical human health exposure, as well as environmental protection. Over the past decade, the EPA has produced an evolving and burgeoning set of risk assessment guidance documents which are intended to lend site-specificity and rigor to the cleanup goal setting ("how-clean-is-clean") process. This set entitled "Risk Assessment Guidance for Superfund" (RAGs), in three volumes, can be accessed on the Superfund Web site <http://www.epa.gov/oerrpage/superfund/programs/risk/ragsa/ci_ra/htm>.

The Administration's 1994 Superfund reauthorization bill contained language calling for a numeric national cleanup goal" and a "national risk protocol." The protocol would have contained standardized exposure scenarios for a range of unrestricted and restricted land uses and standardized formulas for evaluating exposure pathways and developing chemical concentration levels for the 100 contaminants that occur most frequently at Superfund sites *(Environment Reporter,* April 29, 1994, p. 2219). This format, of course, does little to solve the "how-clean-is-clean" dilemma. Viable exposure criteria continue to be absent or unproven for many of the most commonly discarded chemicals and chemical compounds. Without exposure criteria, a health risk assessment format is a hollow one. Failure of the 1994 Superfund reauthorization was regarded by many as a major disappointment, but the "how-clean-is-clean" issues were certain to continue with us, regardless of the 1994 reauthorization outcome.

The EPA also requires that remedies meet "applicable or relevant and appropriate federal and state requirements" (ARARs), such as state mine drainage limits for heavy metals **or** federal limits for PCBs established under the Toxic Substances Control Act (TSCA) authorities.

The introduction of the "Superfund Accelerated Cleanup Model" (SACM) has provided some generalization of cleanup methods in the form of "presumptive remedies and response strategies"[5] discussed later herein. Some states simply impose a blanket requirement that all cleanups achieve background concentrations of waste constituents (*see also*: Staples and Kimerle 1986; EPA 1989; Travis and Doty 1992; Burke 1992; Sims et al. 1996; Sellers 1999, Chapter 2).

ON-SITE REMEDIAL TECHNIQUES

Containment Methods

As the name implies, containment methods are directed toward prevention of migration of liquid hazardous wastes or leachates containing hazardous constituents. Containment usually involves the construction of impermeable barriers to retain liquids within the site, to direct the liquids to collection points for pumping and/or treatment, or to divert ground and surface waters away from the site. Successful application of these methods is usually contingent upon the presence of an impervious layer beneath the material to be contained and the achievement of a good seal at the vertical and horizontal interfaces. Some examples follow.

Slurry Walls. The slurry trench is excavated down to and, if practicable, into an impervious layer. The trench is typically 2 to 5 ft in width. Early applications used a 4 to 7% bentonite clay suspension in water to make up the slurry. The slurry may be mixed with the excavated soil or with other suitable soils to form a very low permeability wall. More recent applications have made use of additives such as polymers to improve the permeability or to protect the slurry from the deleterious effects of leachate. Figure 11.1 shows a trench and soilbentonite slurry wall under construction. The soil removed from the trench is mixed with bentonite clay and replaced in the trench. Figure 11.2 shows a cement-bentonite wall being installed. In this case, the excavated soil is not used. Cement is mixed with the bentonite slurry, which "sets" as a solid wall.

Many variations of the containment wall technique have been developed. The use of high density polyethylene (HDPE) membranes to line the excavated trench or as a curtain in the mid-section of the slurry wall to improve effectiveness is described by Cross. Mitchell and van Court describe and illustrate a geomembrane "envelope," lining the walls of an excavated trench wherein the envelope is filled

---

[5] *See:* Presumptive Response Strategy and *Ex Situ* Treatment Technologies for Contaminated Ground Water at CERCLA Sites, OSWER Directive 9283.1-12.

FIGURE 11.1 Soil-bentonite slurry wall construction. (From Geo-Con Incorporated, 4075 Monroeville Blvd., Suite 400, Monroeville, PA 15146. With permission.)

FIGURE 11.2 Cement-bentonite cut-off wall. (From Geo-Con Incorporated, 4075 Monroeville Blvd., Suite 400, Monroeville, PA 15146. With permission.)

with a sand and water mix to form an impermeable containment wall. Suthersan describes low permeability slurry walls as components of containment systems which direct contaminated groundwater to treatment gates, and permeable reactive trenches using a variety of materials as reactants, or to collect stripped vapors (*see:* EPA 1992, 1998a; Mitchell and van Court 1997; Cross 1996; Suthersan 1997; Pearlman 1999; Sellers 1999, Chapter 3).

Grout Curtains. In somewhat similar fashion, suspension grouts composed of bentonite or Portland cement, or both, may be injected under pressure to form a barrier. The method is most effective when the receiving formation is unconsolidated and porous deposits can be filled by the injection. In other situations, single, double, or triple lines of holes are drilled in staggered positions. Ideally, the grout injected in adjacent holes should penetrate to merge and form a continuous barrier. Chemical grouts are a more recent development and have the advantage of a range of viscosities. Some have viscosities approaching that of water and can be used to seal very fine rock and soil voids (*see:* EPA 1998a; Mitchell and van Court 1997; Cross 1996; Pearlman 1999).

Sheet Piling Cut-Off Walls. Pilings of wood, precast concrete, or steel can be used to form a cut-off wall. Sheet piling of steel is the most effective and has the advantages of great structural strength, it can be driven to depths as great as 100 ft, and it can accommodate irregularly shaped and/or confined areas. It has the disadvantages that it cannot be used effectively in rocky soil, the interlocking joints between the sheet piles must be sealed to prevent leakage,[6] and the steel is subject to attack by the contained corrosive liquids (*see:* EPA 1998a; Sims et al. 1996; Mitchell and van Court 1997; Suthersan 1997, pp. 196–197; Pearlman 1999; Sellers 1999, Chapter 3).

Less frequently used containment techniques include the use of frozen soil barriers and hydraulic barriers (Mitchell and van Court 1997; EPA 1998). Other containment methods make use of surface diversions to route run-off away from the waste deposit and impervious caps to carry rainfall and snowmelt beyond the perimeter of the deposit.

Extraction Methods

Two basic approaches to on-site extraction have gained general acceptance and are effective when properly designed and operated. The methods are pumping of contaminated groundwater to the surface for treatment and discharge or reinjection and active or passive extraction and treatment of soil gases produced in a waste deposit. Uncontaminated groundwater may also be pumped to deny it contact with a waste deposit. In addition, a recognized scientific phenomenon is being employed, in several variations, as the technology *phytoremediation,* with encouraging results. These methods will be briefly overviewed.

Groundwater Pumping. At least three different applications of groundwater pumping are used to control contaminated water beneath a disposal site. These applications are

---

[6] A variety of patented sealant technologies have been developed to seal the joints.

- Pumping to lower a water table
- Pumping to contain a plume
- Groundwater treatment systems

The effect of lowering a water table may be to prevent contaminated water from reaching a surface stream as base flow; to prevent contact with a contamination source; or to prevent migration to another aquifer (Figures 11.3 through 11.6).

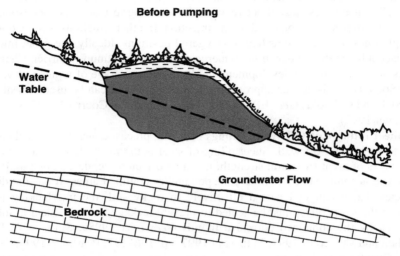

FIGURE 11.3 Lowering a water table to eliminate contact with disposal site (before pumping). (From U.S. Environmental Protection Agency.)

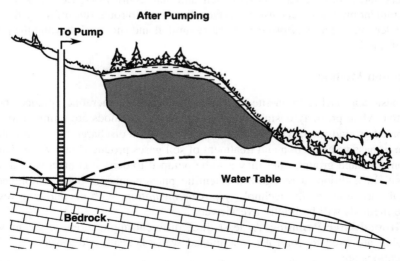

FIGURE 11.4 Lowering a water table to eliminate contact with disposal site (after pumping). (From U.S. Environmental Protection Agency.)

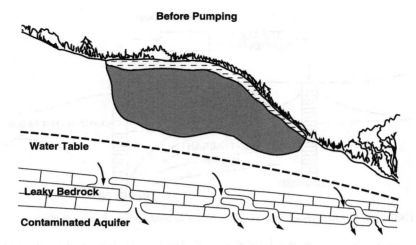

FIGURE 11.5 Lowering a water table to prevent contamination of an underlying aquifer (before pumping). (From U.S. Environmental Protection Agency.)

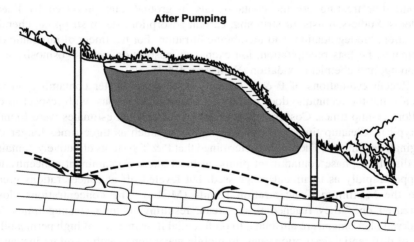

FIGURE 11.6 Lowering a water table to prevent contamination of an underlying aquifer (after pumping). (From U.S. Environmental Protection Agency.)

Extraction wells or combinations of extraction and injection wells may be used to contain a plume and/or alter plume movement to force contaminated groundwater toward collection wells (Figures 11.7 and 11.8). One of the most frequently employed remediation procedures for large plumes of contaminated groundwater is the "pump and treat" (P & T) approach, wherein extraction wells are placed to draw from the plume and prevent or reverse downgradient movement of the plume. The extracted water is treated to remove the pollutant(s) and is then discharged or used on the surface. The treated water may be reinjected at the perimeter of the contaminant plume to create an artificial groundwater mound, thereby assisting in moving

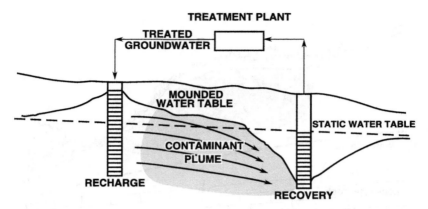

FIGURE 11.7 Reinjection of treated groundwater to contain a contaminant plume.

the contaminants toward the extraction well. The system illustrated in Figure 11.9 employs the ion exchange process for removal of chromium; air stripping of chlorinated solvents; and carbon adsorption to remove stripped organics from the exhaust stream. For treating organic contaminants in groundwater produced by P & T systems, Suthersan lists air stripping, carbon adsorption, steam stripping, chemical oxidation, biodegradation, and membrane filtration. For treatment of inorganic contaminants, he lists precipitation, ion exchange, adsorption, reverse osmosis, steam stripping, and chemical oxidation (Suthersan 1997).

Recent evaluations of P & T projects at 28 groundwater contamination sites reveals that the technique does not always attain expectations, with respect to cost and/or cleanup times. Cost increases of 80% over original estimates were found to be typical. Cleanup times are projected to be as much as three times longer than originally estimated. The studies determined that P & T systems effectively contained the dissolved phase contaminant plume at most sites. Contaminant concentrations dropped rapidly as treatment progressed, but leveled off at concentrations greater than the Maximum Concentration Limits (MCLs). The concentrations slowly decreased once they reached this plateau, resulting in long cleanup times. The observed phenomena are attributed to preferential flow in areas of high permeability; low or differential desorption rates; immobile water zones within soil grains; and/or continuing sources of groundwater contamination. Other referenced material mentions concentrations in remaining groundwater actually rebounding when pumps are shut off. Practitioners are using other aquifer restoration techniques in tandem with P & T technology, or as alternatives, in attempts to achieve more timely cleanup goals (Olsen and Kavanaugh 1993, pp. 42ff; Sellers 1999, Chapter 3; *see also:* EPA 1995, 1999; Keely 1996; Palmer and Fish 1996; Wilson 1997).

Soil Vapor Extraction (SVE). Anaerobic decomposition of organics produces methane gas, which is flammable, can accumulate to explosive concentrations, and is toxic. Deposits of hazardous waste may generate other toxic, flammable, or malodorous vapors. Prevention of dangerous buildups of such vapors is an important aspect of hazardous waste management, in general, and site remediation, in particular. In earlier times, simple venting of such vapors to the atmosphere was widely

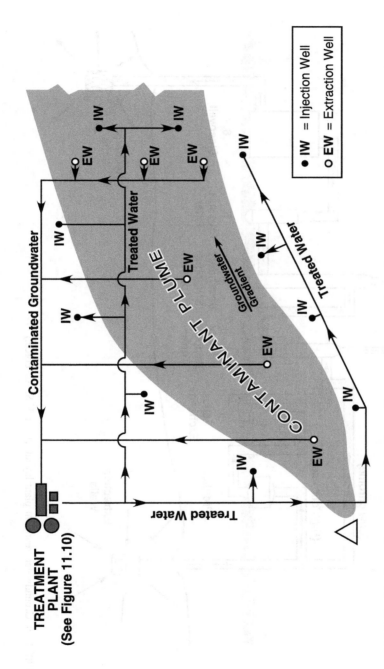

FIGURE 11.8 Combinations of extraction and injection wells to contain a contaminant plume.

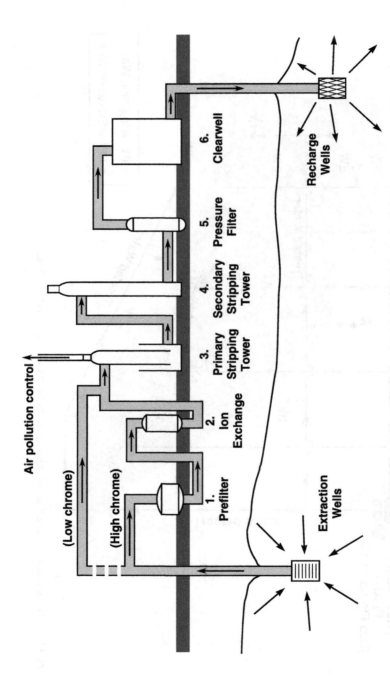

FIGURE 11.9 Groundwater treatment plant layout. (From U.S. Air Force.)

practiced. These primitive practices are now prohibited by most jurisdictions and are generally unacceptable.[7] Elaborate soil vapor collection and treatment systems have been developed to meet site-specific needs, but are not always necessary.

The objective of soil vapor collection and treatment systems is, of course, to prevent hazardous buildups of the gases and to render the collected gases harmless to human health and the environment. Vapors may be vented by passive collection systems, but forced ventilation or vacuum systems are necessary to maintain steady flow to treatment systems. The vapors are collected in pipe wells or trenches by 4- or 6-in. PVC perforated pipe. If a trench or more than one well is necessary, a manifold joins the individual collectors and conveys the vapors to a blower. The blower discharges to a treatment system. Figure 11.10 illustrates some basic configurations.

On-site treatment of extracted vapors is frequently accomplished by granular activated carbon (GAC) adsorption of the organics contained in the removed vapors. The GAC system has the advantages and disadvantages discussed in earlier chapters. The most serious disadvantage is the declining efficiency of carbon adsorption as the adsorptive capacity is approached. Frequent or continuous regeneration or replacement of carbon is necessary to ensure consistent high efficiency.

SVE systems may also be configured to add oxygen to stimulate subsurface aerobic biodegradation processes thereby enhancing removal of subsurface organic contaminants. Effectiveness of SVE systems may also be enhanced with hot air or *in situ* steam extraction. Steam extraction facilitates the removal of moderately volatile residual organics from the vadose zone (Suthersan 1997; Mercer et al. 1997).

On-site destruction of some vapors can be accomplished by flares or afterburners. Supplemental fuel may be necessary to achieve the desired combustion efficiency and/or to sustain combustion (Corbitt 1990, pp. 4.66ff).

Phytoremediation. The ever-intensifying search by legislators, public officials, environmentalists, scientists, regulators, industrial leaders, financiers, and many others for a less costly, less disruptive, less time-consuming means of remediating contaminated sites has spawned or given new life to a variety of technologies. Phytoremediation appears to be a promising means of *in situ* treatment of contaminated soils, sediments, and surface and/or groundwater by direct use of living green plants on sites wherein immediate cleanup is not imperative. The term *phytoremediation* encompasses five subtechnologies, which together or singly perform the following:

- *Phytotransformation* is the uptake of organic and nutrient contaminants from soil and groundwater and the accumulation of metabolites in plant tissue. In site remediation applications, it is important that the metabolites that are accumulated in vegetation be nontoxic or significantly less toxic than the parent compound.
- *Rhizosphere bioremediation* increases soil organic carbon, bacteria, and mycorrhizal fungi, which encourages degradation of organic chemicals in soil. Plants may also release exudates to the soil environment, helping to stimulate the degradation of organic chemicals by inducing enzyme sys-

---

[7] In most situations, vapor releases from RCRA facilities are subject to MACT and/or other standards.

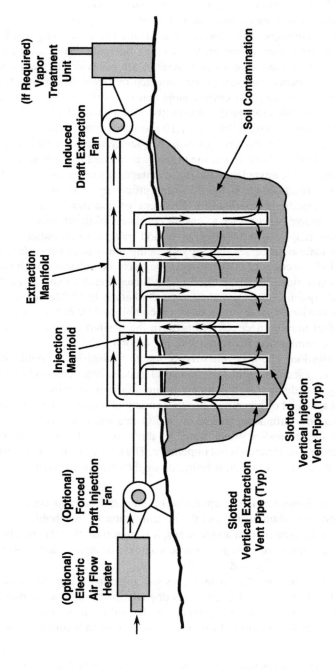

FIGURE 11.10 Soil vapor extraction system. (From U.S. Environmental Protection Agency.)

tems of existing bacterial populations, stimulating growth of new species that are able to degrade the wastes, and/or increasing soluble substrate concentrations for all microorganisms.

- *Phytostabilization* is the holding of contaminated soils and sediments in place by vegetation and immobilization of toxic contaminants in soils. Rooted vegetation prevents or inhibits windblown dust, which is an important source of human exposure from hazardous waste sites. Hydraulic control may be achieved by the transpiration of large volumes of water, thereby preventing migration of leachate toward ground or surface water.
- *Phytoextraction* uses metal-accumulating plants to translocate and concentrate metals from the soil in roots and above-ground shoots or leaves. An important issue is whether the metals can be economically recovered from the plant tissue or whether disposal of the waste is required.
- *Rhizofiltration* uses plant roots to sorb, concentrate, and precipitate metal contaminants from surface or groundwater. Roots of plants are capable of sorbing large quantities of lead and chromium from soil water or from water that has passed through the root zone of densely growing vegetation. The potential for treatment of radionuclide contaminants is being investigated in a Department of Energy pilot project involving uranium wastes and on water from a pond near the Chernobyl nuclear generating plant disaster site (Schnoor 1997).

The advantages of phytoremediation are the low capital costs, aesthetic benefits, minimization of leaching of contaminants, and soil stabilization. The operational cost of phytoremediation is also substantially less and involves mainly fertilization and watering for maintaining plant growth. In the case of heavy metals remediation, operational costs will also include harvesting, disposal of contaminated plant mass, and repeating the plant growth cycle.

The limitations of phytoremediation are that the contaminants below rooting depth will not be extracted and that the plant or tree may not be able to grow in the soil at every contaminated site due to toxicity. In addition, the remediation process can take years for contaminant concentration to reach regulatory levels and thus requires a long-term commitment to maintain the system (Suthersan 1997).

The Interstate Technology and Regulatory Cooperation Workgroup[8] (ITRC) Phytoremediation Work Team has produced a useful decision tree document for determining suitability and effectiveness of phytoremedation at a given site (ITRC 1999). The document can be accessed browsing the EPA Office of Solid Waste and Emergency Response (OSWER) Web site, searching the Phytoremediation Decision Tree. Appendix A provides a summary table showing applications of the five phytoremediation technologies to appropriate media, target contaminants, and suitable plant species (*see also:* EPA 1998b: Sajwan and Ornes 1997; Sellers 1999).

---

[8] The ITRC is a state-led, national coalition of personnel from regulatory and technology programs of states, federal agencies, and tribal, public, and industry stakeholders.

Treatment Methods

On-site treatment of hazardous wastes may be accomplished *in situ* or by excavation, treatment, and replacement (*ex situ*). The EPA recently released the Ninth Edition, Treatment Technologies for Site Cleanup — Annual Status Report, documenting the use of the increasingly numerous treatment technologies to remediate more than 900 contaminated waste sites. In remediating these sites, 32 million cubic yards of soil were treated using *in situ* technology, while 10 million cubic yards were treated *ex situ* (EPA 1999a). *In situ* methods will now be described.

Low Temperature Thermal Desorption. The process uses ambient air, heat, or mechanical agitation to increase the rate of mass transfer of contaminants to the vapor phase. Once in the vapor phase, the contaminants can be further treated by thermal or physical methods. The process can effectively remove halogenated aromatic and aliphatic compounds, volatile nonhalogenated compounds, and semi-volatile nonhalogenated organics (to a limited extent) from the soil matrix (Grasso 1993). Removal efficiencies for this treatment method range to more than 90% and primarily depend on the volatility of the contaminant (Udell 1997). *In situ* desorption of organics may be accomplished by radio frequency or electrical resistance (AC) heating, even in low permeability, clay-rich soils. In sandy, more permeable formations, steam can be injected to create an advancing vapor front which displaces soil, water, and contaminants by vaporization. The organics are transported in vapor-phase to the condensation front, where they can be pumped to the surface. Injection of moderately hot (50EC) water may serve the same purpose, provide easier pumping, and has the added benefit of creating a less harsh environment for beneficial biomass that may enhance removal of residuals (EPA 1994; *see also:* EPA 1995a, 1997; Cook 1996; Udell 1997; Sellers 1999).

Chemical Treatment. Liquid, gaseous, or colloidal reactive chemicals may be applied to, or injected into, a subsurface hazardous waste deposit or a contaminated aquifer by conventional injection wells, by permeable chemical treatment walls, or by deep soil mixing (DSM, discussed later herein). Treatment by these techniques can be oxidative, reductive/precipitative, or desorptive/dissolvable depending upon the character of the wastes to be treated (Yin and Allen 1999). If treatment is to be accomplished by injection or infiltration of an aqueous solution into a contaminated soil or groundwater zone, it must be followed by downgradient extraction of groundwater and elutriate and above-ground treatment and discharge or reinjection. Methods for *in situ* treatment of organics include soil flushing, oxidation, hydrolysis, and polymerization; methods for inorganics include precipitation, soil flushing, oxidation, and reduction (Corbitt 1990, pp. 9.27, 9.28; *see also:* Grasso 1993; Suthersan 1997, pp. 222–224; Rawe 1996; Fountain 1997; Palmer and Fish 1997; Sellers 1999, Chapters 3 and 4; Strbak 2000).

Bioremediation. Bioremediation is a managed or spontaneous process in which microbiological processes are used to degrade or transform contaminants to less toxic or nontoxic forms, thereby mitigating or eliminating environmental contamination. Microorganisms depend on nutrients and carbon to provide the energy needed for their growth and survival. Degradation of natural substances in soils and sediments provides the necessary food for the development of microbial populations in these media. Bioremediation harnesses the natural processes by selecting or promoting the enzymatic products and microbial growth necessary to convert the target contamination to nontoxic end products (van Cawenberghe and Roote 1998).

Organic waste deposits may be seeded with soil microorganisms from other locations or laboratories (exogenous microorganisms) to alter or destroy the wastes. Alternatively, nutrients may be added to an organic waste deposit to enhance naturally occurring (or indigenous) microorganisms and cause them to more actively consume or break down the pollutants. Bioremediation has been widely acclaimed as the hazardous waste treatment technology of the future. Limitations to the feasibility of bioremediation are plentiful, and its successful use requires a thorough understanding of the on-site hydrology, microbiology, and chemical characteristics.

Aerobic biodegradation processes take place in the presence of oxygen and nutrients and result in the formation of carbon dioxide, water, and microbial cell mass. Bioventing[9] may be used to provide subsurface oxygen, in the vadose or unconsolidated zones, by circulating air with or without pumping. In the saturated zone, air sparging[10] may be used to aerate the groundwater. Liquid oxygen, peroxide, or ozone injection can also be used to ensure that aerobic conditions are maintained. The literature reports that aerobic biodegradation has been successfully used to degrade gasoline and other petroleum hydrocarbons, some VOCs, and pesticides. Aerobic treatment schemes for contaminated soils are diagrammed in Figures 11.11 and 11.12.

Anaerobic biodegradation processes take place in the absence of oxygen and result in the formation of methane, carbon dioxide, and cell protein. Experimental work with anaerobes continues, but the practical application thereof is limited. In most cases involving remediation of waste deposits having anaerobic conditions, the approach has been to attempt oxygenation and conversion to aerobic conditions. Alternate electron acceptors such as nitrate or sulfate make use of existing bacterial populations, but in both cases the end products are toxic to humans (van Cauwenberghe and Roote 1998; *see also:* Grasso 1993; Rawe and Meagher-Hartzell 1996; Sims et al. 1996; Sims, Suflita, and Russell 1996; Suthersan 1997, Chapter 5; Ward et al. 1997, pp. 94–95; Sellers 1999, Chapter 3).

Natural Attenuation. As was noted in Chapter 3, chemical transformations of TCA, TCE, and other aliphatics were shown in the early 1980s to occur in groundwater where anaerobic bacteria were present (Vincent 1984). Perhaps the most plentiful example of the viability of natural attenuation can be found in the thousands of leaking underground fuel storage tank sites. It has been clear for a number of years that natural processes, in an obviously anaerobic environment, will achieve remediation of the groundwater and unsaturated zone beneath the tank, once the supply of leakage has stopped. Knowledge of these naturally occurring chemical, biological, and physical processes has continued to grow, giving rise to a *passive*

---

[9] Bioventing uses extraction wells to circulate air with or without pumping.

[10] Air sparging uses injection of air or oxygen under pressure into the saturated zone to transfer volatiles to the unsaturated zone for biodegradation and/or to aerate and oxygenate groundwater to enhance the rate of biological degradation.

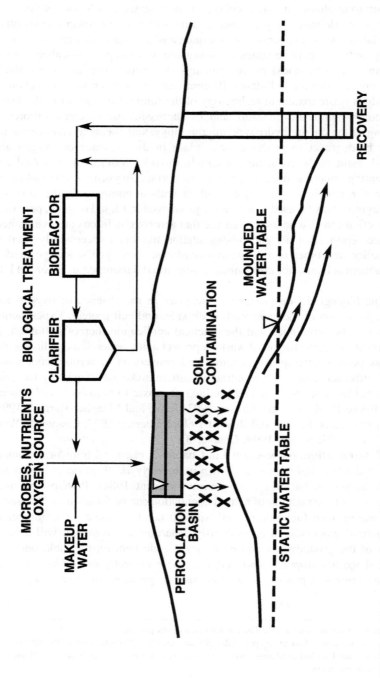

FIGURE 11.11 *In situ* bioreclamation using infiltration. (Adapted from Al W. Bourquin, Bioremediation of hazardous waste, *Hazardous Materials Control*, 2(5), Sept./Oct. 1989.)

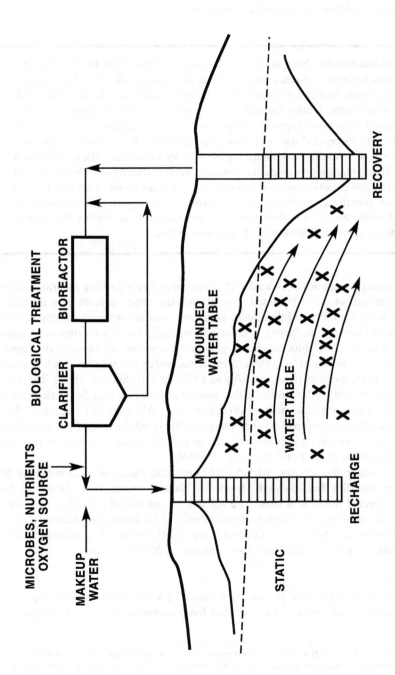

FIGURE 11.12 *In situ* bioreclamation using recharge wells or trenches. (Adapted from Al W. Bourquin, Bioremediation of hazardous waste, *Hazardous Materials Control*, 2(5), Sept./Oct. 1989.)

form of *in situ* remediation, termed variously *intrinsic attenuation, bioattenuation, intrinsic bioremediation, or natural attenuation.*

Natural attenuation has been well documented as a method for treating the fuel components benzene, toluene, ethylbenzine, and xylene (BTEX). Currently, it is not well established as a treatment for most other common classes of groundwater contaminants. Under limited circumstances, it can be applied at sites contaminated with other types if compounds such as chlorinated solvents and metals, but its successful use will depend on attenuation rates, site conditions, and the level of scientific understanding of processes that affect the contaminant … Natural attenuation processes are contaminant specific. Especially significant is the difference between organic and inorganic contaminants. Although natural attenuation reactions can completely convert some organic contaminants to carbon dioxide and water, they can alter the mobility of metals but cannot destroy them (National Academy of Sciences 2000).

The National Academy of Sciences (NAS) conclusion regarding natural attenuation applications where chlorinated solvents are the target contaminant notwithstanding, it is well established that under anaerobic conditions, most common chlorinated solvents undergo reductive dechlorination. Reductive dechlorination results in sequential removal of chlorine atoms, generating a series of intermediate degradation products[11] (Norris et al. 1999). Similar successes are reported with respect to petroleum hydrocarbons (Cho and Wilson 1999; Breedveld et al. 1999). Theoretical and experimental studies indicate that natural attenuation may be workable for other organic compounds as well as mixed plumes (Alleman and Leeson 1999). Some aspects of the technology have advanced to the point that predictive modeling of the fate and transport of chlorinated solvent natural attenuation is possible and functional (Clement et al. 1999; Carey et al. 1999).

As discussed in the section "RCRA and Superfund Remedial Actions" later in this chapter, the EPA has allowed Monitored Natural Attenuation (MNA) to be applied on selected Superfund sites. The Agency has produced a number of guidelines for the technology, including Technical Protocol for Evaluating Natural Attenuation of Chlorinated Solvents in Ground Water (EPA 1998c). The document provides the following advantages and disadvantages of MNA remedies:

Advantages:
- As with any *in situ* process, generation of a lesser volume of remediation wastes reduces the potential for crossmedia transfer of contam-

---

[11] An important, very basic caution, which the newcomer to the topic should have clearly in mind, is the fact that the degradation products of reductive dechlorination may be less, equally, or *more toxic* than the original compound, e.g., PCE-VC.

inants commonly associated with *ex situ* treatment and risk of human exposure to contaminated media.
- There is less intrusion as few surface structures are required.
- There is potential for application to all or part of a given site, depending on site conditions and cleanup objectives.
- These remedies may be used in conjunction with, or as a follow-up to, other (active) remedial measures.
- Overall remediation costs are lower than those associated with active remediation.

Potential disadvantages:
- Longer time frames may be required to achieve remediation objectives, compared to active remediation.
- Site characterization may be more complex and costly.
- Toxicity of transformation products may exceed that of the parent compound.
- Long-term monitoring will generally be necessary.
- Institutional controls may be necessary to ensure long-term protectiveness.
- The potential exists for continued contamination migration and/or cross-media transfer of contaminants.
- Hydrologic and geochemical conditions amenable to natural attenuation are likely to change over time and could result in renewed mobility of previously stabilized contaminants, adversely impacting remedial effectiveness.
- More extensive education and outreach efforts may be required in order to gain public acceptance of monitored natural attenuation.

(*See also:* Sutherson 1997, pp. 149–153; Barker and Wilson 1997; Reinhard et al. 1997; Semprini 1997; EPA 1999b; Sellers 1999, pp. 153–154).

Immobilization. Some types of waste materials may be stabilized or solidified in a matrix by mixing with Portland cement or other pozzolanic material. The hazardous waste constituents are not destroyed, but are immobilized, thereby minimizing leaching to ground or surface waters. Small amounts of waste and solidifying material can be effectively mixed in 55-gal drums which are then landfilled. Larger quantities may be exhumed, mixed in a pugmill or mobile mixing plant, and redeposited in the original or other site. Waste deposits may be mixed *in situ* using a backhoe or other heavy equipment, as illustrated in Figure 11.13. The technology is reported to be most effective for treatment of metal-contaminated soils. Wastes containing oils, chlorinated hydrocarbons, calcium chloride, and organic wastes containing hydroxyl or carboxylic acid functional groups may delay or completely inhibit the solidification of pozzolanic or Portland cement (Wiles 1989, p. 7.93). Volatile organic compounds (VOCS) tend to volatilize during the mixing of soil with stabilization/solidification (S/S) agents and are generally not immobilized (Sellers 1999).

FIGURE 11.13 Solidification by *in situ* mixing using backhoes. (From Geo-Con Incorporated, 4075 Monroeville Blvd., Suite 400, Monroeville, PA 15146. With permission.)

Techniques recently considered "innovative" for *in situ* stabilization have become standard practice. Cementitious stabilization is applicable to a wide range of industrial wastes and results in very stable products. S/S techniques that utilize Portland cement, fly ash, cement kiln dust, quick lime, and slags in various combinations have been used all over the world (Suthersan 1997). Figure 11.14 shows "Deep Soil Mixing" (DSM) equipment capable of mixing chemical reagents with contaminated soil to depths of 150 ft without excavation. The reagents are pumped through the hollow shafts of each auger. Figure 11.15 is a close-up view of the auger-mixing paddle configuration used on the DSM (*see also:* Cartledge et al. 1990; Jones 1990; Fink and Wahl 1996; NAS 1997, pp. 89–91, 98–101).

Destruction Methods

Methods for destruction of hazardous wastes have been adapted to both on-site and off-site applications. Examples of these applications follow.

Incineration. High-temperature incineration is a favored and highly effective means of destruction of as-generated and exhumed hazardous wastes. The wastes are exhumed and incinerated on-site, by mobile/transportable incinerators, with residues redeposited in a secure landfill. Correctly designed and operated high temperature incinerators are capable of very high destruction/removal efficiencies (see Chapter 7), but do not destroy inorganic components such as heavy metals. Mobile/transportable incinerators of both rotary kiln and liquid injection design are used in these applications. These residues must be captured by the emission control system in the incinerator and managed in a secure disposal site. Rotary kilns may be fitted with secondary burners operating at higher temperatures than the kiln, in order to achieve higher efficiencies (adapted from Combs 1989, p. 2ff; and EPA 1998d; *see also:* EPA 1990a; Sellers 1999, Chapter 4).

FIGURE 11.14 Deep soil mixing equipment used *in situ* solidification. (From Geo-Con Incorporated, 4075 Monroeville Blvd., Suite 400, Monroeville, PA 15146. With permission.)

In Situ **Vitrification.** *In situ* vitrification (ISV) uses electrical power to heat and melt soil, sludge, mine tailings, buried wastes, and sediments contaminated with organic, inorganic, and metal-bearing hazardous wastes. The molten material cools to form a hard, monolithic, chemically inert, stable, glass and crystalline product that incorporates and immobilizes the thermally stable inorganics and metals remaining in the mass. The electrical current is applied by a square array of four electrodes driven into the soil or waste mass. The soil melt typically requires a temperature of 2900 to 3600°F. The organic constituents are pyrolized in the melt or migrate to the surface where they combust in the presence of oxygen. Off-gases must be captured and treated. The process is repeated in squares containing up to 1000 tons (Jackson 1996). Since the void spaces initially in the soil are eliminated, ISV results in a volume reduction of 30 to 50% (Sellers 1999; *see also*: Shah et al. 1988; Johnson and Cosmos 1989; Vajda et al. 1995, pp. 294ff; Suthersan 1997, pp. 252, 302).

OFF-SITE TECHNOLOGIES AND PRACTICES

Although there are similarities between some of the technologies and practices employed in the conduct of on-site and off-site remedies, it is important (1) to consider the technologies in each context, (2) to understand some of the differences that prevail, and (3) to know why the differences prevail. The following is an overview of some technologies and practices associated with off-site remedies.

FIGURE 11.15 Deep soil mixing equipment used *in situ* solidification. (From Geo-Con Incorporated, 4075 Monroeville Blvd., Suite 400, Monroeville, PA 15146. With permission.)

FIGURE 11.16 Exposure and recovery of buried drums: "How-Not-To-Do-It." (From the Arizona Department of Environmental Quality.)

Excavation

Excavation of solid wastes in site remediation may simply be the expedient approach or may be necessary when *in situ* methods are not capable of achieving the cleanup objective and the waste to be moved is solid or semisolid. In cases where the applicable technologies provide options for on-site or off-site remedies, costs frequently become the driving factor. Whether an option or a technological necessity, excavation is frequently a major cost factor and may require expert logistical planning. If the soil to be excavated is contaminated with hazardous, radiological, or infectious waste or materials, OSHA worker health and safety planning and implementation requirements[12] become major factors.

Excavation is accomplished using standard or modified earth-moving equipment, however, specialized equipment is required for sites containing buried drums or other containers. Extraordinary care must be exercised to minimize releases from deteriorating containers during excavations on hazardous waste sites. Tedious, one-by-one exposure and recovery of drums is not an unusual necessity in removal actions. Figure 11.16 illustrates the "how-not-to-do-it" problem of damage to the drum and release of its contents. Such operations are most successful when a cable sling can be placed on the drum such that it can be lifted or pulled from the pile. Leaking drums should always be overpacked before movement.

In the general case, waste deposits that are exhumed by excavation are not containerized and are in the solid or semisolid phase. The exhumed wastes may be treated and redeposited on-site, transported to a TSDF, or used for fill material after treatment. Treatment systems that are effective for treatment of solid and/or semisolid hazardous wastes are those that remove (cleanse, desorb, detoxify, or extract) the waste constituents from the soil or other particulate matter or encapsulate or solidify the waste with the soils. Some examples follow.

Thermal Processes. Some form of incineration has been a favored approach to management of exhumed organic waste material. As discussed earlier herein, very high destruction/removal efficiencies (DRE) may be achieved by properly designed and operated thermal destruction units. As before, incineration does not destroy inorganic materials such as heavy metals. The captured solids from the emission control equipment generally require management as a hazardous waste. Incineration is effective with solid, semisolid, or liquid hazardous wastes, but is not *cost* effective for dilute or aqueous wastes. Liquids are usually added, if not present in the waste, to act as a catalyst in the reactions of thermal destruction.

Thermal desorption is gaining in popularity and application, primarily due to lower fuel and operating costs. The mass of the excavated waste is heated to 300 to 1200°F to achieve desorption. The desorbed gases are then raised to destruction temperatures or otherwise managed. The method has the added advantage that metal compounds are not volatilized (Vajda et al. 1995, pp. 282ff). A transportable low-temperature desorption unit is shown in Figure 11.17. Low-temperature desorption units treating soil containing contaminants with relatively low boiling points, such as

---

[12] *See:* Chapter 15.

FIGURE 11.17 Transportable low-temperature thermal desorption unit. (From URS Corporation, 100 California Street, San Francisco, CA 94111-4529. With permission.)

VOCs and total petroleum hydrocarbons (TPHs) operate at temperatures of 300 to 800°F. High-temperature thermal desorption units may treat soils containing contaminants having higher boiling points, such as polycyclic aromatic hydrocarbons (PAHs) and PCBs (Sellers 1999, Chapter 4).

Other applicable thermal processes include pyrolysis, wherein the waste may be transformed into an inert solid by thermal decomposition in the absence of oxygen (*see also:* Brunner 1988; Grasso 1993).

Physical Treatment. Component separation techniques are effective in waste-specific situations. These techniques include soil washing and solvent extraction. The EPA has defined soil washing as a separation process that uses water or water combined with chemical additives and a mechanical process to scrub soils. The technology does not detoxify or significantly alter the contaminant, but transfers the contaminant from the soil into the washing fluid or mechanically concentrates the contaminants into a much smaller soil mass for subsequent treatment (EPA 1997a). Solvent extraction is a physical separation process used in soil remediation to leach waste constituents from the solid matrix to a liquid solution for treatment or other handling. Solvent extraction, with additives such as surfactants or chelating agents, has been shown to be effective in treating sediments, sludges, and soils containing organic contaminants such as PCBs, VOCs, halogenated solvents, and petroleum wastes. In each case, bench-scale testing must precede full-scale operations in order to identify optimum solvents or cleansing media. The end products of these techniques by definition include lesser volumes of increased concentrations of hazardous wastes, which must then be managed.

Chemical Transformation. Although most of the chemical transformation processes are suitable for liquid hazardous wastes and wastewaters having hazardous constituents, they are not widely used on remediation sites. Examples include the use of reagents to remove the chlorine from chlorinated compounds (dehalogenation). Chemical dehalogenation can be an effective process for removing halogens from hazardous organic compounds such as dioxins, furans, PCBs, and chlorinated pesticides. Examples of chemical treatment applications to solid hazardous wastes include chlorination of cyanide wastes and reduction of hexavalent chromium wastes.

Biological Degradation. *Biological treatment* processes have been used most successfully to treat dilute wastes, contaminated groundwater, and wastewaters having hazardous constituents. These processes are generally used in flow-through applications at fixed facilities. *Bioremediation* is frequently used to treat excavated soils and remediation debris in static or recirculating units. On remediation sites, soils are often excavated and debris removed to a bioreactor site, where wastes may be layered with manure or other stimulants to enhance indigenous microbial populations. Alternatively the waste mass may be innoculated with proprietary populations. The unit then circulates the liquid component through the mass, or the mass may be mixed and turned to stimulate the natural biodegration processes (adapted from Vajda et al. 1995, p. 289).

Immobilization/Solidification. Solidification and stabilization techniques employed on-site are adaptable to *ex situ* projects. A wide variety of treatment processes use Portland cement as a binding agent. Pozzolanic materials are frequently added to the cement to react with any free calcium hydroxide and thus improve the strength and chemical resistance of the concrete-like product. Waste/concrete composites can be formed that have exceptional strength and excellent durability and that retain wastes very effectively (Cullinane and Jones 1989). Certain wastes (e.g., oil, grease, chlorinated solvents) interfere with the setting process or facilitate deterioration after setting and are therefore unsuitable for solidification with Portland cement.

Solidification is achieved primarily by adding materials to a waste to produce a solid. It may or may not involve a chemical bonding between the toxic contaminant and the additive. *Stabilization* describes processes that limit the solubility of or detoxify the contaminate. *Encapsulation* is a process involving the complete coating or enclosure of a toxic particle or waste agglomerate with a new substance, e.g., the solidification/stabilization additive or binder. *Microencapsulation* is the encapsulation of individual particles. *Macroencapsulation* is the encapsulation of an agglomeration of waste particles or microencapsulated materials (Wiles 1989).

Thermoplastic microencapsulation has been successfully used in nuclear waste disposal and can be adapted to metals and special industrial wastes. The waste is dried and then dispersed through a heated plastic matrix such as asphalt. The mixture is extruded into and cooled in a fiber or metal drum to give it shape for transport and/or disposal (EPA 1997a). Other materials such as polyethylene, polypropylene, wax, or elemental sulfur can be used for specific wastes where cost is not a factor. The major advantage of asphalt encapsulation, over cement and pozzolan systems, is the ability to solidify very soluble toxic materials. The operation is complex, requiring specialized equipment and a highly trained operating staff (adapted from Cullinane and Jones 1989; *see also:* Russell et al. 1996; Fink and Wahl 1996; EPA 1997b; NAS 1997, Chapter 3).

Mechanical and Hydraulic Dredging

Many of the nation's remedial sites are (or include) impoundments, streams, or estuaries, where contaminated sediments must be removed. These deposits, until removed, may severely disrupt aquatic ecosystems or may threaten public water supplies. Their removal and management pose particular challenges. The sediments may be highly contaminated or the contaminant may be highly toxic (or both). In such cases, special precautions may be necessary to prevent dispersal in the surrounding water or exposure of workers.

Mechanical dredging of contaminated sediments may be appropriate under conditions of low, shallow flow. If the sediments are well consolidated, they may be successfully removed by clamshell, dragline, or backhoe. Stream diversion or diking may be necessary to isolate the area of sediment removal. The physical layout may permit dewatering of the isolated area, followed by mechanical excavation of the sediments to be removed.

Hydraulic dredging is the preferred technique if the sediments have a high liquid content or are unconsolidated or if the contamination is in deep, flowing, or open water where resuspension is a problem. The dredged material may be pumped or barged to shore facilities for further management (adapted from EPA 1982, Chapter 7).

Dredged hazardous wastes are usually in the liquid or slurry form and may have sufficient solids content to justify dewatering. The liquid phase may be suitable for treatment by any of several conventional water treatment processes. The dewatered solids may be managed by the techniques listed earlier (*see:* "Excavation;" *see also:* Dawson and Mercer 1986, Chapters 9 and 10; Wentz 1989, pp. 406–415).

RCRA AND SUPERFUND REMEDIAL ACTIONS

RCRA Corrective Actions

As overviewed in previous chapters, RCRA Sections 3008 and 7003 provide authorities for the EPA to require corrective action whenever there is, or has been, a release of hazardous waste or hazardous waste constituents from a permitted or interim status facility. Moreover, RCRA authorizes the EPA to require corrective action beyond the facility boundary. The EPA interprets the term "corrective action" to cover the full range of possible actions including full cleanups.

The RCRA and Superfund programs follow similar procedures in responding to releases. In both, the first step after discovery of a release is an examination of available data to determine whether or not an emergency action is warranted. In both, short-term measures are authorized to abate the immediate adverse effects of a release. Once an emergency has been addressed, both programs provide for an investigation and formal study of long-term cleanup options. When these analyses are completed, both provide for formal selection of a remedy.

The major procedural difference between the two programs is the ranking of Superfund sites using the Hazard Ranking System (HRS) and the remedial action funding of sites listed on the National Priorities List (NPL). RCRA has neither of these provisions (EPA 1990, pp.VI-12, VI-13).

A recurring theme, in this text, has been (is) the extent to which risk assessment procedures have come to dominate the decision making process in hazardous waste management. The first major step in either RCRA or Superfund remedial actions is a baseline risk assessment. Each following decision step, through selection of the final remedy and development of postremediation monitoring plans, is based upon one or more risk assessment procedures.[13] The imperative felt by regulators, courts, environmentalists, and the public for rationality in standards bids fair for this emphasis to continue and to burgeon.

The facility owner or operator implements RCRA corrective action. A Superfund remedial action may be implemented by the responsible parties, if identified, or by state or EPA contractors. If cleanup is funded by Superfund, the government may seek cost recovery in federal court. There is, of course, no similar recourse for a RCRA corrective action.

Perhaps the greatest area of similarity between the programs is found in procedures and requirements for removal actions initiated under RCRA to achieve "clean closure" [40 CFR 270.1 (c)(5)] and an excavated removal as a Superfund remedy. In both cases, soil is excavated and treated (either on- or off-site) until testing shows no contamination greater than specified. RCRA owners and operators strive to achieve clean closure in order to avoid the procedural burdens of obtaining a post-closure permit and the 30-year groundwater monitoring requirements. Figure 11.18 shows the extent and magnitude of the excavation that may be necessary to achieve clean closure of a former liquid waste impoundment. The exhumed material must be managed by treatment, incineration, or disposal in a secure landfill.

Superfund Remedial Actions

In Chapter 10 we overviewed Subpart F of the National Contingency Plan (NCP) and the highly structured progression of the Superfund process, beginning with the Preliminary Assessment (PA), followed by the Site Inspection (SI), Hazard Ranking Score (HRS), listing on the National Priorities List (NPL), Remedial Investigation (RI), Feasibility Study (FS), and the Record of Decision (ROD). The student desiring greater detail regarding the NCP should consult 40 CFR 300. Practitioners working in Superfund cleanup operations should stay current with the NCP by regular reading of one or more of the EPA/Superfund-oriented

---

[13] The American Society for Testing Materials (ASTM) has published the *Standard Guide for Risk-Based Corrective Action* (Designation E-1739), which is designed to render a consistent decision-making process for the assessment and remediation of petroleum release sites. The complex and highly structured tier process is used to characterize site conditions and risk and evaluate restoration alternatives. As each tier is completed, the process evaluates whether additional site-specific analysis is required or if response actions, or closure, are warranted (Payne 1998). The procedures have been adopted by several states and a provisional RBCA ("Rebecca") for hazardous waste releases is said to be in preparation by ASTM.

FIGURE 11.18 Extent and magnitude of excavation necessary to "clean-close" a former hazardous waste impoundment. (Courtesy of Safety-Kleen Corporation.)

newsletters and periodicals or by accessing the Federal Register Online at <http://www.gpo.gov/su_docs/aces/aces140.html>.

As noted in Chapter 10, the ROD identifies the remedy that the EPA will require and establishes the framework for remedial negotiations between the EPA and the responsible parties. The next step in the Superfund process is the Remedial Design (RD), wherein the selected remedy is translated into an action plan. The RD considers the objectives and technologies overviewed in this chapter, as well as new and emerging technologies, to structure a cost-effective design.

In the Superfund lexicon, "Removal" implies a short-term cleanup action that usually addresses cleanup needs only at the surface of a site. "Removal" actions are conducted in response to an emergency situation (e.g., to avert an explosion, to clean up a spill of hazardous materials, or to stabilize a site until a permanent remedy can be found). "Removal" actions are limited to 12-month duration or $2 million in expenditures, although these limits may be extended.

Again, in the language of Superfund, "Remedial Action" (RA) refers to the final remedy for a site and may include a "Removal." The RA is generally more expensive and of longer duration than a "Removal." The EPA provides an estimate of the average cost of some current treatment remedies at $16 million. Completion of some projects may require up to 10 years, and follow-up monitoring may continue for decades.

If the responsible parties have been identified and have not paid the cost of the cleanup, the EPA initiates cost-recovery in the courts. In fact, cost-recovery may be initiated at any phase of the process. The costs, as noted, can be exceedingly high and are a very persuasive incentive for property owners, plant managers, hazardous waste facility operators, small businesses, and corporations to manage hazardous materials and wastes properly.

Upon completion of the RA, the project enters an Operation and Maintenance (O&M) phase which is designed to ensure that the remedy is operational. O&M costs can be quite high in some cases and are cost-shared by the EPA and the state wherein the site is located. This little-known provision of CERCLA embodies the possibility of wreaking havoc on the budgets of "smaller" states and has been a factor in persuading state legislatures to enact state "Superfund" legislation. The state programs enable funding of the state portion of O&M costs.

Superfund Accelerated Cleanup Model

In 1992, the EPA undertook an effort to streamline the Superfund process by eliminating overlapping site assessments, unproductive and lengthy litigation, and a ponderous remedy selection process. The effort was labeled Superfund Accelerated Cleanup Model (SACM).The model has undergone several iterations and has been adopted for use at all sites. A major change in the process combines PA, SI, the HRS ranking, and the RI/FS in a single site evaluation, yielding a single report. Potentially responsible party (PRP) searches are expedited by beginning searches for "core" groups of PRPs as soon as it is clear that the site will need a remedial response. Regional Decision Teams (RDTs) are formed with experienced and expert personnel from the EPA Regional staff, state officials, on-scene coordinators, remedial project managers, community involvement coordinators, and site and risk assessors. The RDT develops procedural rules, prioritizes sites, decides issues of policy and strategy, signs RODs or action memoranda, etc. The RDT ensures that response actions are fully consistent with CERCLA and NCP requirements [*see:* SACM Regional Decision Teams — Interim Guidance (OSWER Directive 9203.1-051); Payne 1998, pp. 33–58].

Presumptive Remedies and Response Strategies are evolving components of SACM. These tools are based upon a reversal of the earlier belief that each NPL site is unique and requires site-specific remedies. The EPA is taking the position that experience has shown that many sites have similar contamination profiles, waste types, and historical industrial use and will thus require similar remedies. Presumptive remedies are based upon selections of technologies at similar sites that show consistency, while meeting the NCP intent of protecting human health and the environment. The EPA has identified categories of sites where presumptive remedies are said to be appropriate, e.g., municipal solid waste landfills, sites with VOC contamination of soils, sediments and sludges, and wood-treater sites. The agency has developed presumptive response strategies for sites with groundwater contamination and is working on others (EPA 1998e; for a broad, well-reasoned development of strategies for accelerated site cleanup, *see:* Payne 1998).

Brownfields Economic Redevelopment Initiative and Environmental Justice

The term *brownfields* is defined by the EPA as abandoned, idled, or underused industrial and commercial facilities where expansion or redevelopment is complicated by real or perceived contamination. The definition fails to convey the fact that such properties are at the very heart of the spread of urban blight; the sprawling consumption of *greenfields*;[14] the fear of developmental involvement (on the part of lenders and investors); and the perception (or perhaps fact) that poor and minority urban neighborhoods bear disproportionately high and adverse human health and environmental effects from pollution. These closely interrelated environmental, economic, and social detriments are the basis for the EPA's Brownfields Economic Redevelopment Initiative, which is an attempt to address some of urban America's gnawing environmental justice issues. The intent of the initiative was (is) to empower states, communities, and other stakeholders in economic redevelopment to work together to prevent, inventory, assess, safely clean up, and sustainably reuse brownfields.

The EPA implemented the initiative in January 1995 with the Brownfields Action Agenda, which outlined four elements:

- Providing grants (seed money) for brownfields pilot projects
- Removing liability barriers impeding brownfields redevelopment
- Developing partnerships and outreach to all brownfields stakeholders
- Promoting local environmental workforce development and job training

More than 100 commitments from more than 25 organizations and federal agencies are included in the Action Agenda. The commitments, which total $300 million in federal government investments and $165 million in loan guarantees, assist cleanup and redevelopment activities for as many as 5000 properties. The EPA has recently established a Brownfields Technology Support Center which accepts technical support requests from all EPA regions and brownfields localities. The program also provides job training adjunct programs at various community colleges. Federal tax incentives are available to spur the cleanup and redevelopment of brownfields in distressed rural and urban areas. A variety of technical reports, organizational operational directives, and program updates are available at the brownfields Web site <http://www.epa.gov/swerosps/bf.htm> (*see also:* Payne 1998, pp. 302–306).

TOPICS FOR REVIEW OR DISCUSSION

1. What are the distinctions between "biological treatment" and "bioremediation" of hazardous wastes?
2. The text briefly discusses two methods of *in situ* remediation methods that involve heating of the contaminated soil. One is termed _____ and is an example of a _____ method. The other method is referred to as _____ and is an example of a _____ method.
3. Macroencapsulation and microencapsulation are remediation techniques that appear to hold promise. How do they differ?

---

[14] Pristine or undeveloped land.

4. Critics have recently complained of disappointing effectiveness and greater than anticipated time and cost of "pump-and-treat" groundwater remediation technology. Discuss possible alternatives for a generalized groundwater contamination cleanup project.
5. Discuss how a "numeric national cleanup goal" might be formulated. How might it be applied?
6. Discuss developments, changes, impacts of Superfund legislation that may have occurred since preparation of this edition (late 2000).

REFERENCES

Alleman, Bruce C. and Andrea Leeson, Eds. 1999. *Natural Attenuation of Chlorinated Solvents, Petroleum Hydrocarbons, and Other Organic Compounds.* Battelle Press, Columbus, OH.

Barker, J. F. and J. T. Wilson. 1997. Natural Biological Attenuation of Aromatic Hydrocarbons Under Anaerobic Conditions," Chapter 18, in *Subsurface Restoration,* C. H. Ward, J. A. Cherry, and M. R. Scalf, Eds., Ann Arbor Press, Chelsea, MI.

Bourquin, Al W. 1989. "Bioremediation of Hazardous Waste," *Hazardous Materials Control* September/October:16ff.

Breedveld, Gijs D., Magnus Sparrevik, Jenny Aadnanes, and Per Aargaard. 1999. "Natural Attenuation of Jet Fuel Contaminated Run-Off Water in the Unsaturated Zone," in *Natural Attenuation of Chlorinated Solvents, Petroleum Hydrocarbons, and Other Organic Compounds,* Bruce C. Alleman and Andrea Leeson, Eds., Battelle Press, Columbus, OH.

Brunner, Calvin R. 1988. "Industrial Waste Incineration," *Hazardous Materials Control,* July/August:26ff.

Burke, Thomas A. 1992. "Overview: Refining Hazardous Waste Site Policies Through Research," in *Hazardous Waste Site Investigations Toward Better Decisions,* Richard B. Gammage and Barry A. Berven, Eds., CRC Press, Boca Raton, FL.

Carey, Grant R., Paul J. Van Geel, J. Richard Murphy, Edward A. McBean, and Frank A. Rovers. 1999. "Modeling Natural Attenuation at the Plattsburgh Air Force Base," in *Natural Attenuation of Chlorinated Solvents, Petroleum Hydrocarbons, and Other Organic Compounds,* Bruce C. Alleman and Andrea Leeson, Eds., Battelle Press, Columbus, OH.

Cartledge, F. K., H. C. Eaton, and M. E. Tittlebaum. 1990. The Morphology and Microchemistry of Solidified/Stabilized Hazardous Waste Systems. U.S. Environmental Protection Agency, Risk Reduction Engineering Laboratory, Cincinnati, OH, EPA 600-S2-89-056.

Cho, J. S. and J. T. Wilson. 1999. "Hydrocarbon and MTBE Removal Rates During Natural Attenuation Application," in *Natural Attenuation of Chlorinated Solvents, Petroleum Hydrocarbons, and Other Organic Compounds,* Bruce C. Alleman and Andrea Leeson, Eds., Battelle Press, Columbus, OH.

Clement, Prabhakar, Christian D. Johnson, Yunwei Sun, Gary M. Kledka, and Craig Bartlett. 1997. "Modeling Natural Attenuation of Chlorinated Solvent Plumes at the Dover Air Force Base Area-6 Site," in *Natural Attenuation of Chlorinated Solvents, Petroleum Hydrocarbons, and Other Organic Compounds,* Bruce C. Alleman and Andrea Leeson, Eds., Battelle Press, Columbus, OH.

Combs, George D. 1989. *Emerging Treatment Technologies for Hazardous Waste,* Section XV. Environmental Systems Company, Little Rock, AR.

Cook, Kyle. 1996 "*In Situ* Steam Extraction Treatment," in *EPA Environmental Engineering Sourcebook,* J. Russell Boulding, Ed., Ann Arbor Press, Chelsea, MI.

Corbitt, Robert A. 1990. "Hazardous Waste," in *Standard Handbook of Environmental Engineering,* Robert A. Corbitt, Ed., McGraw-Hill, NY.

Cross, Cecil. 1996. "Slurry Walls," Chapter I, in *EPA Environmental Engineering Sourcebook,* J. Russell Boulding, Ed., Ann Arbor Press, Chelsea, MI.

Cullinane, M. John, Jr. and Larry W. Jones. 1989. "Solidification and Stabilization of Hazardous Wastes," *Hazardous Materials Control,* January/February:9ff.

Dawson, Gaynor W. and Basil W. Mercer. 1986. *Hazardous Waste Management.* John Wiley & Sons, NY.

Environment Reporter November 16, 1990, p. 1359, Bureau of National Affairs, Washington, D.C.

Environment Reporter April 29, 1994, p. 2219, Bureau of National Affairs, Washington, D.C.

Fink, Larry and George Wahl. 1996. "Solidification/Stabilization of Organics and Inorganics," Chapter 23, in *EPA Environmental Engineering Sourcebook,* J. Russell Boulding, Ed., Ann Arbor Press, Chelsea, MI.

Fountain, J. C. 1997. "Removal of Nonaqueous Phase Liquids Using Surfactants," Chapter 12, in *Subsurface Restoration,* C. H. Ward, J. A. Cherry, and M. R. Scalf, Eds., Ann Arbor Press, Chelsea, MI.

Grasso, Domenic. 1993. *Hazardous Waste Site Remediation Source Control.* Lewis Publishers, Boca Raton, FL.

IRTC 1999. *Phytoremediation Decision Tree.* The Interstate Technology and Regulatory Cooperation Work Group, Phytoremediation Work Team, Environmental Protection Agency, Washington, D.C.

Jackson, Trevor. 1996. "*In Situ* Vitrification Treatment," Chapter 14, in *EPA Environmental Engineering Sourcebook,* J. Russell Boulding, Ed., Ann Arbor Press, Chelsea, MI.

Johnson, Nancy P. and Michael G. Cosmos. 1989. "Thermal Treatment Technologies for Haz Waste Remediation," *Pollution Engineering,* October:79.

Jones, Larry W. 1990. Interference Mechanisms in Waste Stabilization, I. Solidification Systems. U.S. Environmental Protection Agency, Risk Reduction Engineering Laboratory, Cincinnati, OH, EPA 600-S2-89-067.

Keely, Joseph F. 1996. "Performance Evaluations of Pump-and-Treat Remediations," Chapter 4, in *EPA Environmental Engineering Sourcebook,* J. Russell Boulding, Ed., Ann Arbor Press, Chelsea, MI.

King, Brian J. and Deborah Anne Amidaneau. 1995. "United States Legal and Legislative Framework," in *Accident Prevention Manual for Business and Industry — Environmental Management,* Gary R. Krieger, Ed., National Safety Council, Itasca, IL.

Mercer, J. W., R. M. Parker, and C. P. Spalding. 1997. "Use of Site Characterization Data to Select Applicable Remediation Technologies," Chapter 9, in *Subsurface Restoration,* C. H. Ward, J. A. Cherry, and M. R. Scalf, Eds., Ann Arbor Press, Chelsea, MI.

Mitchell, J. K. and W. A. N. van Court. 1997. "Barrier Design and Installation: Walls and Covers," Chapter 11, in *Subsurface Restoration,* C. H. Ward, J. A. Cherry, and M. R. Scalf, Eds., Ann Arbor Press, Chelsea, MI.

National Academy of Sciences. 1997. *Innovations in Ground Water and Soil Cleanup.* National Academy Press, Washington, D.C.

National Academy of Sciences. 2000. *Natural Attenuation for Groundwater Remediation Executive Summary.* National Academy Press, Washington, D.C.

Norris, Robert D., David J. Wilson, David E. Ellis, and Robert Siegrist. 1999. "Consideration of the Effects of Remediation Technologies on Natural Attenuation," in *Natural Attenuation of Chlorinated Solvents, Petroleum Hydrocarbons, and Other Organic Compounds,* Bruce C. Alleman and Andrea Leeson, Eds., Battelle Press, Columbus, OH.

Olsen, Roger L. and Michael C. Kavanaugh. 1993. "Can Groundwater Restoration Be Achieved?" *Water Environment and Technology* March:42ff.

Palmer, Carl D. and William Fish. 1996. "Chemical Enhancements to Pump-and-Treat Remediation,"Chapter 5, in *EPA Environmental Engineering Sourcebook,* J. Russell Boulding, Ed., Ann Arbor Press, Chelsea MI.

Palmer, C. D. and W. Fish. 1997. "Chemically Enhanced Removal of Metals from the Subsurface," Chapter 14, in *Subsurface Restoration,* C. H. Ward, J. A. Cherry, and M. R. Scalf, Eds., Ann Arbor Press, Chelsea, MI.

Payne, Scott M. 1998. *Strategies for Accelerating Cleanup at Toxic Waste Sites,* CRC Press, Boca Raton, FL.

Pearlman, Leslie. 1999. Subsurface Containment and Monitoring Systems — Barriers and Beyond (Overview Report). U.S. Environmental Protection Agency, Office of Solid Waste and Emergency Response, Technology Innovation Office, Washington, D.C.

RAND. 1989. "Rating Superfund's Progress: In a Word, 'Super-Slow,'" *RAND Research Review* XIII(3), Fall. The RAND Corporation, Santa Monica, CA.

Rawe, Jim. 1996. "*In Situ* Soil Flushing," Chapter 7, in *EPA Environmental Engineering Sourcebook,* J. Russell Boulding, Ed., Ann Arbor Press, Chelsea, MI.

Rawe, Jim and Evelyn Meagher-Hartzell. 1996. "*In Situ* Biodegradation Treatment," Chapter 11, in *EPA Environmental Engineering Sourcebook,* J. Russell Boulding, Ed., Ann Arbor Press, Chelsea, MI.

Reinhard, M., G. P. Curtis, and J. E. Barbash. 1997. "Natural Chemical Attenuation of Halogenated Hydrocarbon Compounds via Dehalogenation," Chapter 24, in *Subsurface Restoration,* C. H. Ward, J. A. Cherry, M. R. Scalf, Eds., Ann Arbor Press, Chelsea, MI.

Russell, Hugh H., John E. Matthews, and Guy W. Sewell. 1996. "TCE Removal from Contaminated Soil and Water," Chapter 6, in *EPA Environmental Engineering Sourcebook,* J. Russell Boulding, Ed., Ann Arbor Press, Chelsea, MI.

Sajwan, Kenneth S. and W. Harold Omes. 1997. "Potential of Mosquito Fem (*Azolla caroliniana* Willd.) Plants as a Biofilter for Cadmium Removal from Wastewater," Chapter 14, in *Emerging Technologies in Hazardous Waste Management* 7, D. William Tedder and Frederick G. Pohland, Eds., Plenum Press, NY.

Schnoor, Jerald L. 1997. *Phytoremediation.* Technology Evaluation Report No. TE-98-01, Ground-Water Remediation Technologies Analysis Center (GWERTAC), Pittsburgh, PA.

Sellers, Kathleen. 1999. *Fundamentals of Hazardous Waste Site Remediation.* CRC Press, Boca Raton, FL.

Semprini, L. 1997. "*In Situ* Transformation of Halogenated Aliphatic Compounds Under Anaerobic Conditions," Chapter 26, in *Subsurface Restoration,* C. H. Ward, J. A. Cherry, and M. R. Scalf, Eds., Ann Arbor Press, Chelsea, MI.

Shah, J. K., T. J. Schultz, and V. R. Daiga. 1988. "Pyrolysis Process," in *Standard Handbook of Hazardous Waste Treatment and Disposal,* Harry M. Freeman, Ed., McGraw-Hill, NY.

Sims, J. L., R. C. Sims, R. R. Dupont, J. E. Matthews, and H. H. Russell. 1996. "*In Situ* Bioremediation of Contaminated Unsaturated Subsurface Soils," Chapter 12, in *EPA Environmental Engineering Sourcebook,* J. Russell Boulding, Ed., Ann Arbor Press, Chelsea, MI.

Sims, J. L., J. M. Sulflita, and H. H. Russell. 1996. "*In Situ* Remediation of Contaminated Groundwater," Chapter 13, in *EPA Environmental Engineering Sourcebook,* J. Russell Boulding, Ed., Ann Arbor Press, Chelsea, MI.

Soundararajan, R., Edwin F. Barth, and J. J. Gibbons. 1990. "Using an Organophilic Clay to Chemically Stabilize Waste Containing Organic Compounds," *Hazardous Materials,* Control January/February:42ff.

Staples, Charles A. and Richard A. Kimerle. 1986. "How Clean is Clean? Site Specific Answers," *Hazardous Substances,* October: pp. 10–12.

Strbak, Lauryn. 2000. In Situ Flushing with Surfactants and Cosolvents. U. S. Environmental Protection Agency, Office of Solid Waste and Emergency Response, Technology Innovation Office, Washington, D.C.

Suthersan, Suthan S. 1997. *Remediation Engineering Design Concepts.* CRC Press, Boca Raton, FL.

Travis, Curtis C. and Carolyn B. Doty. 1992. "Remedial Action Decision Process," in *Hazardous Waste Site Investigations — Toward Better Decisions,* Richard B. Gammage and Barry A. Berven, Eds., CRC Press, Boca Raton, FL.

Udell, K. S. 1997. "Thermally Enhanced Removal of Liquid Hydrocarbon Contaminants from Soils and Groundwater, " Chapter 16, in *Subsurface Restoration,* C. H. Ward, J. A. Cherry, and M. R. Scalf, Eds., Ann Arbor Press, Chelsea, MI.

U.S. Environmental Protection Agency. 1982. Handbook for Remedial Action at Hazardous Waste Sites. Office of Research and Development, Cincinnati, OH.

U.S. Environmental Protection Agency. 1989. Risk Assessment Guidance for Superfund, Volume I, Human Health Evaluation Manual. Office of Emergency and Remedial Response, Washington, D.C., EPA 540-1-89-002.

U.S. Environmental Protection Agency. 1990. RCRA Orientation Manual, 1990 Edition. Superintendent of Documents, U.S. Government Printing Office, Washington, D.C.

U.S. Environmental Protection Agency. 1990a. Mobile/Transportable Incineration Treatment. Office of Reserch and Development, Washington, D.C., EPA 540-2-90-014.

U.S. Environmental Protection Agency. 1992. Engineering Bulletin Slurry Walls. Office of Solid Waste and Emergency Response, Washington, D.C., EPA 540-S-92-008.

U.S. Environmental Protection Agency. 1994. *In Situ* Remediation Technology Status Report: Thermal Enhancements. Office of Solid Waste and Emergency Response, Washington, D.C., EPA 542-K-94-009.

U.S. Environmental Protection Agency. 1995. Pump-and-Treat Ground-Water Remediation, A Guide for Decision Makers and Practitioners. Office of Research and Development, Washington, D.C., EPA 625-R-95-005.

U.S. Environmental Protection Agency. 1995a. Engineering Forum Issue Paper: Thermal Desorption Implementation Issues. Office of Solid Waste and Emergency Response, Washington, D.C., EPA 540-F-95-030.

U.S. Environmental Protection Agency. 1997. Issue Paper: How Heat Can Enhance *In Situ* Soil and Aquifer Remediation. Office of Research and Development, Washington, D.C., EPA 540-S-97-502.

U.S. Environmental Protection Agency. 1997a. Technology Alternatives for the Remediation of Soils Contaminated with As, Cd, Cr, Hg, and Pb. Office of Emergency and Remedial Response and Office of Research and Development, Washington, D.C., EPA 540-S-97-500.

U.S. Environmental Protection Agency. 1997b. Remediation Case Studies: Bioremediation and Vitrification, Volume 5. Federal Remediation Technologies Toundtable, Washington, D.C., EPA 642-R-97-008.

U.S. Environmental Protection Agency. 1998. Introduction to Superfund Liability, Enforcement, and Settlements. Solid Waste and Emergency Response, Washington, D.C., EPA 540-R-98-028.

U.S. Environmental Protection Agency. 1998a. Evaluation of Subsurface Engineered Barriers at Waste Sites. Office of Solid Waste and Emergency Response, Washington, D.C., EPA 542-R-98-005.

U.S. Environmental Protection Agency. 1998b. A Citizen's Guide to Phytoremediation. Office of Solid Waste and Emergency Response, Washington, D.C., EPA 542-F-98-011.

U.S. Environmental Protection Agency. 1998c. Technical Protocol for Evaluating Natural Attenuation of Chlorinated Solvents in Ground Water. Office of Research and Development, Washington, D.C., EPA 600-R-98-128.

U.S. Environmental Protection Agency. 1998d. On-Site Incineration: Overview of Superfund Operating Experience. Office of Solid Waste and Emergency Response, Washington, D.C., EPA 542-R-97-012.

U.S. Environmental Protection Agency. 1998e. Introduction to: Superfund Accelerated Cleanup Model. Office of Solid Waste and Emergency Response, Washington, D.C., EPA 540-R-98-025.

U.S. Environmental Protection Agency. 1999. Hydraulic Optimization Demonstration for Groundwater Pump-and-Treat Systems. Office of Solid Waste and Emergency Response, Washington, D.C., EPA 542-R-99-011.

U.S. Environmental Protection Agency. 1999a. Treatment Technologies for Site Cleanup — Annual Status Report, Ninth Edition. Office of Solid Waste and Emergency Response, Technology Innovation Office, Washington, D.C., EPA 542-R-99-001.

U.S. Environmental Protection Agency. 1999b. Use of Monitored Natural Attenuation At Superfund, RCRA Corrective Action and Underground Storage Tank Sites. OSWER Directive 9200.4-17P, April, 1999, Office of Solid Waste and Emergency Response, Washington, D.C.

U.S. General Accounting Office. 1993. Superfund — Progress, Problems, and Reauthorization Issues. Washington, D.C., GAO/T-RCED-93-27.

U.S. General Accounting Office. 1994a. Superfund — Status, Cost, and Timeliness of Hazardous Waste Site Cleanups. Washington, D.C., GAO/RCED-94-256.

U.S. General Accounting Office. 1994b. Superfund — EPA Has Opportunities to Increase Recovered of Costs. Washington, D.C., GAO/RCED-94-196.

U.S. General Accounting Office. 1999. Superfund — Progress Made by EPA and Other Federal Agencies to Resolve Program Management Issues. Washington, D.C., GAO/RCED-99-111.

Vajda, Gary F., William L. Hall, and Gary R. Krieger. 1995. "Pollution Prevention Approaches and Technologies," in *Accident Prevention Manual for Business & Industry Environmental Management,* Gary R. Krieger, Ed., National Safety Council, Itasca, IL.

Van Cauwenberghe, Liesbet and Diane S. Roote. 1998. *In Situ Bioremediation,* Technology Overview Report No. TO-98-01, Ground-Water Remediation Technologies Analysis Center (GWRTAC), Pittsburgh, PA.

Vincent, J. R. 1984. South Florida Drinking Water Investigation, U.S. Environmental Protection Agency, National Enforcement Investigations Center, Denver, CO, EPA 330-1-84-001.

Ward, C. H., J. A. Cherry, and M. R. Scalf. 1997. *Subsurface Restoration.* Ann Arbor Press, Chelsea, MI.

Wentz, Charles A. 1989. *Hazardous Waste Management.* McGraw-Hill, NY.

Wiles, Carlton C. 1989. "Solidification and Stabilization Technology," in *Standard Handbook of Hazardous Waste Treatment and Disposal,* Harry M. Freeman, Ed., McGraw-Hill Book Company, New York.

Wilson, J. L. 1997. "Removal of Aqueous Phase Dissolved Contamination: Non-Chemically Enhanced Pump-and-Treat," Chapter 17, in *Subsurface Restoration,* C. H. Ward, J. A. Cherry, and M. R. Scalf, Eds., Ann Arbor Press, Chelsea, MI.

Yin, Yujun and Herbert E. Allen. 1999. *In Situ Chemical Treatment.* Technology Evaluation Report No. TE-99-0 1, Ground-Water Remediation Technologies Analysis Center (GWRTAC), Pittsburgh, PA.

APPENDIX A
Typical Plants Used in Various Phytoremediation Applications

Application	Media	Contaminants	Typical Plants
1. Phytotransformation	Soil, groundwater, landfill leachate, land application of wastewater	Herbicides (atrazine, alachlor) Aromatics (BTEX) Chlorinated aliphatics (TCE) Nutrients (NO_3^-, NH_4^+, PO_4^{3-}) Ammunition wastes (TNT, RDX)	Phreatophyte trees (poplar, willow, cottonwood, aspen) Grasses (rye, Bermuda, sorghum, fescue) Legumes (clover, alfalfa, cowpeas)
2. Rhizopshere bioremediation	Soil sediments, land application of wastewater	Organic contaminants (pesticides, aromatics, and polynuclear aromatic hydrocarbons [PAHs])	Phenolics releasers (mulberry, apple, osage orange) Grasses with fibrous roots (rye, fescue, Bermuda) for contaminants 0–3 ft deep Phreatophyte trees for 0–10 ft Aquatic plants for sediments
3. Phytostabilization	Soil, sediments	Metals (Pb, Cd, Zn, As, Cu, Cr, Se, U) Hydrophobic organics (PAHs, PCBs, dioxins, furans, pentachlorophenol, DDT, dieldrin)	Phreatophyte trees to transpire large amounts of water for hydraulic control Grasses with fibrous roots to stabilize soil erosion Dense root systems are needed to sorb/bind contaminants
4. Phytoextraction	Soil, brownfields, sediments	Metals (Pb, Cd, Zn, Ni, Cu) with EDTA addition for Pb selenium (volatilization)	Sunflowers Indian mustard Rape seed plants Barley, hops Crucifers Serpentine plants Nettles, dandelions
5. Phizofiltration	Groundwater, water and wastewater in lagoons or created wetlands	Metals (Pb, Cd, Zn, Ni, Cu) Radionuclides (^{137}Cs, ^{90}Sr, U) Hydrophobic organics	Aquatic plants Emergents (bullrush, cattail, coontail, pondweed, arrowroot, duckweed) Submergents (algae, stonewort, parrot feather, Eurasian water milfoil, Hydrilla)

Source: Schnoor, Gerald L. 1997. Phytoremediation. Technology Evaluation Report No. TE-98-01, Ground-Water Remediation Technologies Analysis Center (GWRTAC), Pittsburgh, PA. (Prepared for the U.S. Environmental Protection Agency through a cooperative agreement with the University of Pittsburgh's Environmental engineering Department.)

12 Medical/Biomedical/Infectious Waste Management

OBJECTIVES

At completion of this chapter, the student should:

- Be familiar with the hazards associated with the traditional "red bag wastes," methods to minimize the hazards, and current criteria for managing the wastes.
- Be familiar with the traditional sanitarian approach to biomedical waste management and the impacts of the AIDS epidemic and the 1988 beach washups on the Atlantic seaboard.
- Understand the regulatory approach of the Subtitle J regulations (40 CFR 259) and the use of the tracking form.
- Be familiar with regulatory developments and trends which are reordering options for management of medical/biomedical/infectious wastes.

INTRODUCTION

For many years, health care workers, hospital administrators, military sanitarians, and other health-related professionals have understood the necessity to protect themselves, their employees/members, and the public from exposure to wastes that might be reservoirs of disease-transmitting organisms. Local ordinances, state and military regulations, and guidelines issued by federal agencies and professional organizations developed around a few simple practices and vice versa. These practices generally included "red-bagging" the solid wastes[1] and isolating them in cool storage, followed by incineration or sterilization and landfilling.

The 1976 enactment of the Resource Conservation and Recovery Act (RCRA) included a definition of hazardous waste which continues as a basis for federal regulation of infectious waste management:

---

[1] The practice of disposing of medical wastes in bright red plastic bags, which distinguish the contents as being distinct from other wastes.

(5) The term "hazardous waste" means a solid waste or combination of solid wastes, which because of its quantity, concentration, or physical, chemical or *infectious* characteristics may —
>   (A) cause, or significantly contribute to an increase in mortality or an increase in serious irreversible, or incapacitating reversible, illness; or
>   (B) pose a substantial present or potential hazard to human health or the environment when improperly treated, stored, transported, or disposed of, or otherwise managed (42 USC 6903).

In 1978, the EPA published proposed regulations for hazardous waste management, which included several classifications of infectious waste. However, the agency did not make a convincing case for the supposed health hazards posed by these wastes and did not include them in the final hazardous waste regulations.

By 1982, the EPA had not promulgated regulations specific to the management of infectious wastes; state and local regulations ranged from nonexistent to overly complex and conflicting; and the agency was under pressure to provide guidance. The agency published the Draft Manual for Infectious Waste Management and, in 1986, published the final version — EPA Guide to Infectious Waste Management. We will borrow heavily from the 1986 Guide in this chapter.

This quiet evolution ended with the nation's growing alarm toward the Acquired Immunodeficiency Syndrome (AIDS) epidemic. Truths, half-truths, and blatant untruths regarding modes of transmission of the Human Immunodeficiency Virus (HIV) caused near panic among some health care workers, in particular, and among the public, in general. Suddenly, landfills began refusing hospital wastes, health care workers began red-bagging *"everything,"* small medical waste incinerators were overwhelmed, and management of infectious waste became a major problem.[2,3]

In May 1988, a garbage slick nearly 1 mi long, surfaced along the Ocean County shore of New Jersey. Needles, syringes, and empty prescription bottles with New York addresses washed up on the shore; 6 weeks later, 10 mi of Long Island beaches closed when medical wastes washed ashore. Throughout the summer of 1988, beaches from Maine to the Gulf of Mexico, along the Great Lakes, and elsewhere experienced washups of medical wastes (Office of Technology Assessment 1988, p. 1). In a similarly disturbing incident, children were found playing with vials of blood they had found in a dumpster (Ostler 1998).

Public and congressional outrage over the closure of beaches and perceived health threats brought about enactment in November 1988 of RCRA Subtitle J, the hastily conceived Medical Waste Tracking Act (MWTA). The EPA rushed the implementing regulations into place in March 1989, reflecting Congress' hope that their

---

[2] This trend abated after some time, but the damage was done. The alarmed reaction among health care workers resulted in large amounts of plastic and paper items being committed to destruction in small incinerators at a time when hospital and municipal waste incineration was generating concern because of emissions of dioxins and furans. The plastics and chlorine-bleached paper were significant sources of these emissions.

[3] In the U.S., the amount (of hospital waste) generated daily is estimated to be between 5 and 7 kg per patient per day, while in Italy, reported amounts are beween 3 and 5 kgs per patient per day (Giroletti and Lodola 1994, p. 161).

impact would prevent beach washups during the summer of 1989 (adapted from Jenkins 1990, p. 55).

The EPA promulgated Subtitle J regulations which were patterned after the RCRA Subtitle C regulations and codified at 40 CFR 259. The regulations included the following elements:

- "Medical Waste" Definition: Medical waste is any solid waste that is generated in the diagnosis, treatment, or immunization of human beings or animals in related research, biologicals production, or testing.
- Medical Waste Generator Requirements: Generators were defined as producers of more than 50 lb of regulated medical waste monthly, managed by shipping off-site. Generators were required to separate, package, label, mark, and track waste according to the regulation.
- Medical Waste Transporter Requirements: Transporters submitted a one-time notification to EPA headquarters, which then issued a medical waste identification number. The ID number was to be used on all tracking forms and reports. Transporters were also required to follow rules regarding transport vehicles; ensure that wastes were properly packaged, labeled, and marked; and comply with rules for tracking, record keeping, and reporting of waste shipments.
- Medical Waste Treatment, Destruction, and Disposal Facility Requirements: These facilities included incinerators, landfills, and treatment operations that grind, steam sterilize, or treat wastes with disinfectants, heat, or radiation. These practices were prescribed, defined, and implemented in similarity to the 1986 Guide.

Subtitle J instructed EPA to develop a 2-year demonstration program to track medical waste in the participating states and to report back to Congress upon completion of the program. Connecticut, New Jersey, New York, and Rhode Island, as well as Puerto Rico, opted to participate. The EPA rendered interim reports in May and December 1990. The latter, EPA 530-SW-90-087B, offers no conclusions regarding effectiveness of the program, and the EPA has published no further evaluation. The program was completed in 1991, and Congress has shown little enthusiasm for an expanded or continued program. The focus for medical waste management regulatory programs has thus reverted to state and local governments. Available guidance includes the 1986 EPA Guide and the more recent "white paper" published by the *Journal of the Air and Waste Management Association*, which will also be quoted herein (*see also:* Reinhardt and Gordon 1991; Drum and Bulley 1994; Turnberg 1996).

MWTA and the 40 CFR 259 Regulations were widely recognized as having minimal effect on the beach washup of medical waste. It was an effort to "do something" about the burgeoning problem of medical waste management and to gather data on the effectiveness of a tracking system patterned somewhat after the "cradle-to-grave" management system for hazardous wastes.

Definition and Characterization of Medical Waste

Disagreement exists between governments, agencies, and practitioners regarding the meanings of the terms "infectious waste" and "medical waste." To avoid unproductively dwelling upon this confusion, we briefly point out the terms used and some indication of their usage. We then adopt a convention for use in this text.

Infectious Waste

In the 1986 guidance document, the EPA defines infectious waste as waste capable of producing an infectious disease. This definition requires a consideration of certain factors necessary for induction of disease. These factors include

- Presence of a pathogen of sufficient virulence
- Dose
- Portal of entry
- Resistance of host

Thus, for a waste to be infectious, it must contain pathogens with sufficient virulence and quantity so that exposure to the waste by a susceptible host could result in an infectious disease. The EPA further recommends categories of waste be designated as infectious waste, as summarized in Table 12.1.

In addition, the EPA has identified an optional infectious waste category which consists of miscellaneous contaminated wastes. The suggestion is that a qualified person or committee should decide whether or not to handle these wastes as "infectious" in specific situations. The optional categories and examples are listed in Table 12.2. The terminology problem is further complicated by the fact that the terms infectious, pathological, biomedical, biohazardous, toxic, and medically hazardous have all been used to describe infectious waste.

Medical Waste

Medical wastes include all infectious waste, hazardous (including low-level radioactive wastes) wastes, and any other wastes that are generated from all types of health care institutions, including hospitals, clinics, doctor (including dental and veterinary) offices, and medical laboratories (Office of Technology Assessment 1988, p. 3).

The terminology confusion is worsened by the EPA's definition of "medical waste" in 40 CFR 259.10 as any solid waste that is generated in the diagnosis, treatment, or immunization of human beings or animals in related research, biologicals production, or testing.

In the Subpart J regulations, the EPA also defined Regulated Medical Wastes as a subset of all medical wastes and included seven distinct categories:

- Cultures and stocks of infectious agents
- Human pathological wastes (e.g., tissues, body parts)
- Human blood and blood products

TABLE 12.1
Categories of Infectious Wastes

Waste Category	Examples[a]
Isolation wastes	Wastes generated by hospitalized patients who are isolated to protect others from communicable diseases
Cultures and stocks of infectious agents	Specimens from medical and pathology agents and associated biologicals laboratories
	Cultures and stocks of infectious agents from clinical, research, and industrial laboratories; disposable culture dishes and devices used to transfer, inoculate, and mix cultures
	Waste from production of biologicals
	Discarded live and attenuated vaccines
Human blood and blood products	Waste blood, serum, plasma, and blood products
Pathological waste	Tissues, organs, body parts, blood, and body fluids removed during surgery, autopsy, and biopsy
Contaminated sharps[b]	Contaminated hypodermic needles, syringes, scalpel blades, Pasteur pipettes, and broken glass
Contaminated animal carcasses, body parts, and bedding[c]	Contaminated animal carcasses, body parts, or bedding of animals that were intentionally exposed to pathogens

[a] These materials are examples of wastes covered by each category. The categories are not limited to these materials. (*Source:* EPA 530-SW-86-014.)

[b] *Note:* Unused sharps that have been improperly managed or discarded should be managed as if contaminated. Both used and unused sharps present the same potential for puncture injuries; testing of improperly disposed sharps to determine the presence of infectious agents is impractical; unused sharps present the same aesthetic degradation of the environment as do used sharps. (*See:* Reinhardt and Gordon 1991, pp. 37–38.)

[c] The descriptor "contaminated" may be superfluous — many landfill authorities do not accept such parts whether exposed/contaminated or not.

TABLE 12.2
Miscellaneous Contaminated Wastes

Miscellaneous Contaminated Wastes	Examples
Wastes from surgery and autopsy	Soiled dressings, sponges, drapes, lavage tubes, drainage sets, underpads, and surgical gloves
Miscellaneous laboratory wastes	Specimen containers, slides and cover slips, disposable gloves, lab coats, and aprons
Dialysis unit wastes	Tubing, filters, disposable sheets, towels, gloves, aprons, and lab coats
Contaminated equipment	Equipment used in patient care, medical laboratories, research, and in the production and testing of certain pharmaceuticals

Source: EPA 530-SW-86-014.

- Sharps (e.g., hypodermic needles and syringes used in animal or patient care)
- Certain animal wastes
- Certain isolation wastes (e.g., wastes from patients with highly communicable diseases)
- Unused sharps (e.g., suture needles, scalpel blades, hypodermic needles)

The similarities between "infectious wastes" as listed in the 1986 Guide and "regulated medical wastes" as listed in 40 CFR 259.10 are obvious. The relevance of the "regulated medical waste" definition is doubtful unless the Subpart J program is resurrected.[4] In hope of some consistency, we will confine the discussion in this chapter to "infectious wastes" except where referenced material makes use of an alternate (*see also*: Office of Technology Assessment 1988, Chapter 1; EPA 1990, 625-7-90-009, Section 2).

INFECTIOUS WASTE MANAGEMENT

The objectives of an effective infectious waste management program should be to provide protection to human health and the environment from hazards posed by the waste. Proper management ensures that infectious waste is handled in accordance with established procedures from the time of generation through treatment of the waste (to render it noninfectious and unrecognizable) and its ultimate disposal.

An infectious waste management system should be documented in a plan[5] and should include the following elements:

- Designation/identification of infectious waste
- Segregation
- Packaging
- Labeling
- Storage
- Transport and handling
- Treatment techniques
- Disposal of treated waste
- Contingency planning
- Staff training

(*See also*: Reinhardt and Gordon 1991, pp. 13ff; Turnberg 1996, pp. 120–126; Ostler 1998, Chapter 9.)

Designation of Infectious Waste

The infectious waste plan should specify which wastes are to be managed as infectious waste. The six categories of Table 12.1 should be included if applicable. A

---

[4] Or the state or locality in which the question arises has medical waste regulations in effect.
[5] Managers or practitioners preparing an Infectious Waste Management Plan (IWMP) should consider combining the IWMP with the Exposure Control Plan required by the Bloodborne Pathogens Standard, if appropriate (*see:* Appendix A to this chapter).

FIGURE 12.1 The universal biohazard symbol.

responsible official or committee should determine which, if any, of the optional categories of Table 12.2 are to be included.

Segregation of Infectious Waste

The 1986 Guide recommends:

- Segregation of infectious waste at the point of origin
- Segregation of infectious waste with multiple hazards as necessary for management and treatment
- Use of distinctive, clearly marked containers or plastic bags for infectious wastes
- Use of the universal biological hazard symbol on infectious waste containers, as appropriate (Figure 12.1)

Also, segregation of infectious wastes assures that the added costs of special handling will not be applied to noninfectious waste.

Packaging of Infectious Waste

Infectious waste should be packaged in order to protect waste handlers and the public from possible injury and disease that may result from exposure to the waste. Accordingly, the 1986 Guide recommends:

- Selection of packaging materials that are appropriate for the type of waste handled
  - Plastic bags for many types of solid or semisolid infectious waste

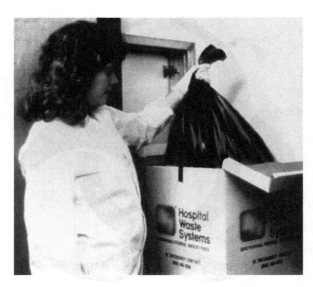

FIGURE 12.2 Red-bagged waste being placed in a rigid container for shipment.

- Puncture-resistant containers for sharps
- Bottles, flasks, or tanks for liquids
- Use of packaging that maintains its integrity during storage and transport
- Closing the top of each bag by folding or tying as appropriate for the treatment or transport
- Placement of liquid wastes in capped or tightly stoppered bottles or flasks
- No compaction of infectious waste or packaged infectious waste before treatment

Shippers of infectious waste are also subject to Department of Transportation regulations for packaging, marking, and labeling of "infectious substances" and "regulated medical waste" if shipped by commercial carriers.[6]

Figure 12.2 shows a red bag containing infectious waste being placed uncompacted in a rigid container for shipment. Figure 12.3 illustrates an infectious waste receptacle in a clinic. Figure 12.4 shows a sharps receptacle receiving a syringe and needle. Figure 12.5 illustrates transfer of sharps for transport. Figure 12.6 demonstrates handling of red-bagged wastes in rigid containers.

Storage of Infectious Waste

Storage temperature and duration are important considerations. Warmer temperatures cause higher rates of microbial growth and putrefaction, resulting in odor problems. The 1986 Guide recommends:

---

[6] *See:* 49 CFR 172.101 Hazardous Materials Table entries "Infectious Substances" and "Regulated Medical Waste;" 172.203 for proper shipping name; 172.432 and Appendix G to Part 173 for labels; 173.134 for definitions; 173.196 and 173.197 for packaging.

FIGURE 12.3 An infectious waste receptacle in a clinic.

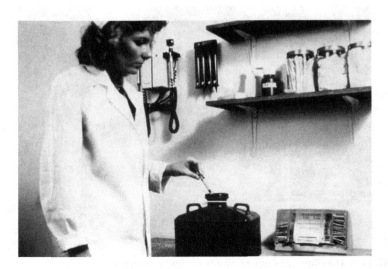

FIGURE 12.4 Sharps receptacle receiving a syringe.

FIGURE 12.5 Transfer of sharps for transport.

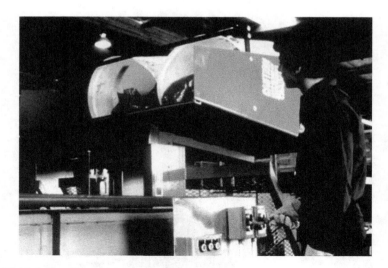

FIGURE 12.6 Handling red-bagged wastes in rigid containers.

- Locating the storage area near the treatment site
- Minimizing storage time
- Proper packing that ensures containment of infectious waste and the exclusion of rodents and vermin
- Limited access to storage area
- Posting of universal biological hazard symbol on storage area door, waste containers, freezers, or refrigerators

Transport of Infectious Waste

The 1986 Guide recommends:

- Avoidance of mechanical loading devices that may rupture packaged wastes
- Frequent disinfection of carts used to transfer wastes within the facility
- Placement of all infectious waste into rigid or semirigid containers before transport off-site
- Transport of infectious waste in closed leakproof trucks or dumpsters
- Use of appropriate hazard symbols in accord with local, state, and federal regulations

Figure 12.7 illustrates a stainless steel conveyor for movement of packaged infectious wastes within a facility. The stainless steel construction provides a smooth surface, which minimizes damage to packages and is easily disinfected in the event of a spill. Cargo space of trucks used for infectious waste transportation should be refrigerated and equipped with a spill containment system.

FIGURE 12.7 Stainless steel conveyor for movement of packaged infectious waste within a facility.

Note: The EPA does not consider the truck as a rigid containment system; rather that it serves only as a transport mechanism. Therefore, all infectious waste should be placed in rigid or semirigid, leakproof containers before being loaded on a truck.

Commercial shipments of infectious wastes are subject to the DOT regulations for a Class 6, Division 6.2 material (49 CFR 173.134). As noted earlier, the shipper is subject to requirements for assigning the correct DOT shipping name, marking, and labeling. The U.S. Postal Service (USPS) regulations entitled "Mailability of Sharps and Other Medical Devices" is published in the USPS Domestic Mail Manual (DMM), Sections 8.1 through 8.10 (39 CFR 111.1). The postal rules are primarily concerned with packaging for sharps, but shippers of infectious wastes should be aware of special requirements for packaging containing dry ice [DMM 8.10(a-c)]. Section 8.5 of the DMM provides instructions for manifesting "Infectious Substances" and a listing of authorized sources of mailing kits for sharps.

TREATMENT OF INFECTIOUS WASTE

In the 1986 Guide, the EPA defined treatment as any method, technique, or process designed to change the biological character or composition of waste. Since landfill operations may cause loss of containment integrity and dispersal of infectious waste, the EPA recommended that all infectious waste be treated prior to disposal. The Guide further recommended:

- Establishing standard operating procedures for each process used for treating infectious waste
- Monitoring of all treatment processes to assure efficient and effective treatment
- Use of biological indicators to monitor treatment (other indicators may be used provided that their effectiveness has been successively demonstrated)
- Treatment for each of the six infectious waste categories per Table 12.3
- The following treatment methods for miscellaneous contaminated wastes (when a decision is made to manage these wastes as infectious):
  - Wastes from surgery and autopsy — incineration or steam sterilization
  - Miscellaneous laboratory wastes — incineration or steam sterilization
  - Dialysis unit wastes — incineration or steam sterilization
  - Contaminated equipment — incineration, steam sterilization, or gas/vapor sterilization

Steam Sterilization

Treatment by steam sterilization is accomplished in either an autoclave or a retort. Both have a chamber in which the waste can be subjected to sterilization by saturated steam at pressures of 15 to 30 psi. The autoclave is the most commonly used steam sterilizer. A variety of designs and capacities are available. The operating and design principle is to subject the waste to the saturated steam, in the absence of air, at prescribed temperature and pressure for a sufficient time to ensure sterilization [adapted from Reinhardt and Gordon (1991, Chapter 6)]. Figure 12.8 illustrates a commercial autoclave for sterilization of infectious wastes. Larger autoclaves are

TABLE 12.3
Recommended Techniques for Treatment of Infectious Waste

Category of Infectious Waste	Recommended Treatment Technique
Isolation wastes	Steam sterilization[a]
	Incineration
Cultures and stocks of infectious agents and associated biologicals	Steam sterilization[a]
	Incineration
	Thermal inactivation[a]
	Chemical disinfection[a]
Human blood and blood products	Steam sterilization[a]
	Incineration
	Chemical disinfection[a]
	Discharge to sanitary sewer[b]
Pathological wastes	Steam sterilization[a]
	Incineration
	Handling by mortician
Contaminated animal carcasses, body parts, and bedding:	
Carcasses and body parts	Steam sterilization[b]
	Incineration
Bedding	Incineration

[a] For aesthetic reasons, steam sterilization (and other treatment methods which do not render the waste unrecognizable) should be followed by incineration, or by grinding with subsequent flushing to sewer system, or by landfilling in accord with state and local regulations.
[b] Provided secondary treatment is online and operating authorities have been notified.

Source: EPA 530-SW-86-014.

found in commercial sterilization facilities, while smaller autoclaves are used by physicians, dentists, clinics, and small hospitals. Autoclaves have the advantage of a live steam jacket which greatly reduces the amount of time needed to reach operating temperature. Disadvantages include the unchanged appearance of sterilized wastes and the unsuitability for treating recognizable body parts (Turnberg 1996). State and local regulatory agencies usually require grinding of autoclaved wastes prior to disposal or discharge to sewers in order to render the wastes unrecognizable (*see also*: EPA 1991, pp. 249ff; Reinhardt and Gordon 1991, Chapter 6; Garvin 1995, p. 118; Ostler 1998, Chapter 9).

Incineration

Incineration continues, at the time of this writing, to be a preferred treatment process for infectious waste management, although estimates of the numbers of hospital/medical/infectious waste incinerators (HMIWI) have significantly declined. In the 1994 Drum and Bulley white paper, the numbers of medical waste incinerators operating in the U.S. was estimated to be 6700, but EPA had consistently reported

FIGURE 12.8 Commercial autoclave for sterilization of infectious waste.

on "… more than 5000…." In 1997, the EPA put the number at 2400. The decline in numbers is due in great part to the promulgation of New Source Performance Standards (NSPS)(MACT) for HMIWI, effective on March 16, 1998, and Emission Guidelines for existing sources, effective on November 17, 1997 (62 FR 48347). The EPA anticipated that the Guidelines would result in the discontinued use of 50 to 80% of the 2400 then operating HMIWI.

Note: Sections 111 and 129 of the Clean Air Act (CAA) requires states with existing HMIWI, subject to the 1997 Emission Guidelines, to submit plans to the EPA that implement and enforce the guidelines. Plan submission and enforcement is optional for Native American Tribes. States having HWIWI subject to the guidelines were to have submitted plans by September 15, 1998. Sections 111(d) and 129 of the CAA require that the EPA develop and implement a Federal plan for HMIWI in jurisdictions that did not submit an approvable plan by September 15, 1999. The EPA promulgated the Federal plan in the August 15, 2000 Federal Register (65 FR 49868) and codified the rule at 40 CFR 62.

At hospitals, where most medical waste is generated, 60% of the waste classified as infectious had been managed by on-site incineration. The on-site option provides many advantages, including sterilization of pathogenic wastes and volume reductions of 90 to 95% prior to ultimate disposal. Most modern medical waste incinerators operate on "controlled-air" using two chambers. The primary chamber, into which the waste is fed, operates with restricted air flow (i.e., "starved air") at 1600 to 1800°F.[7] The waste is pyrolized, and the volatiles move to a secondary chamber where they are combusted at 1800°F or greater temperature. Excess air is provided,

---

[7] Green (1992, p. 110) advocates primary chamber temperatures of 1400 to 1600°F to minimize volatilization of metals in order to minimize the quantities of metals carried out by the fly ash.

in the secondary chamber, to ensure complete combustion. Ash is moved through and exits the primary chamber by the use of hydraulic rams or other feed devices (Reinhardt and Gordon 1991, Chapter 7). Air pollution control equipment collects particulate matter, captures trace metals and organics, and neutralizes acid gases produced in the combustion process (Drum and Bulley 1994, p. 1178).

Figure 12.9 provides a cross-sectional view of an incinerator for infectious wastes. The stack (Figure 12.10), shown producing only faintly visible vapor, should not emit visible smoke. Precision control of incinerator operation is essential (Figure 12.11).

Properly designed and operated infectious waste incinerators, with adequate emission control equipment, can achieve excellent results in terms of destruction of pathogens and organic chemicals, capture and containment of heavy metals, and reduction of volume. However, concern persists regarding atmospheric emissions of highly toxic dioxins and furans by combustion of these wastes at other than optimal feed rates, temperatures, and dwell times (EPA 1991, Chapter 2; Reinhardt and Gordon 1991, Chapter 7; Green 1992, Chapters 3–6; Drum and Bulley 1994, pp. 1177ff). The EPA has developed estimates to the effect that even though dioxin emissions from individual medical waste incinerators are quite small, the collective emissions from more than 5000 facilities[8] were the largest known source of air emissions of dioxin in the nation.

Chemical Disinfection

Disinfection by application of chemicals to contaminated materials has been practiced for many years. Similarly, chemicals have been used as a preventive application for sterilization and for disinfection of infectious or potentially infectious wastes, but the general preference of practitioners and administrators was for hospital waste incinerators. The new emission standards and guidelines have caused a trend toward other methods including chemical disinfection. Current practice usually makes use of chlorine applications such as sodium hypochlorite, accompanied by grinding or shredding and mixing to ensure contact of the disinfectant with all surfaces of the waste particles. The physical destruction of the waste is also necessary[9] to render the waste unrecognizable.

The ability of a chemical disinfectant to kill a targeted organism depends on many factors, including

- Type of microorganism
- Degree of contamination
- Type of disinfectant
- Concentration and quantity of disinfectant
- Contact time between the antimicrobial agent and the targeted organism
- Other relevant factors (e.g., pH, presence of electrolytes, or complex formation and adsorption such as binding to small molecules or ions, macromolecules, or soil) (EPA 1986; Turnberg 1996, Chapter 10).

---

[8] Again, the present number of operating infectious waste incinerators in the U.S. is greatly reduced from this number.

[9] Required by most local codes and state regulations.

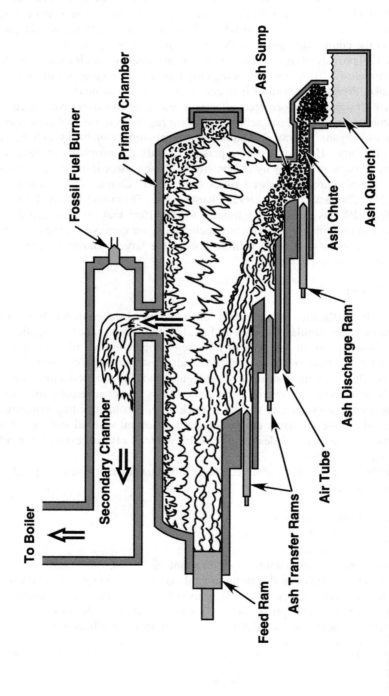

FIGURE 12.9 Cross-section of an incinerator for infectious waste.

FIGURE 12.10 Infectious waste incinerator stack showing no visible smoke.

FIGURE 12.11 Combustion controls for an infectious waste incinerator.

Antimicrobial agents are substances or mixtures of substances that are used to destroy or suppress the growth of harmful microorganisms whether bacteria, viruses, or fungi on inanimate objects and surfaces. Commercial antimicrobial products contain about 300 different active ingredients and are marketed as sprays, liquids, concentrated powders, and gases. More than 8000 antimicrobial products are currently registered[10] with the EPA and sold in the marketplace. Nearly 50% of antimicrobial products are registered to control infectious microorganisms in hospitals and other health care facilities (see also: Reinhart and Gordon 1990, pp. 116ff; Garvin 1995, p. 120; Turnberg 1996, Chapter 10; EPA 1998).

Emerging Treatment Technologies

New or alternative technologies, primarily for sterilization or disinfection of infectious waste, continue to emerge. These include units having microwave, ultraviolet, or plasma arc heating systems, ionizing radiation source material, or newer forms of chemical treatment. Self-contained microwave treatment units have recently become available commercially. The units shred and grind the waste to small, unrecognizable bits that are moistened with high-temperature steam and then pass via screw conveyor tube beneath sequential microwave generators. Temperature is maintained at 200°F during the 30-min passage. The treated material can then be landfilled. Figure 12.12 illustrates the configuration of the microwave unit.

The Stericycle, Inc. patented Electro-Thermal-Deactivation "ETD" process uses low-frequency radio waves to disinfect medical waste. Following grinding of the waste, the system applies the principles of selective absorption of energy, dipolar rotation of liquid molecules, and imposed high-voltage field to achieve *cellular lysis* (rupture) and subsequent bio-burden reduction. The extreme differential in dielectric constants (approximately 20:1) of paper, glass, and plastics vs. organic materials results in the organics selectively absorbing most of the imposed energy. The microbial cell membrane weakens and ruptures, the cell cannot reproduce, and it dies. This occurs at temperatures less than the boiling point of water and reportedly produces no liquid emission due to a recycling water system. The end product can be prepared as a specification grade RDF with or without plastics recycling. Figure 12.13 diagrams the material flow of the proprietary process (Stericycle, Inc. 2000).

Ionizing radiation is considered by the EPA to be a potentially available method for treating medical waste. The process uses a source such as cobalt 60 to destroy infectious agents. The technique has the advantages of minimal use of electrical energy and is suitable for materials that cannot be thermally treated. Disadvantages include complex technology requiring highly trained operating personnel, potential for human exposure, and difficulties associated with disposal of the decayed source material (EPA 1991, pp. 116–117). The technology was used in the U.S. for treatment of biohazardous waste by one company in the early 1990s, although the company

---

[10] As required by the Federal Insecticide, Fungicide, and Rodenticide Act (FIFRA). *Note:* If a purveyor of a chemical used for medical/infectious waste treatment makes an advertised claim of efficacy of treatment (i.e., a level of microbial disinfection, sterilization, etc.) in a specific use, such as hospital waste disinfection (i.e., surface disinfectant), the product is subject to efficacy testing requirements and registration with the EPA Office of Pesticide Registration.

FIGURE 12.12 Microwave disinfection unit. (SANITEC®, Inc., 23 Fairfield Place, West Caldwell, NJ 07006. With permission.)

later abandoned gamma radiation for another technology (Turnberg 1996, Chapter 10; *see also*: Wilson 1992, § III).

Disposal of Treated Waste

Infectious waste that has been effectively treated is no longer biologically hazardous, and may be mixed with and disposed of as ordinary solid waste, provided the waste does not pose other hazards that are subject to federal or state regulations.

The 1986 Guide recommends:

- Contacting state and local governments to identify approved disposal options
- Discharge of treated liquids and pathological wastes (after grinding) to the sewer system (Approval of the local sewer authority must be obtained.)
- Land disposal of treated solids and incinerator ash
- Rendering body parts unrecognizable before land disposal

Some states require that needles and syringes be rendered nonusable before disposal.

Contingency Planning

The infectious waste management plan should include a contingency plan to provide for emergency situations. The plan should include, but not be limited to, procedures to be used under the following circumstances:

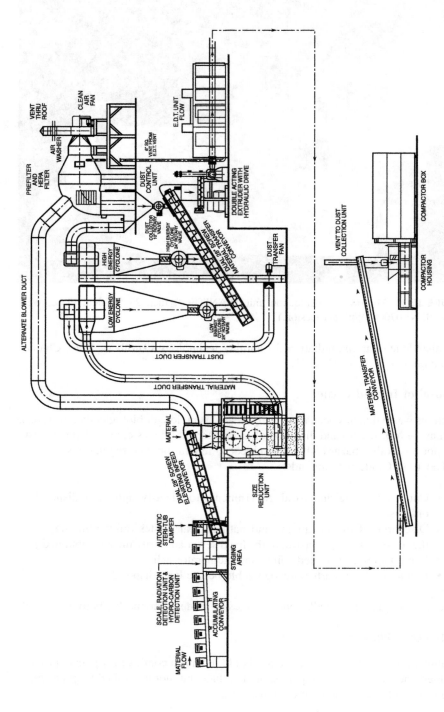

FIGURE 12.13 Electro Thermal Deactivation unit. (Stericycle, Inc., 2816 N. Keith Drive, Lake Forest, IL 60045. With permission.)

- Spills of liquid infectious waste — cleanup procedures, protection of personnel, and disposal of spill residue
- Rupture of plastic bags (or other loss of containment) — cleanup procedures, protection of personnel, and repackaging of waste
- Equipment failure — alternative arrangements for waste storage and treatment (e.g., off-site treatment)

(*See also:* Giroletti and Lodola 1994 concerning hospital waste management in Italy and the European Union.)

Regulatory and Advisory Considerations

As noted earlier, the nonrenewal of the MWTA left formal regulation of infectious waste to the state and local governments. However, the practice of infectious waste management is impinged by a variety of mandatory and discretionary rules, guidelines, standards, and professional practice criteria, in addition to the state and local regulatory structures. A few of the criteria and a reference for each are briefly noted:

Department of Agriculture (DOA), Animal and Plant Health Inspection Service
- Animals and Animal Products (9 CFR 1-199)

Department of Energy (DOE), Nuclear Regulatory Commission (NRC)
- Standards for Protection Against Radiation (10 CFR 20)
- Licensing Requirements for Land Disposal of Radioactive Waste (10 CFR 61)

Department of Defense (DoD), Department of the Army (DA)
- Biological Defense Safety Program (32 CFR 626)
- Biological Defense Safety Program Technical Safety Requirements (DA Pamphlet 385-69)

Environmental Protection Agency (EPA)
- EPA Guide for Infectious Waste Management, 1986 (EPA 530-SW-86-014)
- Emission Guidelines for Existing HMIWI 1997 (40 CFR 60, Subpart C; 62 FR 48348)
- State Plan Requirements for Implementation of Subpart C Guidelines (40 CFR 62; 65 FR 49868)
- General Pretreatment Regulations for Existing and New Sources of Pollution (40 CFR 403)
- New Source Performance Standards for New HMIWI, 1997 (40 CFR 60, Subpart E; 62 FR 48348)

Department of Health and Human Services (HHS), Food and Drug Administration (FDA)[11]

---

[11] The Food Quality Protection Act of 1996 transferred a number of traditional FDA functions having to do with human health protection to the EPA Office of Pesticide Programs.

- Good Laboratory Practice for Nonclinical Laboratory Studies (21 CFR 58)
- Public Health Service (PHS) Interstate Shipment of Etiological Agents (42 CFR 72)
- Recommendations for Prevention of HIV Transmission in Health-Care Settings, 1987, USPHS Centers for Disease Control (CDC)

Department of Labor (DOL), Occupational Health and Safety Administration (OSHA)
- Occupational Exposure to Bloodborne Pathogens Final Standard, Revised 1996 (29 CFR 1910.1030)[12]
- Occupational Exposure to Bloodborne Pathogens — Precautions for Emergency Responders, OSHA 3130 (Revised), 1998

U.S. Postal Service (USPS)
- Mailability of Sharps and Other Medical Devices, 1993, 8.1-8.10, USPS Domestic Mail Manual (DMM) (39 CFR 111.1)

U.S. Department of Transportation (DOT), Research and Special Projects Administration (RSPA)
- Regulations for Transportation of Hazardous Materials (49 CFR 171-178)[13]
- Coast Guard (USCG), Vessels Carrying Oil, Noxious Liquid Substances, Garbage, Municipal or Commercial Waste, and Ballast Water (33 CFR 151)

American National Standards Institute (ANSI)
- Standard for Safety for Alternative Treatment Technologies for the Disposal of Medical Waste, UL 2334 [being developed by the Underwriters Laboratories for ANSI, after consultation with the State and Territorial Association on Alternate Treatment Technologies (STAATT)]

Staff Training

Facilities that generate infectious waste should provide employees with infectious waste management training. This training should include an explanation of the infectious waste management plan and assignment of roles and responsibilities for implementation of the plan. Such education is important for all employees who generate or handle infectious wastes regardless of the employee's role or type of work.

Training programs should be implemented:

- When the infectious waste management plans are first developed and instituted

---

[12] A summary of key provisions of the OSHA Bloodborne Pathogens Final Standard (29 CFR 1910.1030) may be found in Appendix A of this chapter.

[13] Persons responsible for commercial shipment of infectious wastes or for preparation of infectious wastes for commercial shipment are required to successfully complete the HM 181 training described in HM 126(f) (*see:* 49 CFR 172.704).

- When new employees are hired
- Whenever infectious waste management practices or regulatory requirements are changed

Continuing education is also an important part of staff training. Refresher training
aids in maintaining awareness of the potential hazards posed by infectious waste.
Training also serves to reinforce waste management policies and procedures that are
detailed in the infectious waste management plan (EPA 530-SW-86-014, Chapter 3;
see also: EPA 530-SW-86-014, Chapters 4 and 5; Office of Technology Assessment
1988, Chapter 3; EPA 625-7-90/009, Section 2; Boecher et al. 1989; Keene 1989;
Reinhardt and Gordon 1991, Chapter 16; Drum and Bulley 1994; Garvin 1995,
Section III; Turnberg 1996).

TOPICS FOR REVIEW OR DISCUSSION

1. Red-bagged wastes are usually placed in a rigid container for shipment.
   Why is this necessary?
2. How do criteria for ultimate disposal of RCRA hazardous wastes differ
   from those for infectious waste?
3. Why are sharps considered so dangerous to infectious waste handlers?
   Why must unused discarded sharps be managed as if they had been used?
4. The van body of an infectious waste transport truck is not acceptable as
   a "rigid container." Why not?
5. What is meant by "starved air" incineration as in infectious waste incineration? Why is it considered good design for an infectious waste incinerator?
6. What is the major concern regarding atmospheric emissions from infectious waste incinerators?

APPENDIX A
Bloodborne Pathogens Final Standard: Summary of Key Provisions

Fact Sheet No. OSHA 92-46

Purpose: Occupational exposure to blood and other potentially infectious materials is limited since any exposure could result in transmission of bloodborne pathogens which could lead to disease or death.

Scope: All employees who could be "reasonably anticipated" as the result of performing their job duties to **face contact with blood** and other potentially infectious materials are covered. OSHA has not attempted to list all occupations where exposures could occur. "Good Samaritan" acts such as assisting a co-worker with a nosebleed would not be considered occupational exposure. Infectious materials include semen, vaginal secretions, cerebrospinal fluid, synovial fluid, pleural fluid, pericardial fluid, peritoneal fluid, amniotic fluid, saliva in dental procedures, any body fluid visibly contaminated with blood, and all body fluid in situations where it is difficult or impossible to differentiate between body fluids. They also include any unfixed tissue or organ other than intact skin from a human (living or dead) and human immunodeficiency virus (HIV)-containing cell or tissue cultures, organ cultures, and HIV or hepatitis B (HBV)-containing culture medium or other solutions as well as blood, organs, or other tissues from experimental animals infected with HIV or HBV.

Exposure Control Plan: Employers are required to **identify, in writing,** tasks and procedures as well as job classifications **where occupational exposure to blood occurs** — without regard to personal protective clothing and equipment. It must also set forth the **schedule for implementing other provisions** of the standard and specify the procedure for evaluating circumstances surrounding exposure incidents. The plan must be accessible to employees and available to OSHA. Employers must review and update it at least annually — more often, if necessary to accommodate workplace changes.

Methods of Compliance: Universal precautions (treating body fluids/materials as if infectious) **emphasizing engineering and work practice controls** are mandated. The standard stresses handwashing and requires employers to provide facilities and ensure that employees use them following exposure to blood. It sets forth procedures to minimize needlesticks, minimize splashing and spraying of blood, ensure appropriate packaging of specimens and regulated wastes, and to decontaminate equipment or label it as contaminated before shipping to servicing facilities.

Employers must provide, at no cost, and require employees to use appropriate **personal protective equipment** such as gloves, gowns, masks, mouthpieces, and resuscitation bags and must clean, repair, and replace these when necessary. Gloves are not necessarily required for routine phlebotomies in volunteer blood donation centers, but must be made available to employees who want them.

The standard requires a **written schedule for cleaning,** identifying the method of decontamination to be used in addition to cleaning following contact with blood or other potentially infectious materials. It specifies methods for disposing of contaminated sharps and sets forth standards for containers for these items and other regulated waste. Further, the standard includes provisions for handling contaminated laundry to minimize exposures

HIV and HBV Research Laboratories and Production Facilities: These facilities are to follow **standard microbiological practices** and additional practices are specifically intended to minimize exposures of employees working with concentrated viruses and reduce the risk of accidental exposure for other employees at the facility. These facilities must include required containment equipment, and an autoclave for decontamination of regulated waste must be constructed to limit risks and enable easy cleanup. **Additional training and experience requirements** apply to workers in these facilities.

Hepatitis B Vaccination: Vaccinations are required to be made **available to all employees who have occupational exposure to blood** within 10 working days of assignment, at no cost, at a reasonable time and place, under the supervision of licensed physician/licensed health care professional, and according to the latest recommendations of the U.S. Public Health Service (USPHS). **Prescreening may not be required** as a condition of receiving the vaccine. Employees must sign a **declination form** if they choose not to be vaccinated, but may later opt to receive the vaccine at no cost to the employee. Should booster doses later be recommended by the USPHS, employees must be offered them.

APPENDIX A   *(Continued)*
Bloodborne Pathogens Final Standard: Summary of Key Provisions

Post-Exposure Evaluations and Follow-Up: Procedures to be made **available to all employees who have had an exposure incident** are specified. In addition, any laboratory tests must be conducted by an accredited laboratory at no cost to the employee. Follow-up must include a **confidential medical evaluation** documenting the circumstances of exposure, identifying and testing the source individual if feasible, testing the exposed employee's blood if he/she consents, post-exposure prophylaxis, counseling, and evaluation of reported illnesses. Healthcare professionals must be provided specified information to facilitate the evaluation and their written opinion on the need for hepatitis B vaccination following the exposure. Information such as the employee's ability to receive the hepatitis B vaccine must be supplied to the employer. All diagnoses must remain confidential.

Hazard Communication: Warning labels including the **orange or orange-red biohazard symbol** affixed to containers of regulated waste, refrigerators and freezers, and other containers that are used to store or transport blood or other potentially infectious materials are required. **Red bags** or containers **may be used** instead of labeling. When a facility uses universal precautions in its handling of all specimens, labeling is not required within the facility. Likewise, when all laundry is handled with universal precautions, the laundry need not be labeled. Blood that has been tested and found free of HIV or HBV and released for clinical use, and regulated waste that has been decontaminated, need not be labeled. Signs must be used to **identify restricted areas** in HIV and HBV research laboratories and production facilities.

Information and Training: Training within 90 days of effective date, **initially** upon assignment, are mandated and **annually.** Employees who have received appropriate training within the past year need only receive additional training in items not previously covered. Training must include making accessible a copy of the regulatory test of the standard and explanation of its contents, general discussion on bloodborne diseases and their transmission, an exposure control plan, engineering and work practice controls, personal protective equipment, hepatitis B vaccine, response to emergencies involving blood, how to handle exposure incidents, the post-exposure evaluation and follow-up program, and signs/labels/color-coding. There must be **opportunity for questions and answers,** and the **trainer must be knowledgeable** in the subject matter. **Laboratory and production facility workers** must receive **additional specialized initial training.**

Record Keeping: Medical records are to be kept for each employee with occupational exposure for the **duration of employment plus 30 years,** they must be **confidential,** and they must include name and social security number; hepatitis B vaccination status (including dates); results of any examinations, medical testing, and follow-up procedures; a copy of the healthcare professional's written opinion; and a copy of information provided to the healthcare professional. Training records must be maintained for 3 years and must include dates, contents of the training program or a summary, the trainer's name and qualifications, and names and job titles of all persons attending the sessions. Medical records must be made **available to the subject employee,** anyone with written consent of the employee, OSHA, and NIOSH — they are not available to the employer. Disposal of records must be in accord with OSHA's standard covering access to records.

Dates: Important dates are **effective date:** March 6, 1992; **exposure control plan:** May 5, 1992; **information and training requirements and record keeping:** June 4, 1992. The following **other provisions** took effect on July 6, 1992: engineering and work practice controls, personal protective equipment, housekeeping, special provisions covering HIV and HBV research laboratories and production facilities, hepatitis B vaccination and post-exposure evaluation and follow-up, and labels and signs.

REFERENCES

Boecher, Frederick W., David C. Guzewich, and Michael H. Diem. 1989. "Infectious Waste Management at Army Health Care Facilities, Past and Present," *Hazardous Materials Control,* November-December, pp. 73ff.

Drum, Donald A. and Mike Bulley. 1994. Medical Waste Disposal, White Paper, Medical Waste Committee (WT-3), Air & Waste Management Association, *Journal of Air & Waste Management Association,* October, pp. 1176ff.

Garvin, Michael L. 1995. *Infectious Waste Management: A Practical Guide.* CRC Press, Boca Raton, FL.

Giroletti, E. and L. Lodola. 1994. "Medical Waste Treatment," in *Technologies for Environmental Cleanup: Toxic and Hazardous Waste Management,* A. Avogadro and R. C. Ragaini, Eds., Kluwer, Dordrecht, The Netherlands.

Green, Alex E. S. 1992. *Medical Waste Incineration and Pollution Prevention.* Van Nostrand Reinhold, NY.

Jenkins, Pamela R. 1990. *AIDS Infection Control, and the Effective Management of Medical Waste.* Environmental Resource Center, Fayetteville, NC.

Johnson, Jeff. 1995. "Incinerators Targeted by EPA." *Environmental Science and Technology,* January, pp. 33ff.

Keene, John H. 1989. "Medical Waste Management: Public Pressure Versus Sound Medicine." *Hazardous Materials Control,* September-October, pp. 29ff.

Ostler, Neal K. 1998. "Medical Waste Management," Chapter 9, in *Waste Management Concepts,* Neal K. Ostler and John T. Nielsen, Eds., Prentice-Hall, Upper Saddle River, NJ.

Reinhardt, Peter A. and Judith G. Gordon. 1991. *Infectious Medical Waste Management.* Lewis Publishers, Chelsea, MI.

Stericycle, Inc. 2000. Proprietary Material — Patented Electro Thermal Deactivation Process, 2816 North Keither Drive, Lake Forest, IL 60045.

Turnberg, Wayne L. 1996. *Biohazardous Waste Risk Assessment, Policy, and Management.* John Wiley & Sons, NY.

U.S. Congress, Office of Technology Assessment. 1988. Issues in Medical Waste Management — Background Paper, Superintendent of Documents, U.S. Government Printing Office, Washington, D.C., OTA-BP-O-49.

U.S. Department of Labor, Occupational Health and Safety Administration. 1992. Bloodborne Pathogens Final Standard: Summary of Key Provisions, Fact Sheet No. OSHA 92-46, Washington, D.C.

U.S. Environmental Protection Agency. 1986. EPA Guide for Infectious Waste Management, Office of Solid Waste and Emergency Response, Washington, D.C., EPA 530-SW-86-014.

U.S. Environmental Protection Agency. 1990. Guides to Pollution Prevention: Selected Hospital Waste Streams, Center for Environmental Research Information, Cincinnati, OH, EPA 625-7-90-009.

U.S. Environmental Protection Agency. 1990. Medical Waste Management in the United States — Second Interim Report to Congress, Solid Waste and Emergency Response, Washington, D.C.

U.S. Environmental Protection Agency et al. 1991. Medical Waste Management and Disposal, Noyes Data Corporation, Park Ridge, NJ.

U.S. Environmental Protection Agency. 1998. Antimicrobial Pesticides, Office of Pesticide Programs, Washington, D.C., EPA 735-F-98-023.

U.S. Environmental Protection Agency. 1998. Highlights of the Food Quality Protection Act of 1996, Office of Pesticide Programs, Washington, D.C., EPA 735-F-98-029.

Wilson, Michaelle D. 1992. Questionnaire on Hospital Wastes Management, Office of Pesticide Management, U.S. Environmental Protection Agency, Open Document, Washington, D.C.

13 Radioactive Waste Management

OBJECTIVES

At completion of this chapter, the student should:

- Be conversant with basic radioactivity, uses of nuclear energy, and problems of nuclear waste contamination.
- Understand the basic physiological and human health effects of penetrating ionizing radiation and approaches to protection from exposure for workers and the public.
- Have an understanding of the magnitude of the nuclear waste management problem in the U.S., the causes, and the impediments to timely remedy thereof.
- Be conversant with the four separate and distinct types of radioactive wastes and with the management.
- Be similarly conversant with the generation, special problems of management, and regulatory requirements imposed upon handlers of "mixed wastes."

INTRODUCTION

U.S. News and World Report (March 27, 1989) featured an article entitled "Uncle Sam's Toxic Folly." The lead paragraph begins:

Cleaning up radioactive and chemical waste at the nation's nuclear weapons plants and military installations looms as the biggest, toughest and most expensive task of ecological restoration in American history. It presents technical challenges equal to the Apollo moon landing and space shuttle programs, and it will cost roughly as much, about $130 billion ...

A decade later, significant progress has been made in the cleanup of some sites, but the problem and the challenges of restoration of many sites is even greater than envisioned in 1989. The cost estimates are now two and three times greater, and overall progress is disappointing. Nevertheless, new radiological waste management technologies, together with the long-awaited operational status of high-level and transuranic waste disposal sites, are expected to eventually facilitate progress in the cleanup effort. In the following sections, some of the current and planned remediation projects and technologies are overviewed and references to more detailed information are provided. It may be useful to briefly review some of the history, perspectives, and imperatives that brought the nation and the world to the present set of circumstances regarding nuclear energy and radioactive wastes.

In 1895, William Konrad Roentgen, professor of physics at the University of Wurtzburg, showed that the X-rays he had discovered could penetrate matter that was impervious to ordinary light and could produce fluorescence in various substances, such as glass and calcite (Pauling 1958, p. 63). In 1896, the French physicist, Henri Becquerel, discovered that minerals containing uranium gave off rays that were capable of:

- Penetrating black paper and blackening a photographic plate
- Producing fluorescence in certain substances (zinc sulfide and barium platinocide)
- "Ionizing" air and other gases and discharging on an electroscope
- Passing through plates of metal

He called these rays "Becquerel rays" and observed that they were similar to X-rays.

During the next few years, Madame Marie Curie discovered that thorium and its compounds possess properties similar to those of uranium and its compounds. The name "radioactivity" was coined and applied to these extraordinary properties. Madame Curie and her husband, Professor Pierre Curie, working with pitchblende[1] discovered polonium and by 1910 had isolated a new element that was at least 1 million times as active as uranium. They called the new element *radium* (Foster and Alyea 1947, p. 295ff).

Thereafter, the development of nuclear technology progressed through refinement of the X-ray, illumination of timepiece dials, development of medical and industrial tracers, treatment of cancers and allied diseases, fission and fusion weapons, and power generation. These developments brought about unprecedented medical advances, saved thousands of lives by significantly shortening World War II, played a major role in the post-World War II industrial miracles, probably prevented World War III, provided potentially the most environmentally benign source of electrical power, and (to the subject of this chapter) saddled the major world powers with nuclear waste management problems of staggering proportions.

Radioactive waste ("radwaste") management is not a new problem. It began with the Manhattan Project[2] and was recognized in global terms during the first conference on Peaceful Uses of Atomic Energy in Geneva in 1955. However, environmental

---

[1] Pitchblende is an ore containing uraninite and uranium (U_3O_8) which is found in Bohemia, a region and former province of western Czechoslovakia.

[2] Technically, the "Manhattan District," a U.S. Army Corps of Engineers unit, established in 1942, to administer the project that produced the first nuclear bombs.

concerns with nuclear energy and weapons were focused on the anticipated "nuclear winter," which was expected to follow a nuclear war, testing of nuclear weapons, and fears of accidents at nuclear power generation stations. Meanwhile, nations, particularly the U.S. and the Soviet Union, accumulated a massive amount of nuclear waste. In 1989, U.S. Department of Energy (DOE) officials spoke of some radwaste sites being so severely contaminated that abandonment as "national sacrifice zones" might be necessary. In following years, DOE representatives were unwilling to discuss such a concept, but have recently openly discussed sensitive topics such as levels of cleanup criteria linked to "restricted use" of contaminated sites.

In March 1995, the DOE released the Environmental Management 1995 report and an Executive Summary entitled Estimating the Cold War Mortgage. These documents spoke clearly of sites which could be remediated only partially. In several cases, the technology did not and does not exist to achieve even that end. Some of these sites will remain closed to public access; others will be suitable only for restricted use; a few have radioactive contaminants that cannot presently be removed from groundwater. By 1995, the entire operation was expected to require 75 years, and the "mid-range" cost estimate had grown to $230 billion (DOE 1995a,b). Now, in the new century, even more pessimistic estimates are seen and heard.

During the past, present, and next decades, many thousands of engineers, scientists, technicians, administrators, and project managers have been and will be engaged in the radioactive waste cleanup effort. Accordingly, space is devoted here to provide an overview of the problem and the effort to manage it. The topic and subtopics occupy millions (probably billions) of pages in the technical, scientific, professional, diplomatic, defense, and electric power industry literature. It is obviously impossible to deal with the material in depth, however, references are provided in areas critical to the hazwaste/radwaste practitioner.

BACKGROUND

How is it possible that the U.S. could let such a problem grow to such proportions? The history is complex and should not be oversimplified, but four factors seem to stand out.

1. The Defense/Energy establishment's first priority, for three decades, was the development, production, and modernization of nuclear weapons. The work was carried out behind a wall of secrecy that made it possible for environmental considerations to be "postponed," while weapons imperatives were pursued.
2. The nature of the waste is that it is not biodegradable. It is not destroyed by incineration or other conventional treatment techniques. It began accumulating during times in which burial was considered good waste management.

During the first three decades of the nuclear era, scientists, regulators, and promoters of nuclear power tended to view waste management as a technical problem for which modern technology would provide a solution … It was not until the late 1970s that the federal government allocated substantial funds and personnel to develop a plan for the long-term management of nuclear wastes (League of Women Voters 1985).

---

3. During the 1970s, President Carter, expressing proliferation concerns, imposed bans on commercial reprocessing of spent nuclear fuel rods. The ban had the effect of causing significantly more high-level radwaste to accumulate than would have been the case without the ban. In 1981, President Reagan lifted the ban, but the industry has declined to make the commitment to reprocessing because (a) of uncertainty about future governmental policies regarding reprocessing and (b) uranium had become so cheap that industry could not afford to reprocess spent fuel rods.
4. As will be discussed, attempts to develop permanent or temporary repositories are hamstrung by public outcry, political power, court decisions, and technical difficulties. Each delay is compounded in further delays, exponentially higher costs, and ever-increasing frustration on the part of all involved.

With only very limited reprocessing capability, and no long-term storage and disposal for high-level waste yet available, the nation's effort to manage high-level waste and spent nuclear fuel (SNF) is near chaos.[3] After years of wrangling and frustration, the WIPP facility[4] is finally receiving transuranic waste. The nation must soon create additional storage space for spent nuclear fuel or begin shutting down nuclear-power generating facilities. Cleanup operations at weapons facilities cannot proceed without repositories for high-level waste. The most promising scenario for keeping former Soviet Union warheads and weapons materials out of terrorist hands is to bring the warheads to the U.S. for demilitarization, but facilities are overburdened with U.S. weapons slated for demilitarization. The numbers and complexities of the issues that attend radwaste management are such that public attitudes and political postures are shaped by fears — legitimate and unfounded. For an excellent overview of nuclear waste management and the attendant cross-currents and uncertainties, see *Understanding Radioactive Waste, Fourth Edition* (Murray 1994).

THE NATURE, EFFECTS, AND MEASUREMENT OF RADIOACTIVITY

Radioactivity

Some atoms are unstable (radioactive) and undergo a spontaneous decay process, emitting radiation until they reach a stable form. Such atoms are called radioisotopes.

---

[3] The DOE projects that cumulative spent fuel discharges will exceed reactor pool capacity by 2333 metric tons in 2000. Aditional storage needs will total 10,686 metric tons by 2010, double by 2020, and plateau at 25,244 metric tons in 2039 (Congressional Research Service 1996).
[4] *See:* Later section on the Waste Isolation Pilot Plant (WIPP).

TABLE 13.1
Half-lives of Some Radioisotopes

Radioisotope	Half-life
Iodine-132	2.4 hours
Rhodium-105	36.0 hours
Xenon-133	5.3 days
Barium-140	12.8 days
Cerium-144	284 days
Cesium-137	30 years
Carbon-14	5,730 years
Uranium-234	250,000 years
Uranium-235	704,000,000 years[a]
Uranium-238	4,470,000,000 years[a]
Helium-4	12,500,000,000 years

[a] Tang and Saling (1990, p. 20).

Source: Adapted from Enger et al. (1989).

The decay process may last from a fraction of a second to billions of years depending upon the type of atom. The rate of radioactive decay is measured in half-lives, the time required for half the atoms in a sample to spontaneously decay to another form. Each isotope has its own half-life, i.e., the amount of time required for half of the atoms in a sample to decay. Table 13.1 shows the variability in half-lives of some atoms.

Nuclear energy is released by the processes of fission and fusion. During fission, the nucleus of an atom is split into two smaller nuclei, called fission products, releasing neutrons, radiation, and heat in the process. The released neutrons can cause nearby atoms to split, and if sufficient fissionable material is present, a chain reaction can begin. Such a chain reaction generates heat from the fission process and from the decay of radioactive products. An uncontrolled nuclear chain reaction can progress to an atomic explosion (adapted from Office of Technology Assessment 1985). The fission process is illustrated in Figure 13.1.

The fusion process involves the combination of small atomic nuclei to form more massive nuclei, instability of one or both nuclei, and the simultaneous release of energy. Figure 13.2 illustrates three possible types of fusion.

The fission process is harnessed to provide intense heat for steam generation, which in turn powers turbine generators in a nuclear power facility (Figure 13.3). Both fusion and fission processes are employed in nuclear weapons to create massive uncontrolled chain reactions.

Types of Radiation

The radioactive isotopes found in radwaste emit three types of penetrating ionizing radiation — alpha (α) and beta (β) particles and gamma (γ) rays. Radioactivity is a process in which a nucleus spontaneously disintegrates or "decays," resulting in a

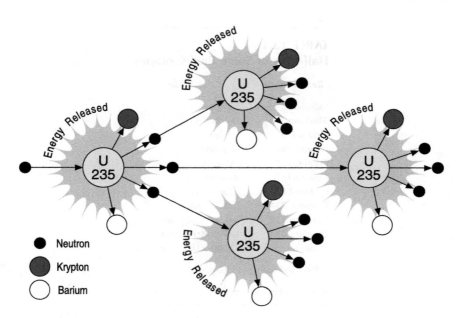

FIGURE 13.1 The nuclear fission process. (Adapted from Eldon D. Enger, J. Richard Kormelink, Bradley F. Smith, and Rodney J. Smith, *Environmental Sciences: The Study of Interrelationships*, Wm. C. Brown Publishers, Dubuque, IA. With permission of the McGraw-Hill Companies.)

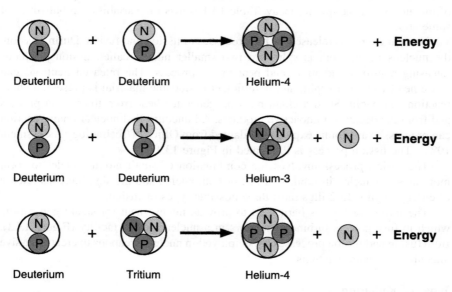

FIGURE 13.2 The nuclear fusion process. (Adapted from Eldon D. Enger, J. Richard Kormelink, Bradley F. Smith, and Rodney J. Smith, *Environmental Sciences: The Study of Interrelationships*, Wm. C. Brown Publishers, Dubuque, IA. With permission of the McGraw-Hill Companies.)

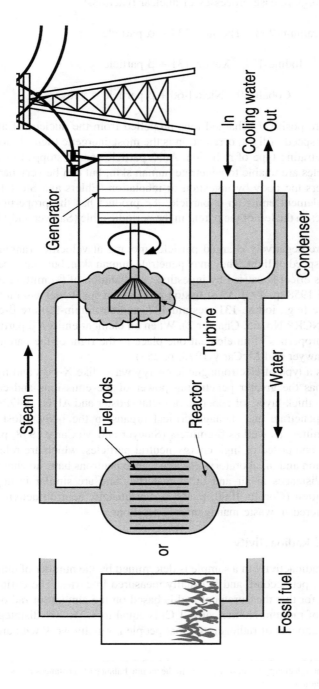

FIGURE 13.3 Nuclear power generation. (Adapted from Eldon D. Enger, J. Richard Kormelink, Bradley F. Smith, and Rodney J. Smith, *Environmental Sciences: The Study of Interrelationships*, Wm. C. Brown Publishers, Dubuque, IA. With permission of the McGraw-Hill Companies.)

release of one or more types of ionizing radiation.[5] The following are examples of the decay and energy release processes of nuclear reactions:

Uranium-238   Thorium-234 + α particle

Iodine-131   Xenon-131 + β particle

Cobalt-60   Nickel-60 + β particle + γ rays

α *particles* are positively charged ions propelled from the nucleus of atoms at about 10% of the speed of light. α radiation is the most energetic (densely ionizing) but the least penetrating type of radiation. An α particle can be stopped by a sheet of paper. α particles are unable to penetrate human skin, but can be very harmful if an α-emitter enters the body by ingestion or inhalation (Ehlers and Steel 1958, p. 490). When an element emits an α particle, the product has the properties of an element two places to the left of the parent in the periodic table (Sawyer and McCarty 1978, p. 258).

β *particles* are negatively charged particles moving at velocities ranging from 30 to 99% of the speed of light. They may penetrate human skin, but like α particles, their most serious effect is caused by ingestion or inhalation of β-emitting isotopes (Ehlers and Steel 1958, p. 490). Most fission products in spent-fuel assemblies and reprocessed waste (e.g., iodine-131, cesium-137, and strontium-90) are β-emitters (NCRPM 1979, NCRP No. 65, Chapter 2). When an element emits a β particle, the product has the properties of an element one place to the right of the parent in the periodic table (Sawyer and McCarty 1978, p. 258).

γ *radiation* is a type of electromagnetic energy wave, like X-rays and light and radio waves. It has the greater penetrating power of the emissions and can pass through relatively thick layers of concrete or metal (Foster and Alyea 1947, p. 299). γ radiation can penetrate and damage critical organs in the body. Most fission products are γ-emitters as well as β-emitters (Sawyer and McCarty 1978, p. 266).

Neutrons are composed of high-energy neutral particles, which are released in a nuclear detonation and in laboratory research. Since neutrons have no charge, they can travel long distances in air and other materials and are similar in degree of hazard to γ radiation (Corbitt 1990, p. 9.87). Fortunately, neutron activity is not normally encountered in waste management operations.

Measurement of Radioactivity

The intensity of radioactivity in a sample is determined by the number of emissions, or disintegrations, per second, and is usually measured in curies. The curie (Ci) is the standard unit for this measurement and is based on the amount of radioactivity contained in 1 g of radium. Numerically, 1 Ci is equal to 3.7×10^{10} disintegrations per second. The amounts of radioactivity that people normally work with are in the

---

[5] Radiation that has enough energy to cause a change in the atomic balance of substances it passes through is called ionizing radiation.

millicurie (1 thousandth of a curie) or microcurie (1 millionth of a curie) range. Elements with shorter half-lives (e.g., thorium-234 at 24.1 days) are more radioactive than those with longer half-lives (e.g., uranium-238 at 4.5 billion years).

Radiation exposure is measured in rems,[6] a unit that indicates the amount of radiation received and the biological implications of the exposure. In 1 year, the average person in the U.S. is exposed to approximately 160 mrems (thousandths of a rem) of radiation, two thirds of which comes from natural background sources such as mineral ores, cosmic radiation from outer space, and the radioactive carbon and potassium found in most living things. Slightly less than one third of this annual exposure comes from medical sources (i.e., X-rays) (adapted from Office of Technology Assessment 1985, pp. 21–23).

Human Health Effects of Exposure to Radiation

Radiation is converted to other forms of energy when it is absorbed by matter. Because of this energy conversion, damage occurs at a cellular, tissue, organ, or organism level when organisms are irradiated. Radiation effects on man are classified as *somatic* or *genetic*. Somatic effects are those that cause damage to the exposed individual and include anemia, fatigue, loss of hair, cataracts, skin damage, and cancer. Genetic effects include inheritable changes resulting from mutations in reproductive cells (Sawyer and McCarty 1978, p. 266). The degree and kind of damage varies with the kind and amount of radiation, the duration of the exposure, and the particular type of cells irradiated.

Ingestion and inhalation are frequent forms of chronic exposure. Several historic cases provide classic examples.

- Madame Curie was felled by cancer, at the age of 47, due to her exposure while working with radioactive substances. Irene Joliet-Curie, who continued her mother's research, also died with cancer at age 59.
- Workers in clock factories, during the period of 1915 to 1935, painted the numerals and hands on clocks with fluorescent radium to give them night visibility. The workers twirled the paint brushes on their tongues to provide a very sharp point. The workers experienced very high incidences of bone sarcoma and carcinomas of the head and paranasal sinuses (NAS 1972, p. 126ff; *see also:* Martland and Humphries 1929).
- The high incidence of lung cancer among uranium miners has been widely reported and documented. The incidence is most strongly associated with miners who have the highest exposure to radon and radon daughters and who are also cigarette smokers (NAS 1972, p. 146).

---

[6] *Rem:* 10 CFR 20.1004 defines the rem as "… a measure of the dose of any ionizing radiation to body tissues in terms of its estimated biological effect relative to a dose of one roentgen of X-rays." In more practical terms, a rem is the amount of radiation that is required to produce the same biological effect as one roentgen of gamma or X-radiation. (For further discussion, *see:* Sawyer and McCarty 1978, p. 259; Corbitt 1990, pp. 9.87ff; Meyer 1989, p. 480; Murray 1994, pp. 20–21).

An acute radiation dose — 50 rems or more over a 24-h period — results in radiation sickness within 1 h to several weeks. The chance of death is nearly 100% from a dose greater than 1000 rems; 90 to 100% from 600 to 1000 rems; and 50% from 400 rems. Survival is almost certain if the dose is 200 rems or less.

Other consequences range from gastrointestinal and circulatory system disorders to long-term effects such as cancer, birth abnormalities, genetic defects, and poor general health. Long-term effects also result from chronic exposure to low-level radiation. In radioactive waste disposal, the concern centers on the possibility of such chronic low-level exposure caused by releases of radioactive waste (adapted from Office of Technology Assessment 1985, p. 21; *see also:* Miller and Majumdar 1985; Tang and Saling 1990, Chapter 2; Murray 1994, Chapter 5).

RADIATION PROTECTION

A basic understanding of the nature of radioactivity and protection of human health and the environment from adverse impacts of radioactivity is a major thrust of this chapter. However, we devote only limited time and space under this heading to the topic of protection. As shown in the next section of this chapter, there are actually *four* types of radioactive waste and four separate and distinct radioactive waste management problems. Each has its own set of standards, regulations, and practices, including waste management. A few concepts and principles apply to all or most of the four types. We now take up those concepts.

Permissible Dose Concepts and Applications

The International Commission on Radiological Protection (ICRP)[7] has since 1928 issued numbered publications attempting to define "permissible dose" in workable terms. The task was hampered initially by the absence of thorough understanding of the effects of radioactivity and later by controversy in the interpretation of research findings. The Commission has consistently held that a linear, no threshold dose response[8] relation is likely to be a good approximation of the true condition at low doses. In other words, not even the smallest dose of radiation is regarded as entirely safe. The 1990 recommendations, contained in ICRP Publication 60, provide the following:

> ... reflected new data that indicated a probably higher risk of stochastic late harm per unit dose than previously assumed. Because of that increased risk estimate, the 1990 Recommendations also reduced the dose limits from 50 to 20 mSv for workers and from 5 to 1 mSv for members of the public (both 5-year averages).[9] It is important that reduced dose limits were not regarded as a tool to reduce doses in general; it just meant that the border line between barely tolerable and always unacceptable moved down. Instead, reduction of doses in general (much below the dose limits) came about because of the increased risk figures. They meant that measures that had previously

---

[7] The ICRP is an independent international network of specialists in various fields of radiological protection. At any one time, about 100 eminent scientists are actively involved in the work of the ICRP (Valentin 2000).

[8] *See:* Dose-Response Evaluation, Chapter 4.

[9] 1 sievert (Sv) = 100 rems; 1 millisievert (mSv) = 0.1 rem (*see:* Glossary).

appeared disproportionately expensive now became reasonable alternatives in optimisation (Valentin 2000).

The work of the ICRP, as well as that of the U.S. counterpart, the National Council on Radiation Protection and Measurement (NCRP), together with the National Research Council (NRC), the National Academy of Sciences (NAS), the U.S. Nuclear Regulatory Commission (USNRC), and the DOE, are reflected in the regulations for occupational exposure to ionizing radiation. Occupational groups are limited by 10 CFR 20 to permissible doses per year, as follows:

- Total effective dose equivalent (TEDE)   5 rems
- Any organ other than the lens of the eye   50 rems
- Lens of the eye   15 rems
- Hands and forearms; feet and ankles   50 rems
- Skin of the whole body   50 rems

(10 CFR 20.1201). The general population is provided a higher level of protection by an annual limit of 0.1 rem (10 CFR 20.1301).

The ALARA Concept

In 1975, the USNRC published Regulatory Guide 8.8 entitled Information Relevant to Assuring That Occupational Radiation Exposures at Nuclear Power Plants Will Be as Low as Reasonably Achievable or ALARA. The concept is based upon the assumption that the relationship between dose and biological effect is linear and that no threshold effect is involved. The 1977 revision provides the ALARA philosophy as follows:

1. Merely controlling the maximum dose to the individual is not sufficient; the collective dose to the group (measured in person-rems) must be kept as low as is reasonably achievable.
2. "Reasonably achievable" is judged by considering the state of technology and the economics of improvement in relation to all of the benefits from these improvements.
3. Under the linear, nonthreshold concept, restricting the doses to individuals at a fraction of the applicable limit would be inappropriate if such action would result in the exposure of more persons to radiation and would increase the total person-rem dose (adapted from Tang and Saling 1990, p. 41).

(*See also:* Sabo 1985; Berlin and Stanton 1989, pp. 79–84; Murray 1994, p. 31).

Pathways of Dispersion and Human Exposure

To the extent that radioactive particles enter the human body by ingestion (i.e., eating, drinking, and breathing), it is necessary to isolate the source of radioactivity or to render it harmless. Account must be taken of all pathways to humans including ingestion of water, food crops, milk, and fauna (e.g., livestock, fish); direct and indirect exposure to radioactive materials; and background (natural) exposure.

The impact potential of a release of radioactive materials is measured in terms of the concentration and release rate to a dispersion pathway. The seriousness of a release depends upon a number of factors:

- The initial concentration of radionuclides in the source material: greater concentrations can be subjected to greater dispersion and remain at dangerous levels.
- The physical form of the matrix in the waste stream in which the radionuclides are bound: considerable variation in emission rate occurs due to moisture content, density, permeability, particle size, etc.
- The nature and intensity of the release mechanism: atmospheric releases tend to be more mobile and to be dispersed rapidly. Surface water releases may be channeled and diluted or impounded. Groundwater releases may be slowly but broadly dispersed.

The Nuclear Fuel Cycle, diagrammed in Figure 13.4, illustrates a dispersion pathway by which radionuclides may endanger human populations. Each process, movement, or use of the material embodies actual or potential release(s) of radioisotopes to the environment. The schematic also illustrates, graphically, the potential reduction in handling and environmental releases that could be achieved by reprocessing spent nuclear fuel. Figure 13.5 illustrates some of the many pathways by which radioactive wastes may reach biota and man (*see:* Berlin and Stanton 1989, Chapter 4, or Murray 1994, Chapter 14, for a detailed discussion of the mobilization and dispersion of radioactive waste sources).

Physical Protection

Radioactive material having a short half-life may be stored, if properly shielded and isolated, until it decays to an acceptable level. If in solution, the material may be diluted below the maximum permissible concentration (MPC) and treated or stored. Materials having longer half-lives require long-term storage in deep repositories. These requirements are overviewed in subsequent sections of this chapter.

Protection against external radiation exposure of the body requires consideration of three factors: distance, time, and shielding. Distance between the source and receptor humans should be maximized; time of exposure should be minimized; and shielding should be provided by the greatest density and thickness of shielding material that is practicable.

Radiological Monitoring Programs

Radiological monitoring is conducted at radwaste facilities to verify that there is no unacceptable migration of pollutants through pathways that could lead to man. The monitoring is necessary to ensure that safe working conditions are maintained for

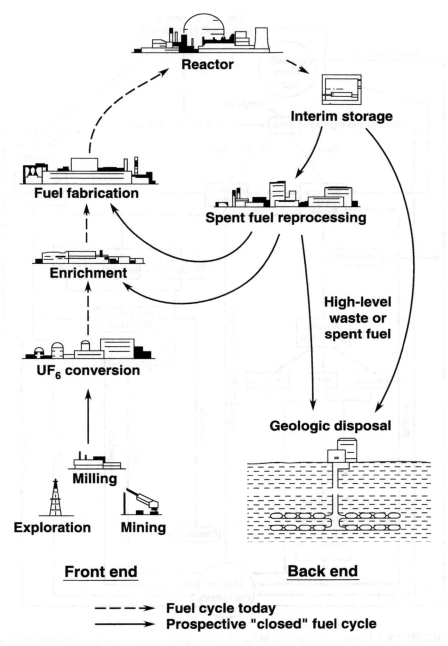

FIGURE 13.4 The nuclear fuel cycle. (From the U.S. Environmental Protection Agency.)

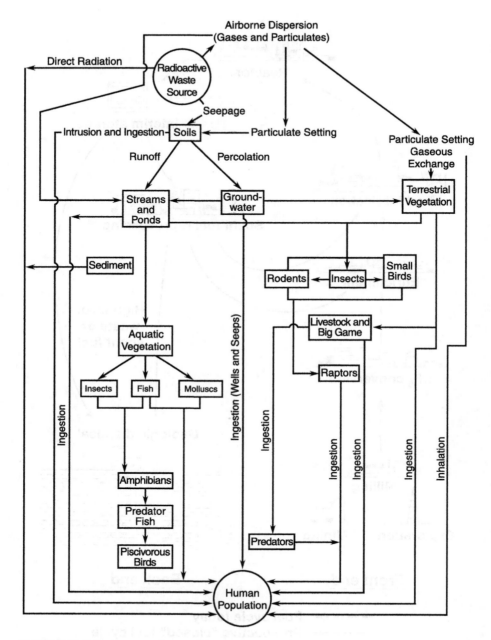

FIGURE 13.5 Pathway analysis to biota and man: generation and disposal locations on a common site. (From Robert E. Berlin and Catherine S. Stanton, *Radioactive Waste Management*, by permission from John Wiley & Sons, NY.)

on-site employees and that safe environmental conditions are maintained for the general public. Monitoring programs are also designed and operated to:

- Assess the level of impact of site operations on the environment and the public.
- Determine if the site is in compliance with applicable regulations, standards, and performance objectives.
- Enable timely detection of migration.
- Enable long-term predictions of waste isolation capabilities that will be important in the post-closure period for the site.
- Establish a database for use in the design of future waste disposal sites and monitoring programs.
- Evaluate the effectiveness of effluent control measures and equipment.

A facility radiological monitoring program has three components:

1. Baseline monitoring prior to facility construction is conducted to document the concentration of radionuclides in the soil, water, air, and biota, prior to disturbance of the land.
2. Environmental monitoring at the site boundaries and adjacent to the site is conducted to detect any changes from baseline conditions. Monitoring parameters are those pertinent to the radioactive materials, chemically toxic substances, and leachate indicators managed at the site.
3. Effluent monitoring is designed to determine concentrations and release rates of the gaseous and liquid effluents that are released by the facility. The data are used to assess (a) the effectiveness of control systems and engineered barriers in preventing releases and (b) compliance with environmental regulations.

In addition, radiation safety monitoring programs, including external radiation surveys, airborne radiation surveys, and internal radiation monitoring, are conducted to measure radiation rates and exposure indicators throughout the facility and among workers (Berlin and Stanton 1989, Chapter 4; *see also:* Jester and Yu 1985; Tang and Saling 1990, Chapter 2).

Limits on *emissions* of radioactivity include a USNRC limit on total emissions, except tritium and dissolved gases, of 0.185 TBq (terabecquerel) [5 curies (Ci)] per year from any nuclear reactor. The EPA limits emissions from the entire nuclear fuel cycle, per electrical gigawatt-year of output, to 1850 TBq (50,000 Ci) of krypton-85, 185 MBq (5 mCi) of iodine-129, and 18.5 Bq (0.5 mCi) of transuranics.[10] The emissions limits have been designed to ensure that the dose limits are met (NAS 1995, pp. 94–95).

---

[10] One becquerel (Bq) is the amount of radioactivity that yields one nuclear transformation per second (*see:* Glossary).

REGULATORY STRUCTURES

Historical Development of Policies and Statutes

In the previous section, we noted that four different classifications of radioactive wastes present separate and distinct management problems and are regulated accordingly. The four are

1. High-level waste (HLW)
2. Transuranic waste (TRU)
3. Low-level waste (LLW)
4. Uranium mine and mill tailings

Although the four "levels" of radwaste are regulated separately, a general view regarding responsibility prevailed in Congress and among the states during the early years of the nuclear era. That view was to the effect that due to the long half-lives of some radwaste constituents, only the most durable and permanent of institutions can be counted upon to carry out the continuing responsibilities associated with radwaste management. Accordingly, the task was assigned to governments — federal and state (Gehr 1990). That view was reflected in the Nuclear Waste Policy Act (NWPA, 1982) assignment of responsibility for management of HLW to the federal agencies. Hindsight, and the experience of other countries, raises substantial questions regarding the wisdom of that policy as it pertains to SNF:

> Foreign nuclear utilities generally have more responsibility for waste disposal than their American counterparts. Proponents of this approach believe that placing the burden of implementing waste disposal solutions on the waste producers may encourage better managerial and financial accountability for the program.

(GAO 1994, p. 12).

Nevertheless, the original concerns regarding security of materials that can be converted to weaponry remain as valid now as in 1982. It remains to be seen how effectively the other countries' sense of responsibility and security systems will perform with respect to controlling materials adaptable to weaponry or terrorism.

Much of the TRU is generated by military programs. The management of TRU has thus evolved as a federal government program.

Management of LLW was originally assigned to the Atomic Energy Commission (AEC), but growing concerns on the part of the states wherein storage was taking place moved Congress in 1959 to amend the basic legislation. The new legislation authorized states to enter into agreements with the AEC to regulate LLW under regulations and standards set by AEC. As a result, existing low-level waste disposal sites are licensed and regulated by host states, since all are in "agreement states." Any proposed site in nonagreement states would be regulated by the federal government.

In 1980, Congress passed new low-level waste legislation mandating decentralized responsibility, making the disposal of commercial low-level waste a state responsibility. States were free to build their own dump sites or could form regional compacts with the approval of Congress to establish burial sites. The legislation provided for refusal by regional groups to accept waste from noncompact states after 1985 (Friedman 1985).

After years of no management, Congress in 1978 assigned responsibility for management of uranium mill tailings piles to the DOE. These responsibilities are discussed further in the following sections.

Statutory and Regulatory Framework

The NWPA, as amended in 1987 by the Nuclear Waste Policy Act Amendments (NWPAA), establishes the framework and assigns responsibility for management of HLW. The two acts:

- Assign responsibility for accepting and disposing of waste from privately owned reactors, in the U.S. to the DOE.
- Establish a schedule for the siting, construction, and operation of a high-level waste repository.
- Authorize the DOE to site, construct, and operate one monitored retrievable storage (MRS) facility.
- Authorize the DOE to develop a system for transporting high-level nuclear waste to an MRS facility and repository.
- Define the working and decision-making relationships between the federal and state governments and the Indian tribes.
- Require the establishment of a fund to cover nuclear waste disposal costs.

(Adapted from GAO 1994.)

The original provisions for HLW repositories contained in NWPA were amended to consider only the Yucca Mountain, NV site as the first geologic repository. The original act also called for a monitored retrievable storage (MRS) facility at Oak Ridge, TN. The amendment canceled the Oak Ridge siting proposal and established an MRS Review Commission to evaluate the need for the MRS (adapted from Tang and Saling 1990, Chapter 1). These delays in bringing both temporary and permanent disposal facilities to operational status and the resultant accumulations of HLW have been the source of great concern and controversy.

The EPA is charged with responsibility to develop and promulgate environmental standards for protection of public health and the environment from radioactive materials. Such standards may include limits on radiation exposures to workers and members of the general public and concentrations or quantities of radioactive materials in uncontrolled areas (Berlin and Stanton 1989, pp. 79–80).

The USNRC and the DOE regulate radiation control within and in areas affected by facilities which they license or operate. The DOE regulates the activities of contractors at the weapons-related facilities. Regulatory responsibilities for radwaste are shared by the USNRC, EPA, and Department of Transportation (DOT). The USNRC regulates and licenses all waste handling/processing and disposal activities. The EPA sets standards for exposure of the general public to radiation and reviews impact statements for major projects. The DOT establishes packaging, marking and labeling standards, sets qualifications for carrier personnel, and monitors transportation [*see also:* Office of Technology Assessment (1985, Chapters 4 and 5)].

Department of Energy Management of Cleanup Programs

The Energy Reorganization Act of 1974 abolished the AEC and transferred the agency's waste management and remedial action functions to the DOE. The DOE retained many of AEC's personnel, policies, priorities, and attitudes. By 1988, the DOE was generally discredited with regard to attitudes and progress toward environmental responsibilities. Although the focus was upon the major weapons facilities, all DOE operations having to do with cleanup and waste management were under Congressional scrutiny. In January 1989, President Bush nominated retired Admiral James D. Watkins as Secretary of Energy. Watkins, a veteran of Admiral Hyman Rickover's nuclear navy programs, brought organizational and technical skills to bear and began the difficult process of refocusing the DOE from weapons production to cleanup of the sites. The following years were characterized by top to bottom overhaul and shakeup of the agency. Although the Environmental Management program has focused and achieved some order and problem solving ability, budget, administrative, security, technical, political, and public relations problems continue to hamstring the cleanup program, particularly the repository projects.

HIGH-LEVEL RADIOACTIVE WASTE MANAGEMENT

HLW Defined and Described

The USNRC description of HLW includes spent fuel from reactors in civilian power-generating plants, naval propulsion units, and obsolete nuclear weapons; the highly concentrated wastes from reprocessing fuel rods; and the solids generated in fuel reprocessing. This classification is frequently divided into "commercial" and "defense" subcategories. However, Goranson (1978) asserts: "Separation of high-level wastes into 'commercial' and 'defense' has meaning only to the technologist concerned about specific chemical composition (acid vs. neutralized waste) and heat generation. The public and the media use the terms interchangeably."

HLW Treatment and Disposal

From the earliest days of the nuclear era, development and implementation of suitable treatment and disposal for radwastes have lagged their production. The evolution of scientific and engineering knowledge has brought about several alternative concepts, including deep geologic repositories, ocean disposal, disposal in thick Antarctic ice, disposal in deep space, and highly theoretical transmutation schemes. Prior to 1970, the U.S. disposed of much of the generated HLW in the ocean. The U.S. and other nations had placed 90,000 barrels of radioactive waste on the ocean floor when a moratorium halted this practice in 1970.

Other nations have continued the practice in disregard of several international agreements and conventions (see Chapter 7).[11] Most notable among these activities is the disposal of radioactive materials, nuclear fuel elements, and entire "hot" reactors by the former Soviet Union in the sea off Novaya Zemla and the Kola Peninsula. A partial inventory of the Russian dumping includes four reactor compartments with three damaged cores, one complete submarine with two reactors containing fuel, and three reactor tanks (one with fuel)[12] (Olgaard 1997).

Reprocessing has been proven technically feasible for reduction of the quantities of HLW that are otherwise destined for disposal. As noted earlier, reprocessing of commercial SNF was halted in 1977 by President Carter, primarily because of concerns regarding control of plutonium and the proliferation of nuclear weapons. Reprocessing of military SNF continues at three sites. Reprocessing is not, however, a panacea. It is very costly, produces large volumes of TRU, and is a major source of HLW (Murray 1994, Chapter 14). Commercial facilities in France, Britain, and Japan have proceeded with reprocessing of SNF, however, a data manipulation scandal at a British reprocessing facility has cast doubt on the future of reprocessing in Europe. Meanwhile, France is reported to be moving ahead with plans for a deep underground disposal facility (*Environment Reporter,* February 18, 2000).

As noted earlier, the NWPA, enacted in 1982, and the 1987 amendments have focused on the design, site selection, and construction of a HLW repository. The 1987 amendments added provision for a monitored retrievable storage facility (MRS), but conditioned construction of the MRS upon NRC authorization for construction of the permanent repository. As noted in the section on HLW management, the Yucca Mountain HLW facility has been seriously delayed. The Congress, in the conditions imposed by the 1987 amendment, effectively prevented construction of the MRS, and the attention of nuclear utilities has focused on shifting of spent fuel from storage pools to dry storage casks, which could be stored at the reactor sites [Congressional Research Service (CRS) 1996]. A tally of U.S. nuclear power facilities' spent fuel pool storage *capacities*, used and remaining, can be accessed at: <http://www.nrc.gov/OPA/drycask/sfdata.htm>.

The Yucca Mountain Repository (YMR) construction was expected to require 6 years, but public opposition, disapproval of the project by the state of Nevada and subsequent legal maneuvering, a large number of geotechnical, regulatory, and policy issues, and inadequate funding have delayed the project. In 1989, the DOE

---

[11] "Under the 1972 Convention on the Prevention of Marine Pollution by Dumping Wastes and Other Matter in the Ocean (London Ocean Dumping Convention), the International Atomic Energy Agency (IAEA) was charged with developing regulations to restrict dumping of radioactivity into the oceans to levels that pose 'no unacceptable degree of hazard to humans and their environment.' The resulting IAEA guidelines are based on the proposition that additions of radionuclides to the oceans should not exceed rates that, if continued for 1,000 years, would lead eventually to doses exceeding 1mSv (100 mrem) per year to the most exposed individuals (IAEA 1986). The models used by the IAEA to estimate these rates consider a variety of pathways by which humans could be exposed to radionuclides from seawater, including ingestion of fish, shellfish, seaweed, plankton, desalinated seawater, and sea salt; inhalation of evaporated seawater and airborne particulates originating from ocean sediments; and external irradiation from swimming and onshore sediments" (NAS 1995, Chapter 3).

[12] For a detailed overview of the scope of Russian submarine nuclear fuel disposition, *see:* Kirk 1997.

announced a schedule to begin operating the MRS in 1998 and operating the repository in 2010. The DOE schedule incorrectly assumed that a site would be found for the MRS and that the NRC license would be issued in time to maintain that schedule. The General Accounting Office in a 1993 report estimated that it was unlikely that the repository would begin operation until 2007 and possibly not before 2014 (GAO 1993, p. 28). The DOE now consistently forecasts operational status for YMR in 2010 (NRC 1999). Figure 13.6 is a composite of graphic artists' renditions of the configuration of the YMR, based upon Department of Energy and U.S. Geological Survey materials.

Meanwhile, HLW must be stored in safe secure facilities. The DOE is attempting to safely store 396,000 m^3 of previously generated HLW.[13] Liquid HLW is stored in tanks and buried in drums at various DOE facilities, and leakage has reached disastrous proportions at some DOE sites. The amount of waste generated will increase as weapons are dismantled, facilities are disassembled and remediated, and contaminated sites are restored (DOE 1994a, pp. 1–5). The Department's Environmental Management (EM) program is responsible for identifying and reducing risks and managing wastes on 350 projects at 48 sites. The DOE now aims to clean up most of its contaminated facilities by 2006 (Ware 1999).

All of the operating nuclear power plants in the U.S. are storing spent fuel rods that have been removed from their reactor cores in water pools at the power generation sites (USNRC 1999, p. 45). Water is a convenient storage medium because it is inexpensive, available, can cool by natural circulation, provides shielding from radiation, and provides visibility for handling. Most utilities are at or nearing storage capacity and are insisting that the DOE had a statutory and contractual responsibility to accept the wastes in 1998. Indeed, on August 30, 2000, the U.S. Court of Appeals for the Federal Circuit ruled that a lawsuit by three electric utilities, seeking monetary damages for the DOE's alleged breach of contractual obligations to dispose of the companies' SNF beginning no later than January 1, 1998, may proceed.

The electric power industry and the DOE are examining and debating technologies, options, advantages, and disadvantages for operating above-ground temporary storage facilities. The debate revolves around the issues of temporary storage at the reactor sites vs. a central storage facility or facilities. The focus is upon a variety of designs for dry storage canisters and casks, including multipurpose canisters for transportation, storage, and disposal of spent fuel rods (Figure 13.7). The power industry appears to be resigned to the necessity to provide additional on-site storage. A likely scenario for power industry SNF assemblies is cooling in water pools, removal, disassembly, and above-ground **storage** in canisters or casks as the industry awaits operational access to YMR or another permanent HLW **disposal** site (CRS 1996).

HLW, including spent fuel rods, must be rendered immobile and insoluble prior to **disposal**. Liquid HLW may be subjected to one of a number of calcination processes that produce a reduced-volume, stable, dry solid. The calcined material may then be incorporated into a molten glass mixture and solidified. Other wastes may be immobilized in the borosilicate glass as well (DOE 1994b, pp. 23ff).

---

[13] HLW other than SNF.

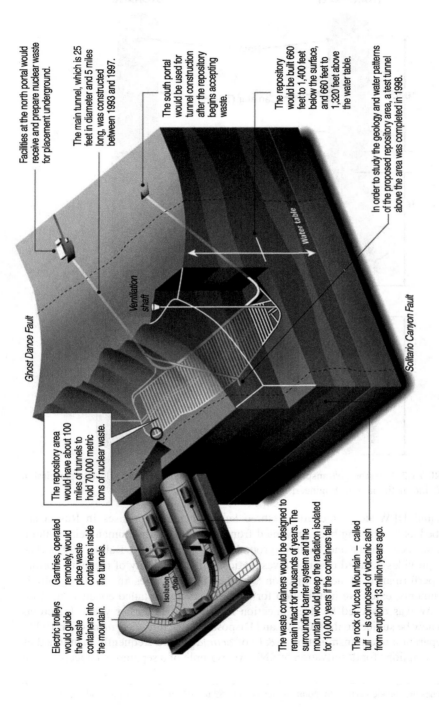

Facilities at the north portal would receive and prepare nuclear waste for placement underground.

The main tunnel, which is 25 feet in diameter and 5 miles long, was constructed between 1993 and 1997.

The south portal would be used for tunnel construction after the repository begins accepting waste.

The repository would be built 660 feet to 1,400 feet below the surface, and 660 feet to 1,320 feet above the water table.

In order to study the geology and water patterns of the proposed repository area, a test tunnel above the area was completed in 1998.

Ghost Dance Fault

Solitario Canyon Fault

Ventilation shaft

Water table

The repository area would have about 100 miles of tunnels to hold 70,000 metric tons of nuclear waste.

Electric trolleys would guide the waste containers into the mountain.

Gantries, operated remotely, would place waste containers inside the tunnels.

Isolation door

The waste containers would be designed to remain intact for thousands of years. The surrounding barrier system and the mountain would keep the radiation isolated for 10,000 years if the containers fail.

The rock of Yucca Mountain – called tuff – is composed of volcanic ash from eruptions 13 million years ago.

FIGURE 13.6 Configuration of the Yucca Mountain High-Level Waste Repository. (Adapted from graphics by Jay Carr of the *Dallas Morning News* and the U.S. Environmental Protection Agency. Source material: U.S. Department of energy and U.S. Geological Survey.)

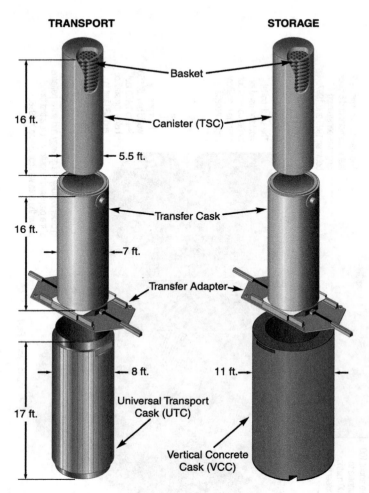

FIGURE 13.7 Dry storage/transportation system for spent nuclear fuel rods. (By permission of NAC International, 655 Engineering Drive, Norcross, GA 30092.)

Liquid HLW deep injection at three large nuclear enterprises in Russia are described as "… having been prevented from affecting nearby human beings, flora, fauna, surface waters and shallow groundwaters" (Foley and Ballou 1998). The source provides detailed research investigation of the suitability of receiving formations, performance of deep injection of other liquid wastes in similar geologic environments, extensive justification for the practice, and detailed conclusions. Liquid HLW was disposed by deep injection in the U.S., in earlier years. Injections would now be subject to the RCRA Land Disposal Restrictions (*see also:* Gibb 1999).

Separation and transmutation (S&T) of *actinides*[14] are frequently mentioned as available options for deactivation of SNF. An example of a separation concept would

---

[14] Actinides are the elements with atomic numbers from 89 to 103, inclusive. All are radioactive.

use a combination of PUREX solvent extraction of actinides up to plutonium, followed by a TRU extraction of the actinides above plutonium. Transmutation techniques would use a light water reactor (LWR), an advanced liquid-metal reactor (ALMR), or an accelerator-driven subcritical nuclear reactor for producing neutrons for a concept called the accelerator transmutation of waste (ATW). Other proposals for S&T approaches are under study, and there are numerous, highly theoretical concepts described in the literature. The NRC has produced extensive literature on these processes and has concluded that given the present state of knowledge, the time required to develop the concepts and the cost uncertainties of implementing an operational scale S&T facility renders these technologies infeasible at present (*see:* NAS 1995; NRC 1996; Murray 1994, Chapter 21; *see also:* Organization for Economic Co-operation and Development 1984; Office of Technology Assessment 1985, Chapter 3; League of Women Voters 1993, pp. 21–23; Nebel and Wright 1993, Chapter 22; GAO 1994; IAEA 1994; NRC 1995; Easterling and Kunreuther 1995; Diakov 1997; Foley and Ballou 1998; NRC 1999).

TRANSURANIC WASTE MANAGEMENT

TRU Defined and Described

Transuranic elements are those having atomic numbers greater than 92 (i.e., having more protons than uranium). TRU waste is defined in the U.S. as radwaste that is not classified as HLW, but contains an activity of more than 100 nCi/g from α-emitting TRU isotopes having half-lives greater than 20 years (Tang and Saling 1990, p. 173). TRU wastes are discarded materials that have been generated from nuclear weapons production, research, and development since the 1940s. The waste typically includes metal tools, gloves, lab coats, rags, scrap, equipment, debris, etc. that are contaminated with plutonium during laboratory and facility operations (DOE 1994c, pp. 36P). TRU is not generally found outside the DOE complex and is mainly produced from the reprocessing of nuclear fuel, nuclear weapons production, and reactor fuel assembly. Approximately 55% of the DOE's TRU waste is MTRU.[15] The waste is currently stored at 11 federal facilities around the country. Prior to 1970, most of the TRU was buried in shallow land disposal sites. Since 1970, the waste has been stored in a variety of containers, mainly 55-gal drums and wooden and metal boxes. Congress in NWPAA directed the DOE to dispose of these wastes at the Waste Isolation Pilot Plant (WIPP) near Carlsbad, NM.

TRU waste may contain sufficiently high concentrations of γ-emitting nuclides that remote handling is necessary.[16] Most TRU wastes contain primarily α-emitters and, when packaged, are safe for contact handling.[16] Nevertheless, great care must be taken to avoid damage to containers because of the dangers of ingestion of spilled or leaked α-emitters. Since TRU wastes, by definition, have long half-lives, the most suitable method for disposal is isolation in geologic repositories.

---

[15] Mixed transuranic waste is a mixture of transuranic waste and a RCRA hazardous waste (*see:* Glossary).
[16] This dichotomy has given rise to still another classification: contact-handled (CH) TRU has sufficient restrictions on radioactive content and packaging that personnel can work in the immediate vicinity without shielding; remote-handled (RH) TRU requires nearby personnel to be shielded.

TRU Disposal

The Waste Isolation Pilot Plant. In 1979 Congress authorized studies and development operations preliminary to construction of the WIPP. The facility was to be constructed in the Salado Formation, near Carlsbad, NM, if the early investigative and planning phases showed that the project was feasible. This 3000-ft-thick rock salt formation is in a seismically stable area and is devoid of circulating groundwater. The facility was designated "... a research and development facility to demonstrate the safe disposal of radwaste resulting from the defense activities and programs of the U.S. exempted from regulation by the NRC." The intent was to provide a laboratory for demonstration and validation of disposal of defense radwastes in salt formations. The project was to have begun radioactive waste operations in October 1988 (Khareis 1990).

As the operational date approached, a variety of delays caused the 1988 startup date to be missed, and the schedule continued to recede further into the future. As noted earlier, the refocusing effort at DOE is having a positive effect, but problems continue to impede progress. Some examples at WIPP are

1. The TRU wastes, having biodegradable content, including organic solvents and refuse, are RCRA "mixed waste"[17] and must meet the requirements of 40 CFR 191. The deposited wastes are to be contained in 55-gal drums; they will be backfilled by "blown-in" granular salt to absorb any gases that may be released from the drums. A number of delays in the permitting process, including protests by environmentalists, the issue of venting the gasses, restrictions on placement of mixed waste, a permit requirement for characterization of both mixed and non-mixed waste, and a prohibition of remote-handled waste, have caused continuing delays.

2. Several aspects of the project required resolution by Congress. The Bureau of Land Management (BLM) land had to be transferred to the DOE. Funding for state construction of roads and other facilities was to be appropriated. A bill to accomplish those ends failed in 1989 was not brought to the floor in 1990. Congress passed the Waste Isolation Plant Land Withdrawal Act (WIPP LWA) in 1992. In addition to accomplishing the necessary land withdrawal, the Act required the EPA to promulgate specific criteria for determination if the facility complies with EPA's generic high-level and transuranic waste disposal standards (*Environment Reporter*, January 20, 1995, p. 1798). The DOE has authority to dispose of nonmixed TRU waste, but the Part B RCRA permit for disposal of mixed waste was issued by the New Mexico Environment Department (NMED), which has RCRA regulatory authority in the state. The first shipment of defense-generated transuranic radioactive waste arrived at the facility on March 26, 1999. By July 14, 1999, 63 shipments of TRU waste had been received at WIPP (DOE 1999).

---

[17] NRC 1999a, pp. 35–36.

The DOE has proposed a modification to its Hazardous Waste Facility Permit for the Waste Isolation Pilot Plant that would support a centralized waste characterization facility. The agency states that the increase in storage capacity and permitted storage areas will provide WIPP the ability to characterize wastes on-site, reducing transportation, handling, time, and resources that are required to accomplish the characterizations and audits at multiple facilities (DOE 2000).

If the project is allowed to continue in the present format, the DOE hopes to demonstrate that TRU waste can be safely stored in a deep-bedded salt formation 2150 ft below ground surface. If the demonstration is successful, the WIPP will be operated as a repository for an additional 20 years. Figure 13.8 illustrates the configuration of the shafts, tunnels, and storage areas of the WIPP. The scientific knowledge that is expected to result from WIPP will greatly enhance knowledge of management technology for safe handling and storage of radioactive wastes (*see also:* Berlin and Stanton 1989, pp. 107–110; DOE 1994a, pp. 35–37; DOE 1994c, pp. 0036P; Murray 1994, Chapter 21; NRC 1999a).

LOW-LEVEL WASTE MANAGEMENT

LLW Defined and Classified

Low-level wastes are defined by the Low Level Radioactive Policy Act (LLRPA) as "radioactive waste not classified as high-level waste, transuranic waste, spent nuclear fuel, or mill tailings." LLW often contains small amounts of radioactivity dispersed in large amounts of material. It is generated by uranium enrichment processes, reactor operations, isotope production, and medical and research activities. The 1980 LLRPA assigned responsibility for management of this category of waste to the states, authorized states to enter into compacts for the development of regional disposal facilities, and provided statutory authority for states to refuse acceptance of wastes generated outside their regional borders after 1986. By 1984 it became evident that no new disposal facilities would be available by 1986. In an effort to establish a sense of urgency among states, Congress enacted the Low-Level Radioactive Waste Policy Amendments Act of 1985, requiring states and compacts to comply with strict timetables for establishing LLW disposal sites (DOE 1994c, pp. 186P).

LLW is defined so broadly that some waste streams may meet the definition, but contain some radionuclides which may not be suitable for disposal in near-surface facilities. Accordingly, USNRC has developed a classification scheme, which is implemented by 10 CFR 61. The classifications are summarized in Table 13.2.

Treatment and Disposal of LLW

The general nature of LLW includes items and materials incidental to, and contaminated during, radwaste handling, including dry trash, plastics, paper, glass, clothing, discarded tools and equipment, wet sludges, and organic liquids (Tang and Saling 1990, p. 195). Some forms of LLW may be concentrated by evaporation, crystallization, and drying. Some may be amenable to incineration, calcination, or compaction. Much of the waste goes directly to near-surface land disposal. Class B and C wastes receive deeper burial, more cover, and/or incremental protection.

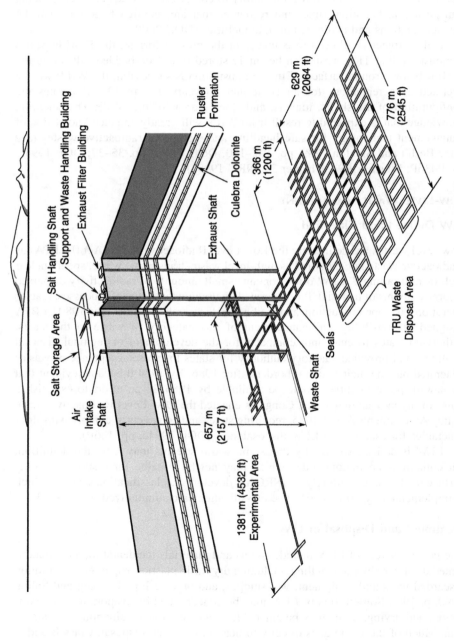

FIGURE 13.8 Configuration of the Waste Isolation Pilot Plant. (From the U.S. Department of Energy, Washington, D.C.)

TABLE 13.2
USNRC Waste Classification System Maximum Radionuclide
Concentration (μCi/cc or Ci/m³)

Radionuclide	Class A	Class B	Class C
Hydrogen-3	40	MC	—
Colbalt-60	700	MC	—
Nickel-63	3.5	70	700
Nickel-63 (in activated metal)	35	700	7,000
Strontium-90	0.04	150	7,000
Cesium-137	1	44	4,600
Carbon-14	0.8	—	8
Carbon-14 (in activated metal)	8	—	80
Nickel-59 (in activated metal)	22	—	220
Niobium-94 (in activated metal)	0.02	—	0.2
Technetium-99	0.3	—	3
Iodine-129	0.008	—	0.08
Total all radionuclides with <5 year half-life	700	MC	—
α-emitting transuranic wastes with half-life >5 years	10[a]		100[a]
Plutonium-241	350[a]		3,500[a]
Curium-242	2,000[a]		20,000[a]

Note: MC = maximum concentration — all waste above Class A limit is Class B; — = no limit is applicable for this class.

[a] Units are nanocuries/gram.

Source: Adapted from Dornsife (1985) and 10 CFR 61.55.

By 1994, only two sites — Barnwell, SC, and Richland, WA — were accepting LLW, and in June 1994, Barnwell closed to contributors other than the Southeast Compact states. However, in 1995, Barnwell withdrew from the Southeast Compact because North Carolina failed to open the new regional facility, which was to replace the Barnwell operation. Barnwell then began accepting wastes from many states and, by late 1999, was receiving LLW from 39 states, reducing its operating life to about 12 years (*Environment Reporter*, January 14, 2000, p. 66). In July 2000, Barnwell joined Connecticut and New Jersey in a newly formed Atlantic Compact and restricted disposal to the new compact, thereby extending the life of the facility by at least 50 years. The compacts, member states, and host states are presently aligned as indicated in Table 13.3.

Siting of LLW disposal facilities is proving to be as difficult as the HLW and TRU siting has been. The Ward Valley site in southeastern California is a case in point. The site was intended to serve the Southwest Compact for 30 years. In July 1992, California, host state for the Southwest Compact, applied to then Interior Secretary Manuel Lujan to purchase the site from the Bureau of Land Management. In August 1993 newly appointed Secretary Bruce Babbitt notified Governor Pete Wilson that he proposed selling the land to California based upon the outcome of

TABLE 13.3
Low-Level Radioactive Waste Disposal
Compact Membership

Compact	Host State	Member State
Atlantic	South Carolina	Connecticut
		New Jersey
Appalachia	Pennsylvania	West Virginia
		Maryland
		Delaware
Southeast	North Carolina	Florida
		Georgia
		Tennessee
		Alabama
		Mississippi
		Virginia
Central States	Nebraska	Arkansas
		Louisiana
		Kansas
		Oklahoma
Midwest	Ohio	Wisconsin
		Indiana
		Iowa
		Minnesota
		Missouri
Central Midwest	Illinois	Kentucky
Rocky Mountain[a]		Nevada
		Colorado
		New Mexico
Southwest	California	Arizona
		North Dakota
		South Dakota
Northwest	Washington	Idaho
		Oregon
		Utah
		Alaska
		Hawaii
		Montana
		Wyoming
Texas	Texas	Maine
		Vermont

Unaligned: New Hampshire, New York, Massachusetts, Rhode Island, Puerto Rico, District of Columbia, Michigan.

[a] Northwest accepts Rocky Mountain LLW per agreement between compacts.

Source: DOE 1999, pp. 43–44.

narrowly focused public hearings. In September 1993, the California Department of Health Services (DHS) issued a disposal facility operating license to U.S. Ecology, the contractor retained by the state for initial work on the site. This would normally have triggered construction of the site.

A month after the license was issued, Senator Barbara Boxer of California announced that an unreviewed report (the "Wilshire report") by three U.S. Geological Survey (USGS) geologists "... found 'significant potential' for radioactive contamination of the groundwater and eventual contamination of the Colorado River." Both the DHS and U.S. Ecology took strong issue with the report, noting that it had not been subjected to the normal internal USGS review process and that the authors had relied upon faulty and incomplete information to support their conclusions.

In October 1993, opponents filed two lawsuits claiming that improper procedures had been followed in the issuance of the operating license and that the Wilshire report provided significant new evidence showing the unsuitability of the Ward Valley site. Shortly thereafter, Secretary Babbitt notified the Governor that he was "postponing further action pending final resolution of the litigation." In February 1994, Senator Boxer reiterated her request to Secretary Babbitt that "the work of the three USGS Geologists be expanded and subjected to a thorough and objective scientific review." The DHS retained environmental consultants who reviewed the technical data and concluded that contamination of the Colorado River by releases from the Ward Valley site would be "impossible." Secretary Babbitt then asked the National Academy of Sciences (NAS) to review the issues raised in the Wilshire report. The NAS review was expected to be completed in the spring of 1995.

Meanwhile, in May 1994, the Superior Court combined the two lawsuits and dismissed all of the allegations except one, ruling that DHS should re-examine the licensing decision in the light of the Wilshire report. The matter then awaited a California Supreme Court Decision.

Further impediments grew from concerns for the Desert Tortoise, a threatened species. U.S. Ecology and a Desert Tortoise Task Force submitted a plan to the U.S. Fish and Wildlife Service (USFWS) in 1990 to mitigate impacts of the site on the tortoise. The plan included installation of several miles of tortoise-proof fencing along Interstate 40 to eliminate road kills, the leading cause of mortality for tortoises in the northern portion of Ward Valley. The plan was expected to more than compensate for the loss of the 80 acres that would be used for the LLW facility. The USFWS received the mitigation plan favorably, but two organizations sued the agency to force creation of a critical habitat for the tortoise. In February 1994, USFWS designated a critical habitat that included land in which the proposed disposal site was located. The decision would have required DHS to seek another biological opinion from USFWS. As a result of the designation — the site was then located within a critical habitat — the Environmental Protection Agency and several other Federal agencies had to be consulted.

The NAS, which had been asked by Secretary Babbitt to review the *geological* issues raised in the Wilshire report, entered objections that the *habitat* should not be fragmented by projects like the Ward Valley disposal facility. Further litigation ensued, and in June 1999, Governor Gray Davis announced that he would not appeal a U.S. District Court decision supporting the Clinton administration's refusal to transfer the federal land needed for construction of the Ward Valley facility. Govenor Davis stated that he would create an advisory group to investigate environmentally sound alternatives for disposal of California's LLW (*Environment Reporter*, June 11, 1999, p. 256).

Appendix A to this chapter provides a summary of commercial LLW options available to the states. In summary, at the time of this writing there are (were) two LLW operating facilities and another licensed only for Class A wastes. A recent GAO report[18] states: "States acting alone or in compacts had collectively spent almost $600 million attempting to develop about ten new disposal facilities. None of these efforts have been successful." Thus, 11 states dispose of their LLW at Hanford, WA, 3 states can access the Barnwell site, and an unknown number generating no Class B or C wastes may access the Envirocare facility in Tooele County, UT.

MIXED WASTE MANAGEMENT

Mixed Waste Defined and Described

"Mixed waste" is defined as a waste mixture that contains both radioactive materials subject to the Atomic Energy Act (AEA) and a hazardous waste component regulated under RCRA. The hazardous waste (i.e., the non-AEA material) can be either a listed hazardous waste in Subpart D of 40 CFR 261 or a waste that exhibits any of the hazardous waste characteristics identified in Subpart C of 40 CFR Part 261 (EPA 1998). The radioactive component may be TRU or LLW (designated MTRU or LLMW). High level radioactive waste is also a mixed waste because it has highly corrosive components or has organics or heavy metals that are regulated under RCRA. HLW may include other highly radioactive material that USNRC determines by rule requires permanent isolation.

Almost all of the commercially generated (non-DOE) mixed waste is composed of LLW and RCRA hazardous waste, i.e., LLMW. Commercially generated LLMW is produced in all 50 states at industrial, hospital, and nuclear power plant facilities. Radioactive and hazardous materials are used in a number of processes such as medical diagnostic testing and research, pharmaceutical and biotechnology development, and pesticide research, as well as nuclear power plant operations. Approximately 4000 m^3 of LLMW were generated in the U.S. in 1990. Of this amount, 2840 m^3 (or 71%) was liquid scintillation cocktail.[19] Organic solvents, corrosive organics, and waste oil made up 18% and toxic metals were 3% of the LLMW total (EPA 1998a).

Once a waste is determined to be a mixed waste, the waste manager must comply with both AEA and RCRA statutes and regulations. The requirements of RCRA and AEA are generally consistent and compatible. Where provisions of the two acts are found to be inconsistent, the AEA takes precedence (EPA 1998a). At the time of this writing, the EPA had issued LDR standards for four categories of mixed wastes:

---

[18] U.S. GAO 1999.

[19] The liquid scintillation cocktail is a fluid used in medical laboratories to analyze DNA and proteins. It often uses radioactive tracers and materials such as toluene and xylene.

1. Radioactive lead solids with a BDAT treatment standard of macro-encapsulation
2. Radioactive elemental mercury with a BDAT treatment standard of amalgamation
3. Radioactive hydraulic oil contaminated with mercury with BDAT standard incineration
4. Radioactive high level wastes generated during the reprocessing of fuel rods with BDAT standard vitrification

The remaining mixed wastes are subject to the promulgated standards that apply to the hazardous portion of the waste unless the EPA publishes specific standards for mixed waste treatability groups in the future. The EPA has issued a variety of extensions, variances, and policy statements regarding mixed waste treatment, storage, and disposal (EPA 1998). *The practitioner contemplating handling of mixed waste is advised to seek current regulatory requirements from the state regulatory agency or EPA regional office regarding the specific mixed waste to be managed* (*see also:* Wagner 1997; NRC 1999a; Rothfuss 1999; EPA 1999).

URANIUM MINE AND MILL TAILINGS MANAGEMENT

Tailings Defined, Described, and Characterized

Uranium **mine** tailings usually consist of waste rock and low-grade ores which may be piled near or in the mine or may be used in construction of the **mill** tailings pond(s). The uranium **mine** tailings generally contain low levels of radioactive materials, are considered to be subject to the Bevill Amendment,[20] and have not been brought under RCRA control.

Uranium **mill** tailings are the sandy residue of the uranium extraction processes. Much of the ore contains less than 1% uranium, so that extraction produces large volumes of bulky wastes. Estimates place the waste-to-product ratio at 1300:1. After extraction of the uranium, the tailings contain other natural radionuclides such as thorium-230, radium-226, and radon-222. The tailings are discharged, in a slurry, to a basin or impoundment where the solids are retained behind a man-made dam. The liquid overflows or is pumped to a waste treatment facility where the radium is coprecipitated with barium sulfate (Hare and Aikin 1984).

The tailings have accumulated in large piles in several uranium-producing areas of the U.S., Canada, and elsewhere. A recent fact sheet issued by the Australian environmental agency estimates that there are more than 500 million tonnes (551 million tons) of uranium mill tailings located in 18 countries around the world (Environment Australia 1998). The piles have been poorly controlled, and where control has not been established they constitute a health and environmental hazard. Where control has not been established, the remaining radionuclides are subject to dispersion in windblown tailings and the remaining radium is leached from the piles by rainfall. The effectiveness of remediation efforts has not been convincingly demonstrated. Seepage from the piles may contain radium in concentrations exceeding criteria.

---

[20] *See:* Bevill Amendment in Glossary.

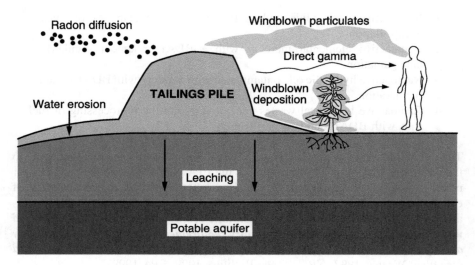

FIGURE 13.9 Potential exposure pathways originating in uranium mill tailings piles. (Adapted from Eldon D. Enger, J. Richard Kormelink, Bradley F. Smith, and Rodney J. Smith, *Environmental Sciences: The Study of Interrelationships*, Wm. C. Brown Publishers, Dubuque, IA. With permission of the McGraw-Hill Companies.)

Radium is easily taken up by plants and ingested by animals. The pathway to humans via these introductions to the food chain can be direct or short.[21] Figure 13.9 illustrates pathways of migration from a uranium mill tailings pile.

Tailings were used in construction and as backfill around building foundations in several Colorado, Utah, New Mexico, and Idaho communities. Persons living and working in the buildings were exposed to radon gas concentrations exceeding maximum allowable levels for uranium miners. Some 13,555 potential cleanup sites were originally surveyed to determine if they were contaminated. More than 5000 of the 5199 properties found to be contaminated had been remediated by the end of 1993.

Uranium mill tailings were dispersed in other ways that endangered human health. Several disastrous failures of tailings dams allowed tailings to be washed into streams. Mine and mill owners were required to retrieve the tailings, but the effectiveness of the cleanup is questionable.

Treatment and Control

The uranium industry grew rapidly, during the early days of the nuclear era, and large numbers of tailings piles were created. However, the expected growth in nuclear power generation failed to materialize, and many of the mills closed. At the peak

---

[21] A detailed analysis issued by The Uranium Institute of the health threat to humans by radon gases from uranium mill tailings discounts that threat. A key finding states: "The radon concentrations associated with the tailings emissions are extremely small on both a relative (compared to typical background levels) and absolute (in terms of dose and risk) level. In the authors' view the individual risk of cancer associated with the predicted concentrations is below a level that can be considered completely insignificant and trivial, i.e., *de minimis*" (Chambers et al. 1998).

of the domestic industry, 28 uranium mills were operating. All are now inoperative and many were abandoned. Abandoned (and uncontrolled) tailings piles became numerous. When present in populated areas, the tailings present a potential long-term health hazard. The tailings contain low concentrations of radium which decays to radon a radioactive gas which has been linked to increased incidence of lung cancer when present in confined areas such as homes or mines. The radium in tailings will not decay entirely for thousands of years (USNRC 1999). As noted earlier, the tailings also contain other radioactive contaminants that can pollute groundwater (DOE 1994c). During the nearly three decades of the existence of the Atomic Energy Commission (AEC), the officials of the Commission claimed that the AEC had no jurisdiction to assert control over tailings management.

The Uranium Mill Tailings Radiation Control Act of 1978 (UMTRCA) made the Department of Energy responsible for management of the tailings piles at 24 inactive sites, in 10 states, where uranium was produced for the national defense programs. Remedial actions taken at these sites are site-specific and hinge upon the requirement that the site and surrounding areas comply with EPA standards. Actions taken range from stabilization on-site, to removal of the tailings for stabilization on-site, to removal of the tailings for stabilization elsewhere. Surface cleanup has been completed at all sites except Grand Junction, CO. That site will remain open until 2023 to accept additional waste. The DOE groundwater cleanup phase was initiated in 1991. Groundwater cleanup has been estimated by the DOE to cost more than $310 million (USNRC 1999a, p. 4).

Stabilized uranium mill tailings piles must have a cover designed to control radiological hazards for a minimum of 200 years and for 1000 years to the greatest extent reasonably achievable. The cover must also limit radon releases to 20 pCi/m² averaged over the disposal area. Radon release limitation requirements apply to any portion of the tailings disposal site unless radium contrates do not exceed 5 pCi/g in the first 15 cm below the surface and 15 pCi/g in layers more than 15 cm below the surface (DOE 1995; *see also:* Rustum 1982; Tang and Saling 1990, Chapter 7; League of Women Voters 1993; DOE 1994a; DOE 1997, Chapter 5; USNRC 1999a).

TOPICS FOR REVIEW OR DISCUSSION

1. The text discusses four types of radiation, three of which may be of concern in managing radioactive waste. One type probably will not be encountered on a radioactive waste site. Discuss.
2. If α-particles are so easily attenuated (i.e., by a paper barrier), why is this type radiation of concern to hazardous or radioactive waste workers?
3. Physiological effects of radiation on man are classified as *somatic* or *genetic*. Explain the terms and their implications.
4. Occupational exposure to radioactive materials is limited by 40 CFR 20 to 50 rems per year to each of several organs, but only 5 rems total body exposure. Explain!
5. Three radioactive repository/storage sites are in various stages of development in the U.S. Identify them and the types of wastes to be managed at each.

6. What is meant by RCRA "mixed waste"? Why is it a major problem (a) from a technical standpoint and (b) from a regulatory standpoint?

7. Most TRU wastes are relatively mild in terms of radioactivity. Why, then, is TRU to be managed in deep underground repositories? What special management problems attend TRU wastes?

8. Some of the LLW disposal compact memberships are not noteworthy for geographical contiguity nor proximity. What problems do you foresee in this regard?

9. Considering overall history of development, what problems do the Yucca Mountain, WIPP, and Ward Valley sites have in common? As a future policy-maker/implementer, how would you develop and implement plans for a future site in a manner that would avoid the controversies that have attended the three planned sites?

APPENDIX A
Commercial Low-Level Radioactive Waste Disposal Options

	Restrictions	Other Information

Facilities Currently Available to Accept Commercial LLRW

Location
Barnwell County, SC

Operator
Chem-Nuclear Systems, LLC

LLRW Accepted
Classes A, B, C
Subject to waste acceptance criteria

South Carolina is a member of the congressionally approved Atlantic Compact, known in statute as the Northeast Compact. The compact commission has authority over importation of commercial LLRW into the compact region for disposal and has authorized South Carolina to import LLRW consistent with state law.
Under South Carolina legislation enacted in June 2000, importation of waste from states other than members of the Atlantic Compact (SC, CT, NJ) may not result in the facility's acceptance of more than specified total volumes of waste, in accordance with the following schedule:

FY 2001 (7/1/00–6/30/01)	160,000 ft³
FY 2002	80,000 ft³
FY 2003	70,000 ft³
FY 2004	60,000 ft³
FY 2005	50,000 ft³
FY 2006	45,000 ft³
FY 2007	40,000 ft³
FY 2008	35,000 ft³

After June 30, 2008, acceptance of waste from non-Atlantic Compact states is prohibited.

The South Carolina Budget and Control Board is expected to set a policy for allocation of the reduced capacity among non-Atlantic Compact generators. Board staff have shown an early preference for a system allocating future capacity to noncompact generators based on the volumes of waste they shipped to Barnwell from July 1, 2000, through June 30, 2001.
New States may join the Atlantic Compact only if they volunteer to host a regional LLRW disposal facility and meet other conditions.

Location
Tooele County, UT

Operator
Envirocare of Utah, Inc.

LLRW Accepted
Up to class A limits for most radionuclides and mixed waste of the same radioactive content
Subject to waste acceptance criteria

Utah is a member of the congressionally approved Northwest Compact, which grants the compact committee authority over importation of commercial LLRW into the compact region for disposal. The committee's policy is to allow access only for waste approved for export by the originating compact or, to the extent possible, by the originating unaffiliated state.

Currently accepts mainly high-volume, low-activity wastes. The company's existing cell for class A LLRW is close to capacity, but in August 2000 state regulators made an initial decision, subject to public comment, to approve license amendments and permit modifications requested by Envirocare of Utah to allow construction of a new disposal cell for class A LLRW.
In addition, Envirocare of Utah has had a license application pending since November 1999 to accept class B and C LLRW as well as mixed waste of the same radioactive content. If licensed by the state regulatory agency to accept class B and C waste, Envirocare must still obtain approval from the Utah legislature and Governor.

Location
Occupies part of the Hanford federal reservation near Richland, WA; land is leased from the U.S. Department of Energy to the state and subleased to the operator

Washington is a member of the congressionally approved Northwest Compact, which grants the compact committee authority over importation of LLRW into the compact region for disposal. The committee's policy is to allow access only for LLRW generated

Operator's lease expires in 2005 but may be renewed.
No additional states are eligible to join the Northwest Compact.

Continued.

APPENDIX A   (Continued)
Commercial Low-Level Radioactive Waste Disposal Options

	Restrictions	Other Information
Operator US Ecology, Inc. **LLRW Accepted** Cflasses A, B, C Subject to waste acceptance criteria	within a member state of the Northwest Compact (AK, HI, ID, MT, OR, UT, WA, WY) or the Rocky Mountain Compact (CO, NV, NM), which has a contract with the Northwest Compact and Washington. Correspondence from the Washington Department of Ecology in December 1998 states that Washington "does not anticipate providing disposal access to additional states."	

Facilities with Plans to Provide Long-Term Management of Commercial LLRW

Location Andrews County, TX[a] Lea County, NM[b] **Operator** Waste Control Specialists, LLC (WCS) **Licensed for Disposal of** Solid and hazardous waste, polychlorinated biphenyls (PCBs), those naturally occurring radioactive materials (NORM) that are exempt from state licensing requirements at Andrews County site	WCS has offered to work with Texas regulators in evaluating the Andrews County site as the location for a disposal facility for the Texas Compact. State legislative action is required to pursue that option, however, and the Texas legislature is not due to reconvene until January 2001. In addition, in July 1999, WCS made a presentation to the Lea County Commission in New Mexico concerning a proposal to site an LLRW facility in that state. As proposed, the facility would accept class A, B, and C LLRW and mixed waste from commercial generators and from the U.S. Department of Energy. To date, WCS has not submitted an application for a New Mexico facility.	Texas state regulators have issued to WCS a Class III radioactive waste license — a waste processing license that includes storage within the definition of processing. It currently contains an activity and volume cap of approximately 200,000 Ci and 300,000 ft³ for all waste at the facility, regardless of whether the waste is there for processing or storage. Since the main focus of WCS' operation in Andrews County is to process radioactive waste and either ship it to a disposal facility or return it to the generator, the activity and volume cap effectively preclude the company from storing large volumes of waste for long periods of time.
Location Ward and Andrews Counties, TX[a] **Operator** Envirocare of Texas, Inc. **Licensed for Disposal of** Not applicable	In November 1999, Envirocare of Texas submitted a license application for a Class III facility for processing and storage of commercial LLRW in Ward County. The application proposes the acceptance of waste with an activity cap of approximately 1 million curies for 40 years, followed by another 500 years of active monitoring and maintenance. In May 2000, Texas regulators notified Envirocare of amendments and corrections that need to be made to the application before a more detailed technical review can be conducted. In a November 1999 press release about its Class III facility application, Envirocare referred to the proposed storage method as "assured isolation" and stated that the proposed facility seeks to allow Texas to meet its obligations under the Texas Compact. Texas regulators, however, have questioned their own authority to issue a license for assured isolation or for	In December 1996, Envirocare filed an application for a Class II radioactive waste processing and storage facility in Andrews County. At that time, Envirocare also applied to the state for a hazardous waste processing permit for the Andrews County site. This permit, combined with a Class II license, would allow the facility to process mixed waste. However, when Envirocare later filed its Class III facility application for Ward County, the company requested that the Class II application be placed on hold. Only recently did Envirocare submit additional information requested by state regulators for the Class II application. Although both Class II and III licenses allow the holder to process and store radioactive waste, a Class III license allows the holder to process and store more curies. In addition, a Class II

Continued.

APPENDIX A *(Continued)*
Commercial Low-Level Radioactive Waste Disposal Options

Restrictions	Other Information
long-term storage facility. In response, Envirocare has notified the regulators that the application is for storage and processing, with the process proposed being long-term radioactive decay.	applicant need not conduct an independent environmental assessment, whereas a Class III applicant is required to do so. Texas regulators issued an initial draft hazardous waste processing permit for the Andrews County site on June 15, 2000. Regulatory staff are currently working on the final draft. Issuance of the final draft will be followed by a 40-day public comment period and potentially by an administrative hearing.

[a] Texas is a member of the congressionally approved Texas Compact, which grants the Texas Compact Commission authority over importation of LLRW into the compact region. However, the commission has not yet been fully appointed and has not convened to establish any policies.

[b] New Mexico is a member of the congressionally approved Rocky Mountain Compact, which grants the Rocky Mountain Board authority over importation of LLRW into the compact region. The board exercises this authority.

Source: Afton Associates, Inc. for the Midwest Compact, September 2000.

REFERENCES

Berlin, Robert E. and Catherine C. Stanton. 1989. *Radioactive Waste Management.* John Wiley & Sons, NY.

Chambers, Douglas B., Leo M. Lowe, and Ronald H. Stager. 1998. *Long Term Population Dose Due to Radon from Uranium Mill Tailings.* The Uranium Institute, Twenty Third Annual Symposium, 1998.

Corbitt, Robert A. 1990. "Hazardous Waste," in *Standard Handbook of Environmental Engineering,* Robert A. Corbitt, Ed., McGraw-Hill, NY.

Diakov, Anatoly S. 1997. "Disposition of Weapons-Grade Plutonium in Russia: Evaluation of Different Options," in *Dismantlement and Destruction of Chemical, Nuclear and Conventional Weapons,* Nancy Turtle Schulte, Ed., Kluwer, Dordrecht, The Netherlands.

Dornsife, William P. 1985. "Classification of Radioactive Materials and Wastes," in *Management of Radioactive Materials and Wastes,* Pennsylvania Academy of Science, Easton, PA.

Easterling, Douglas and Howard Kunreuther. 1995. *The Dilemma of Siting a High-Level Nuclear Waste Repository.* Kluwer, Boston.

Ehlers, Victor M. and Ernest W. Steel. 1958. *Municipal and Rural Sanitation.* McGraw-Hill, NY.

Enger, Eldon D., J. Richard Kormelink, Bradley F. Smith, and Rodney J. Smith. 1989. *Environmental Science: The Study of Interrelationships.* Wm. C. Brown, Dubuque, IA.

Environment Australia. 1998. *Uranium Mill Tailings Disposal,* Canberra ACT 2601, Australia.

Environment Reporter, January 20, 1995, p. 1795. Bureau of National Affairs, Washington, D.C.

Environment Reporter, June 11, 1999, p. 256. Bureau of National Affairs, Washington, D.C.

Environment Reporter, January 14, 2000, p. 66. Bureau of National Affairs, Washington, D.C.

Environment Reporter, February 18, 2000, p. 287. Bureau of National Affairs, Washington, D.C.

Foley, Michael G. and Lisa M. G. Ballou, Eds. 1998. *Deep Injection Disposal of Liquid Radioactive Waste in Russia.* Battelle Press, Columbus, OH.

Foster, William and Hubert N. Alyea. 1947. *An Introduction to General Chemistry.* D. Van Nostrand Company, NY.

Friedman, Robert S. 1985. "Political Considerations of Nuclear Waste Disposal Policy," in *Management of Radioactive Materials and Wastes,* Pennsylvania Academy of Science, Easton, PA.

Gehr, Arthur C., Esq., December 4, 1990. Partner, Snell and Wilmer. Personal communication, Phoenix, AZ.

Gibb, Fergus G. F. 1999. "High-Temperature, Very Deep, Geological Disposal: A Safer Alternative for High-Level Radioactive Waste?," *Waste Management* 19(3), pp. 207–211, Pergamon/Elsevier, London.

Goranson, Richard B. 1978. "Long-Term Management of Defense High-Level Waste," in *Waste Management and Fuel Cycles '78,* Proceedings of the Symposium on Waste Management, March 6–8, Tucson, AZ.

Hare, F. Kenneth and A. M. Aikin. 1984. "Nuclear Waste Disposal Technology and Environmental Hazards," in *Nuclear Power — Assessing and Managing Hazardous Technology,* Martin J. Pasqualetti and K. David Pijawka, Eds., Westview Press, Boulder, CO.

International Atomic Energy Agency. 1986. *Definition and Recommendations for the Convention on the Prevention of Marine Pollution by Dumping of Wastes and Other Matter in the Ocean, 1972.* Safety Series No. 78, Vienna.

International Atomic Energy Agency. 1994. *Safety and Engineering Aspects of Spent Fuel Storage.* Proceedings of the International Symposium on Safety and Engineering Aspects of Spent Fuel Storage, Vienna, October 10–14, 1994.

Jester, William A. and Charley Yu. 1985. "Environmental Monitoring of Low-Level Radioactive Materials," in *Management of Radioactive Materials and Wastes.* Pennsylvania Academy of Science, Easton, PA.

Khareis, Tarek. 1990. U.S. Department of Energy, WIPP Project. Personal communication. July 31.

Kirk, Elizabeth J. 1997. *Decommissioned Submarines in the Russian Northwest Assessing and Eliminating Risks.* Elizabeth J. Kirk, Ed., Kluwer, Dordrecht, The Netherlands.

League of Women Voters. 1985. *The Nuclear Waste Primer.* Nick Lyons Books, NY. (1993 edition published by League of Women Voters Education Fund).

Martland, H. S. and R. E. Humphries. 1929. "Osteogenic Sarcoma in Dial Painters Using Luminous Paint," in *Archival Pathology,* Vol. 7.

Meyer, Eugene, 1989. *Chemistry of Hazardous Materials, Second Edition.* Prentice-Hall, Englewood Cliffs, NJ.

Miller, E. Willard and Shyamal K. Majumdar. 1985. "Environmental and Biological Effects of Ionizing Radiation," in *Management of Radioactive Materials and Wastes.* Pennsylvania Academy of Science, Easton, PA.

Murray, Raymond L. 1994. *Understanding Radioactive Waste, Fourth Edition.* Battelle Press, Columbus, OH.

National Academy of Sciences. 1972. *The Effects on Populations of Exposure to Low Levels of Ionizing Radiation.* Report of the Advisory Committee on the Biological Effects of Ionizing Radiation (BEIR Report), Washington, D.C.

National Academy of Sciences. 1995. *Management and Disposition of Excess Weapons Plutonium.* Committee on International Security and Arms Control, National Academy Press, Washington, D.C.

National Council on Radiation Protection and Measurements. 1979. *Management of Persons Accidentally Contaminated with Radionuclides.* NCRP Report No. 65, Washington, D.C.

National Research Council. 1995. *Technical Bases for Yucca Mountain Standards.* Committee on Technical Bases for Yucca Mountain Standards, National Academy Press, Washington, D.C.

National Research Council. 1996. *Nuclear Waste Technologies for Separations and Transmutation.* Committee on Separations Technology and Transmutation Systems, National Academy Press, Washington, D.C.

National Research Council. 1999. *Alternative High-Level Waste Treatments at the Idaho National Engineering and Environmental Laboratory.* Committee on Idaho National Engineering and Environmental Laboratory (INEEL), National Academy Press, Washington, D.C.

National Research Council. 1999a. *The State of Development of Waste Forms for Mixed Wastes.* Committee on Mixed Wastes, National Academy Press, Washington, D.C.

Nebel, Bernard J. and Richard T. Wright. 1993. *Environmental Science.* Prentice-Hall, Englewood Cliffs, NJ.

Olgaard, Povl L. 1997. "Worldwide Decommissioning of Nuclear Submarines: Plans and Problems," in *Decommissioned Submarines in the Russian Northwest Assessing and Eliminating Risk,* Elizabeth J. Kirk, Ed., Kluwer, Dordrecht, The Netherlands.

Organization for Economic Co-operation and Development. 1982. *Uranium Mill Tailings Management.* Nuclear Energy Agency, Paris.

Organization for Economic Co-operation and Development. 1984. *Geological Disposal of Radioactive Waste.* Nuclear Energy Agency, Paris.

Pauling, Linus. 1958. *General Chemistry.* W. H. Freeman, San Francisco, CA.

Rothfuss, Heather. 1999. Genetic Engineering of A Radiation Resistant Bacterium for Biodegradation of Mixed Waste. U.S Environmental Protection Agency, Office of Research and Development, National Center for Environmental Research and Quality Assurance, Washington, D.C.

Rustum, Roy. 1982. *Radioactive Waste Disposal.* Pergamon Press, New York.

Sabo, A. T. 1985. "Radiation Protection Standards and Radiation Risks," in *Management of Radioactive Materials and Waste.* Pennsylvania Academy of Science, Easton, PA.

Sawyer, Clair N. and Perry L. McCarty. 1978. *Chemistry for Environmental Engineers, Third Edition.* McGraw-Hill, NY.

Tang, Y. S. and James H. Saling. 1990. *Radioactive Waste Management.* Hemisphere, NY.

U.S. Congressional Research Service. 1996. Report for Congress — Civilian Nuclear Spent Fuel Temporary Storage Options — Need for Additional Storage Capacity. The Committee for the National Institute for the Environment, Washington, D.C.

U.S. Department of Energy. 1994a. Environmental Management 1994. Office of Environmental Management Information, Washington, D.C., DOE/EM-0119.

U.S. Department of Energy. 1994b. Committed to Results: DOE's Environmental Management Program. Office of Environmental Restoration, Washington, D.C., DOE/EM-0152P.

U.S. Department of Energy. 1994c. Environmental Fact Sheets. Office of Environmental Management, Washington, D.C.

U.S. Department of Energy. 1995. Decommissioning of U.S. Uranium Production Facilities, Washington, D.C., DOE-EIA-0592.

U.S. Department of Energy. 1995a. Environmental Management 1995. Center for Environmental Management Information, Washington, D.C., DOE/EM-0228.

U.S. Department of Energy. 1995b. Estimating the Cold War Mortgage — The 1995 Baseline Environmental Management Report — Executive Summary. Office of Environmental Management, Washington, D.C.

U.S. Department of Energy. 1997. Uranium Mill Tailings, Integrated Data Base, Chapter 5. Office of Environmental Management, Washington, D.C.

U.S. Department of Energy. 1999. Waste Management FY 1999 Progress Report. Center for Environmental Management Information, Washington, D.C.

U.S. Department of Energy. 2000. Request for RCRA Class 2 Permit Modification in Accordance with 20.4.1.900 NMAC (incorporating 40 CFR Part 270). Waste Isolation Pilot Plant, Carlsbad, New Mexico.

U.S. Environmental Protection Agency. 1998. Effects of RCRA LDR Regulations on Mixed Waste Management. U.S. EPA Mixed Waste Team, Washington, D.C.

U.S. Environmental Protection Agency. 1998a. An Overview of Mixed Waste. U.S. EPA Mixed Waste Team, Washington, D.C.

U.S. Environmental Protection Agency. 1999. Proposed Rule for Storage, Treatment, Transportation, and Disposal of Mixed Waste. Solid Waste and Emergency Response, Washington, D.C., EPA 530-F-99-045.

U.S. General Accounting Office. 1993. NUCLEAR WASTE — Yucca Mountain Project Behind Schedule and Facing Major Scientific Uncertainties. Washington, D.C., GAO/RCED-93-124.

U.S. General Accounting Office. 1994. NUCLEAR WASTE — Foreign Countries' Approaches to High-Level Waste Storage and Disposal. Washington, D.C., Report No. GAO/RCED-94-172.

U.S. General Accounting Office. 1999. Low Level Radioactive Wastes: States Are Not Developing Disposal Facilities. Washington, D.C., GAO/RCED-99-238.

U.S. Nuclear Regulatory Commission. 1999. Information Digest, 1999 Edition (NUREG 1350, Vol. 11). Office of the Chief Financial Officer, Washington, D.C.

U.S. Nuclear Regulatory Commission. 1999a. Uranium Mill Tailings. Technical Issues Paper No. 19, Washington, D.C.

U.S. Office of Technology Assessment. 1985. Managing the Nation's Commercial High-Level Radioactive Waste. U.S. Congress, Superintendent of Documents, U.S. Government Printing Office, Washington, D.C.

Valentin, Jack. 2000. ICRP — The International Commission on Radiological Protection, Stockholm, Sweden.

Wagner, Julie. 1997. New and Innovative Technologies for Mixed Waste Treatment. U.S. Environmental Protection Agency, Office of Solid Waste, Permits and State Programs Division, Washington, D.C.

Ware, Patricia. 1999. ADOE Aims to Clean Up Most Sites by 2006, Environmental Official Tells Senate Panel, Environment Reporter, March 19, 1999, Bureau of National Affairs, Inc. Washington, D.C.

14 Underground Storage Tank Management

OBJECTIVES

At completion of this chapter, the student should:

- Understand the nature and magnitude of the environmental threat of leaking underground storage tanks.
- Understand the causes of underground storage tank and piping failures.
- Be familiar with the theories and practice of internal tank testing and external monitoring for leaks.
- Be familiar with remediation measures, tank rehabilitation procedures, and requirements for new tank and piping installations.
- Be conversant on the RCRA Subtitle I requirements for underground storage tank management.
- Be conversant on the distinctions between migration of subsurface release of MTBE and releases of other gasoline components and know where to find current information on the problem.

INTRODUCTION

Leaking underground fuel storage tanks (USTs) can cause fires or explosions and/or contaminate groundwater. More than 50% of the population in the U.S. depends upon groundwater for domestic use. Petroleum products, including gasoline, are highly mobile as they float on a sloped or flowing groundwater surface. Flammable liquids and/or vapors seeping into basements or other subterranean spaces can create explosive conditions, inhalation exposure hazards, or both. Thus, leaking USTs have been and are a major threat to the public health and safety and to the environment.

In the Hazardous and Solid Waste Amendments of 1984 (HSWA), Congress added a new Subtitle I to the Resource Conservation and Recovery Act (RCRA) to address the problem of leaking underground tanks used for storage of petroleum and hazardous *substances*. The implementing federal regulations are found in 40 CFR 280 and 281 (53 FR 37082). (Tanks used for storing hazardous *wastes* are regulated by 40 CFR 264 and 265.)

In 1994, the U.S. Environmental Protection Agency (EPA) estimated that about 1.2 million tanks at more than 500,000 sites were subject to federal regulation (EPA

FIGURE 14.1 Galvanic corrosion of an unprotected steel underground storage tank. (From Environmental Technology, Inc., 2541 E. University, Phoenix, AZ. With permission.)

1994d). Tens of thousands of these tanks, including their piping, had leaked or were then leaking. The EPA reported that in the 10 years since the Subtitle I regulatory program was authorized, the number of confirmed releases had reached 262,000.[1] Moreover, the agency expected the total number of releases to reach 400,000 during the next few years (EPA 1994c). By December 1998, more than 1 million substandard USTs that had been in service in 1988 had been taken out of operation, thereby removing them as sources of leaks. Of the 892,000 USTs then in operation, the EPA estimated that approximately 500,000 met the Part 280 and 281 standards (EPA 1998).

Typical condition of steel tanks being removed in remediation or replacement activity is shown in Figures 14.1 and 14.2. Many older tanks, and the associated piping, are of unprotected steel construction and can be expected to develop leaks unless they are removed or rehabilitated. In this chapter we will overview the nature and causes of the problem, the related technologies, and the regulatory structure.

Leaking Underground Storage Tanks — Problems and Causes

As noted, large numbers of the older USTs are of "bare" steel construction. Older tanks, especially those more than 10 years old and/or unprotected from corrosion, are likely to develop leaks. A leak from an UST, if undetected or ignored, can cause very large amounts of petroleum product to be lost to the subsurface. In a recent case, a tank at a city-owned vehicle maintenance facility lost an estimated half-million gallons of gasoline to a producing aquifer. In another case, a major oil company found it necessary to buy and vacate all of the residences on a city block

---

[1] Another EPA publication puts the figure at 341,000 in September 1997. About 30,000 new releases are reported each year (EPA 1998a). The June 2000 Report to Congress, referenced later herein, states that by September 1999, 400,000 releases had been reported (EPA 2000a, p. 5).

FIGURE 14.2 Corroded underground storage tank after removal. (From Environmental Technology, Inc., 2541 E. University, Phoenix, AZ. With permission.)

adjacent to a company-owned service station. Leaking gasoline had migrated from the underground tanks at the station, and liquid gasoline and vapors entered basements on the block. Water supply wells adjacent to older service stations are frequently contaminated with gasoline or other petroleum products.

Underground storage tanks usually release contaminants into the subsurface environment as a result of one or more of four factors: corrosion, faulty installation, piping failure, or spills and overfills. Galvanic corrosion, or the breakdown of hard refined steel, is the most common cause of release from bare steel UST systems. Because the majority of older UST systems are of bare steel, corrosion is believed to be the leading cause of releases (EPA 1998a, p. IV-2). This may not be true, however, in areas of the arid southwestern U.S.

Galvanic Corrosion

The rate and severity of corrosion varies depending upon a number of site-specific factors (e.g., soil moisture, conductivity) that are almost always present when bare steel is placed underground. Steel is, by definition, an alloy of iron and carbon, containing other constituents such as manganese, chromium, nickel, molybdenum, copper, tungsten, or cobalt. These metals have differing electromotive activities and the more active metals tend to displace the less active. Dissimilar metals may be present in the soil surrounding a steel tank. Most commonly, part of the tank becomes negatively charged with respect to the surrounding area. The negatively charged part of the UST acts as a negative electrode and begins to corrode at a rate proportional to the intensity of the current (adapted from EPA 1990, p. IV-2). Galvanic corrosion always occurs at a specific point on a tank or pipe where the current exits. As the current passes through this point, the hard steel is transformed into soft ore, a small hole forms, and the leak occurs. The hole, so formed, is usually small (Figure 14.3),

FIGURE 14.3 Typical pinhole leak caused by galvanic activity. (From U.S. Environmental Protection Agency.)

but large quantities of liquid may be released (*see:* Cole 1992, Appendix A, for a thorough discussion of the galvanic corrosion of steel underground tanks).

Faulty Installation

Installation failure encompasses a wide variety of problems such as inadequate backfill, allowing movement of the tank, and separation of pipe joints. Mishandling of the tank during installation can cause structural failure of fiberglass reinforced plastic (FRP) tanks or damage to steel tank coatings and cathodic protection. Cole lists the causes of failure that are related to backfill:

- Improper, inhomogeneous (sic) backfill material
- Inadequate or improper compaction
- Rocks or debris left in excavation
- Voids left under tank
- Failure to prevent migration of backfill
- Placing a tank directly on a concrete slab or hard native soil

Cole (1992, p. 49).

Piping Failures

The underground piping which connects tanks to each other, to delivery pumps, and to fill drops is even more frequently of unprotected steel (Figure 14.4). EPA studies indicate that piping failure accounts for 50 to 80% of leaks at UST facilities. The piping failures are nearly all caused by poor workmanship and/or corrosion. Threading of galvanized steel pipe exposes electrically active metal and creates a strong tendency to corrode if not coated and cathodically protected. The problem is com-

FIGURE 14.4 Typical corroded piping at an underground storage tank replacement site. Note hole in pipe nipple between elbows. (From Environmental Technology, Inc., 2541 E. University, Phoenix, AZ. With permission.)

pounded if the fittings and valves used in the system are of dissimilar metals. Improper layout of piping runs, incomplete tightening of joints, inadequate cover pad construction, and construction accidents can lead to failure of delivery piping (adapted from Cole 1992, p. 2; Munter et al. 1995, p. 190). Figure 14.5 illustrates a typical service station tank and piping layout.

Spills and Overfills

Spills and overfills, usually caused by human error, contribute to the release problem at UST facilities. In addition to the direct contamination effect, repeated spills of petroleum products or hazardous wastes can intensify the corrosiveness of soils. Spills and overfills are almost totally attributable to human error. These mistakes can be avoided by following correct tank filling practices and by providing spill and overfill protection. The EPA regulations require catchment basins to contain spills and the installation of automatic shutoff devices, overfill alarms, or ball float valves (EPA 1994a; EPA 1998a, p. IV-3).

Compatibility of UST and Contents

Another possible cause of tank failure has become of concern in areas of the nation that are experimenting with additives, blends, and alternative (automotive) fuels in the hope of achieving improved air quality. The rush to replace steel tanks has enhanced the popularity of fiberglass-reinforced plastic (FRP) tanks and tank liners, to the end that large numbers of them have been put into service. Some FRP tanks or liners may not be compatible with some methanol-blended (and possibly some ethanol-blended) fuels or with additives such as methyl tertiary butyl ether (MTBE).

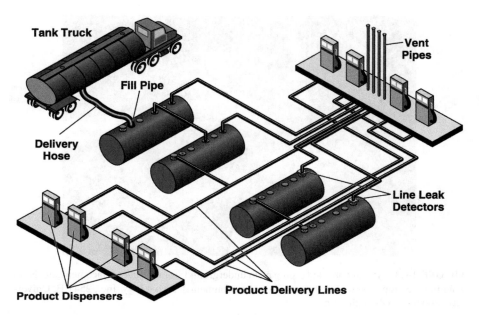

FIGURE 14.5 Typical service station tank and piping layout. (From U.S. Environmental Protection Agency.)

Compatibility for tanks means that the fuel components would not change the physical or mechanical properties of the tank. Compatibility for liners requires that the fuel components not cause blistering, underfilm corrosion, or internal stress or cracking. Owners/operators of FRP-constructed or lined tanks should consult the appropriate standards of the American Petroleum Institute (API) (adapted from Leiter 1989, p. 55).

Mobility of Leaked Hydrocarbon Fuels

Motor fuels, when leaked, are acted upon by gravitational forces which act to draw the fluid downward. Other forces act to retain the fuel, which is either adsorbed to soil particles or trapped in soil pores. The amount of fuel retained in the soil is of primary importance, as it will determine both the degree of contamination and the likelihood of subsequent contaminant transport to groundwater (Bauman 1989, p. 3). Upon reaching the saturated zone, some of the lighter components may dissolve in water, but large quantities can float on the water surface, sliding downgradient over great distances. This mobility frequently causes remediation of leaking UST sites to be costly, involving many recovery wells and large-scale separation of pumped water and recovered product. MTBE, a gasoline additive (see box), is water soluble, does not partition with the gasoline, and is transported with the groundwater flow.

Methyl tertiary butyl ether (MTBE), a synthetic chemical oxygenate, is blended with gasoline to improve combustion and reduce carbon monoxide and ozone emissions from automobile exhaust systems. MTBE may comprise up to 10%, by volume, of gasoline (Robinson et al. 1993). This additive chemical does not behave in the manner of gasoline, or other additives, when released to the subsurface. In a recent well-reasoned paper, a developer/marketer of biomerme-dation products[2] summarized the factors contributing to the complexity of reme-diating properties contaminated with fuels containing the chemical:

- MTBE degrades very slowly under aerobic conditions.
- MTBE is not recognized as an anaerobically degradable compound.
- Unlike BTEX, MTBE is highly soluble and does not retard on the aquifer matrix. The compound is therefore capable of rapid and pervasive disper-sion in groundwater.
- The toxicity and carcinogenicity of MTBE have not been established.
- Taste and odor thresholds for MTBE are very low.
- Although MTBE is extremely volatile, when dissolved in water, it is dif-ficult to strip which complicates sparging and pumping options. In the latter case, pumped water may have to be treated in bioreactors (Regenesis 1999).

The Regenesis paper, which reports on experiments with oxygen release com-pounds (ORC®) in wells, suggests that it may be possible to achieve some deg-radation of MTBE in groundwater by enhancing conditions for aerobic activity. The paper can be accessed at <http://www.regenesis.com/ORCtech/Tb2231.htm>.

The impacts and issues generated by the use and release of MTBE are many faceted and conflicting. An EPA publication reports that experiments with lab-oratory microcosms constructed with material from an MTBE-contaminated aquifer indicate that significant reductions in MTBE were achieved under meth-anogenic conditions (EPA 2000). A study by the Lawrence Livermore National Laboratory concluded that MTBE is a "frequent and widespread contaminant" in groundwater throughout California and does not degrade significantly once it is there. The study estimates that MTBE has contaminated groundwater at over 10,000 shallow monitoring stations in California. The California Depart-ment of Health Services (DHS) adopted a primary (health-based) drinking water standard of 5 ppb in April 2000 (ACWA 2000).

The EPA issued an Advance Notice of Proposed Rulemaking (ANPRM) to issue a rule under the Toxic Substances Control Act (TSCA)(40 CFR 755) to Control MTBE in Gasoline. The ANPRM states that the outcome of the rule-making could be a total ban on the use of MTBE as an additive or several lesser limitations (65 FR 16093, March 24, 2000). On July 12, 2000, the EPA issued a Notice of Proposed Rulemaking (40 CFR 80), which adjusted the Clean Air Act regulations to "... increase the flexibility available to refiners to formulate

---

[2] Regenesis Bioremediation Products, Inc.

RFG[3] without MTBE while still realizing ozone benefits that are similar to those ... (existing)" (65 FR 42920). On August 4, 2000, the EPA issued a Notice of Proposed Rulemaking (40 CFR 80 and 86) to "... develop a framework to construct a national mobile source air toxics program ... and make a commitment to revisit the issue of mobile source air toxics controls in a 2004 rulemaking." The action creates a list of 21 Mobile Source Air Toxics (MSATs), which includes MTBE; however, the only MSAT for which immediate action is proposed is benzene (65 FR 48057).

It is unclear what, if any, perturbations the MTBE concerns hold for practitioners; however, it is clear that UST-related program managers, consultants, technicians, contractors, manufacturers, financial interests, and insurers must take steps to remain informed and focused on the subject.

As noted, leaked or spilled gasoline percolates to the groundwater surface, then floats on that surface, traveling downgradient at rates determined largely by the physical characteristics of the geologic materials, which make up that portion of the vadose zone and by the slope of the water table. The highly soluble and miscible MTBE readily mixes with a moving groundwater plume and may move ahead of the gasoline plume (Weaver et al.1999; *see also:* Swain 2000; EPA 2000; Cater et al. 2000).

PROTECTION OF TANKS AND PIPING FROM CORROSION

Galvanic corrosion is the most common cause of corrosion and subsequent release from bare steel UST systems. Steel tanks and piping can be protected from corrosion by coating them with a corrosion-resistant coating and by using "cathodic" protection. Cathodic protection reverses the electric current that causes corrosion and can be applied in the form of sacrificial anodes or as an impressed current.

Protection by Sacrificial Anode

Sacrificial anodes are pieces of metal that are more electrically active than steel in the UST to which they are attached. Because the anodes are more active, the electric current will exit from them rather than from the steel tank. Thus the tank becomes the cathode and is protected from corrosion while the attached anode is sacrificed. Depleted anodes must be replaced in order to achieve continuous protection of the UST (EPA 1998b).

Protection by Impressed Current

An impressed current protection system uses a rectifier to convert alternating current to direct current. The current is sent through an insulated wire to the anodes, which are metal bars buried in the soil near the UST. The current flows through the soil

---

[3] Reformulated gasoline.

FIGURE 14.6 Composite steel-fiberglass reinforced plastic tanks.

to the UST system and returns to the rectifier through an insulated wire attached to the UST. Since the electric current flowing from these anodes to the tank system is greater than the corrosive current attempting to flow from it, the UST is protected from corrosion (EPA 1998b; *see also:* Cole 1992, Appendix A).

Protection by Cladding or Dielectric Coating

Steel-fiberglass-reinforced-plastic composite tanks are adequately protected from corrosion by the thick outside layer (or cladding) of FRP (Figure 14.6). Cathodic protection is not needed with this method of protection (40 CFR 280.20). New steel tanks for petroleum storage must be coated with a dielectric coating (asphalt or paint) and cathodically protected (40 CFR 280.20). Care must be taken during installation to protect the coating from damage. Any separation ("holiday") of the coating from the tank tends to focus the galvanic forces, accelerates corrosion, and may cause a release (Leiter 1989, p. 22).

Protection of Piping

The UST regulations require that piping in contact with the ground be constructed entirely of fiberglass-reinforced plastic or if of steel be cathodically protected by:

- Coating with suitable dielectric material
- Field-installed cathodic protection system designed by a corrosion expert
- Impressed current system
- Cathodic protection conforming with listed codes and standards (40 CFR 280.20)

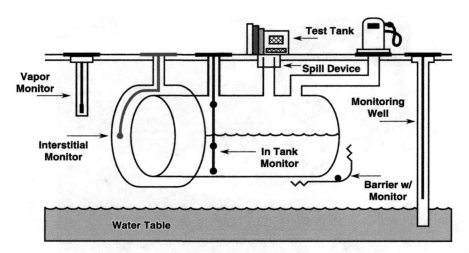

FIGURE 14.7 Underground storage tank leak detection alternatives. (From U.S. Environmental Protection Agency.)

DETECTION OF LEAKS FROM UNDERGROUND STORAGE TANK SYSTEMS

Seven general methods of leak detection are used for underground storage tanks. Some of the methods have many variations. Practitioners and tank testing companies vigorously argue the merits of particular methods and the supporting technologies. The Subtitle I regulations allow owners or operators of UST facilities to choose between leak detection methods and impose specific requirements on the use of each method. The student should refer to Figure 14.7 as the methods are briefly described.

1. *Automatic Tank Gauging.* This method uses probes which are permanently installed in the tank and an external control device to monitor product level and temperature. These systems automatically calculate the changes in product volume that can indicate a leaking tank (EPA 1998a; *see also:* Leiter 1989, p. 174; Wilcox 1990, pp. 119ff).
2. *Groundwater Monitoring.* This method is used to detect the presence of gasoline or other liquid product floating on the groundwater. Monitoring wells are placed at strategic locations in the ground near the tank and piping runs. The wells may be sampled periodically by hand or continuously with permanently installed equipment. The method is effective only at sites where groundwater is within 20 ft of the surface (EPA 1998a).
3. *Soil Vapor Monitoring.* Leaked petroleum product releases vapors into the soil surrounding the UST. Vapor monitoring around the tank and piping senses the presence of vapors from leaked product. The method requires that tanks be backfilled with porous soils and that monitoring locations be carefully planned. Vapor monitoring can be performed manually, on a prescribed frequency, or continuously, using permanently installed equipment (EPA 1998a).

4. *Secondary Containment and Interstitial Monitoring.* Secondary containment is achieved by placing a barrier between the UST and the environment. The barrier may be a vault, liner, or double-walled structure. Leaked product from the UST is detected by monitoring of the space between the tank and the barrier. Alternatively, tanks can be equipped with inner bladders to provide secondary containment. If product escapes from the inner tank or piping, it will then be directed toward an interstitial monitor located between the walls (EPA 1998a).

5. *Statistical Inventory Reconciliation.* This method uses sophisticated computer software to determine whether a tank system is leaking. The computer conducts a statistical analysis of inventory, delivery, and dispensing data. These data are then analyzed to determine if any product has been released (EPA 1998a).

6. *Manual Tank Gauging.* Manual gauging can be used only on tanks of 2000-gal capacity or smaller. The method requires taking the tank out of service for at least 36 hr each week to take measurements of the tank's contents. Tanks of not more than 1000-gal capacity may use this method alone. Tanks of 1001- to 2000-gal capacity must also use periodic tank tightness testing and for only 10 years after installation or upgrade of the UST. After 10 years, these USTs must use one of the other detection methods listed in 1 through 5 above (EPA 1994b).

7. *Tank Tightness Testing with Inventory Control.* The method combines monthly inventory control information (measured daily and compiled monthly) with periodic tank tightness testing. Inventory control involves taking measurements of tank contents, recording the amount of product pumped each operating day, and reconciling these data at least once monthly. Tank tightness testing includes a variety of methods used to determine if a tank is leaking; most of these methods involve monitoring changes in product level or volume in a tank over a period of several hours (EPA 1998a).

Detection of Leaks in Pressurized Underground Piping

An automatic line leak detector is required. The automatic line leak detector uses combinations of flow restrictors and flow shutoffs to monitor pressure in a line. Automatic line leak detection must be accompanied by one of the following methods: groundwater monitoring, vapor monitoring, secondary containment and interstitial monitoring, or an annual tightness testing of the piping (EPA 1998a).

Detection of Leaks in Underground Suction Piping

Leak detection is not required if the suction piping meets the following basic design requirements:

- Below-grade piping operating at less than atmospheric pressure is sloped so that the contents of the piping will drain back into the storage tank if suction pressure is released, *and*

- Only one check valve is included in each suction line and it is located directly below the suction pump in the dispensing unit.

Suction piping that does not meet the above requirements must be subjected to one of the following:

- Line tightness tests every 3 years
- Groundwater monitoring
- Soil vapor monitoring
- Secondary containment and interstitial monitoring

(EPA 1998a).

RCRA Subtitle I Regulations and Requirements

Background

In recognition of the leaking underground storage tank problem, Congress included the original Subtitle I in the 1984 Hazardous and Solid Waste Amendments (HSWA). Subtitle I contained provisions prohibiting installation of new tanks that were not designed to prevent releases due to corrosion, structural failure, or incompatibility and imposed notification requirements upon owners of UST facilities.

The implementing regulations (40 CFR 280) were significantly broadened and strengthened in 1988. The new regulations established (1) the technical standards that **new** and **existing** UST facilities were/are required to meet and (2) financial assurance requirements that required owners or operators to demonstrate that they could pay for cleanup of leaks from their UST facilities. The December 22, 1998 deadline for **existing** UST owners/operators to replace, upgrade, or close substandard tanks (and piping) has passed. Nevertheless, many tanks and/or subsurface piping systems are not in compliance, as discussed herein.

The goals of the UST regulations are

- To prevent leaks and spills
- To detect leaks and spills if and when they occur
- To ensure that owners and operators can pay for correction of problems created by leaks that may occur
- To ensure that state regulatory programs for underground storage tanks impose regulations that are as strict or more strict than the federal regulations (EPA 1988b)

Implementation Schedule

The regulations impose differing requirements upon owners or operators of **new** and **existing** UST systems. **New** UST systems are those that are installed after December 1988. Existing systems are those installed before December 1988 (40 CFR 280.12).

Requirements for New Petroleum UST Systems

Owners or operators of UST systems that were installed after December 1988 must meet five technical requirements. The requirements paraphrased[4] are

- Tank and piping must be protected from corrosion in accord with a code of practice developed by a nationally recognized association or independent testing laboratory [§ 280.20(a)].
- The piping that routinely contains regulated substances and is in contact with the ground must be properly designed, constructed, and protected from corrosion in accord with a code of practice developed by a nationally recognized association or independent testing laboratory [§ 280.20(b)].
- Owners and operators must install specified equipment to prevent spilling and overfilling associated with product transfer to the UST [§ 280.20(c)].
- All tanks and piping must be properly installed in accord with a code of practice developed by a nationally recognized association or independent testing laboratory [§ 280.20(d)].
- All owners and operators must ensure that certification, testing, or inspection is used to demonstrate compliance with paragraph (d) of this section by providing a certification of compliance on the UST notification form in accord with § 280.22 [§ 280.20(e)].

Requirements for Existing UST Systems

As noted above, an existing UST system is one that was installed prior to December 1988. The implementation schedule for existing systems required immediate adoption of tank-filling procedures that will prevent spills and overfills. By December 1998 (10 years following promulgation of the UST regulations), all **existing** UST systems were required to comply with one of the following:

- **New** UST system performance standards of § 280.20
- The **upgrading** requirements of paragraphs (b) through (d), below
- Closure requirements of subpart G of part 280, including applicable requirements for corrective action of subpart F [§ 280.21(a)]
  - Tank upgrading requirements: Steel tanks must be upgraded to meet one of the following requirements in accord with a code of practice developed by a nationally recognized association or independent testing laboratory:
    Interior lining
    Cathodic protection
    Internal lining combined with cathodic protection [§ 280.21(b)]
  - Piping upgrading requirements: Metal piping that routinely contains regulated substances and is in contact with the ground must be cathod-

---

[4] The five standards include extensive detail, particularly with respect to the codes of practice and options. The practitioner involved in this work should study Section 280 carefully.

ically protected in accord with a code of practice developed by a nationally recognized association or independent testing laboratory, and must meet the requirements of § 280.20(b)(2) [§ 280.21(c)].

- Spill and overfill prevention equipment: To prevent spilling and overfilling associated with product transfer to the UST system, all existing UST systems must comply with new UST system spill and overfill prevention equipment requirements specified in § 280.20(c) [§ 280.21(d)].

Corrective Action Requirements

The release investigation and corrective action requirements pertaining to UST releases are found at 40 CFR 280, Subparts E and F. The EPA regulations are carefully worded to defer to the implementing (state or local) agency, but also establish minimum standards for release responses. The standards, paraphrased,[5] require owners and operators to:

- Report to the implementing agency, within 24 hr, discovery of a release, unusual operating conditions, or monitoring results that indicate that a release may have occurred.
- Conduct appropriate tests, repair, replace, or upgrade the defective unit, and begin corrective action as required by Subpart F.
- Contain and immediately clean up spills and overfills.
- Report releases equal to or in excess of reportable quantities of hazardous substances to the National Response Center.
- Investigate the extent of the release, monitor, and mitigate fire and safety hazards.
- Remedy hazards posed by contaminated soil and groundwater.
- Conduct free product removal and abate free product migration.

(*See:* 40 CFR 280, Subparts E and F).

The § 280, Subparts E and F, referenced above do not provide guidance nor requirements regarding the release reporting, investigation, confirmation, response, or corrective action of an MTBE release. Considering the aforementioned indications that MTBE may not behave as and/or partition with other fuel components, the regulatory agencies can be expected to issue directives for contaminant-specific management of such releases. Although it is not possible to predict the form or content of future directives, the practitioner should maintain currency with the technical literature pertaining to the topic and take appropriate steps to minimize the subsurface transport and environmental impact of any release of the material for which he/she might have the opportunity to (1) minimize the health or environmental impacts of or (2) be held responsible for. Some beginning references and Web sites are

---

[5] The Subpart E and F requirements are condensed here for overview knowledge. They are extensive and subject to augmentation by the implementing agency. The practitioner should become fully familiar with the content of Subparts E and F, as well as applicable state and/or local codes.

- Papers presented at the Fifth International *In Situ* and On-Site Bioremediation Symposium in San Diego, California, April 19–22, 1999 (collected and published in Alleman and Leeson 1999: presented or authored by Reid et al.; Hurt et al.; Jong and Wilson; Anthony et al.; Zenker, Borden, and Barlaz; McLinn; Miller; and Edwards, Hayer, and Krueger)
- EPA 1998d; EPA 1999; EPA 2000
- http://www.epa.gov/swerffrr/petrol.htm
  http://www.epa.gov/otaq/fuels.htm
  http://www.epa.gov/safewater/mtbe.html
  http://www.epa.gov/OUST/fedlaws.htm
  http://www.epa.gov/ada/reports.html

Financial Responsibility Requirements

Subpart H (40 CFR 280.90 through 280.111) establishes extensive and complex financial assurance requirements that either the owner or operator of a UST system must meet. The intent of these regulations is to ensure that money is available to pay for cleanup of releases of petroleum product and to compensate third parties for bodily injury and property damage resulting from a release.

The financial responsibility regulations require "per occurrence" coverage, as follows:

- Petroleum marketers — $1 million per occurrence
- Petroleum nonmarketer, having monthly average petroleum product throughput greater than 10,000 gal — $1 million per occurrence
- Petroleum nonmarketer, having monthly average petroleum product throughput less than 10,000 gal — $500,000 per occurrence

Annual aggregate coverage is the total amount of financial responsibility coverage required to pay for the costs of all leaks that might occur in 1 year. Owners and operators of more than 100 USTs must demonstrate annual aggregate coverage of at least $2 million; owners of 100 or fewer tanks must demonstrate at least $1 million in annual aggregate coverage (EPA 1998, p. IV-13; 40 CFR 280.93).

The required coverage may be shown in one of the following ways:

- State assurance funds
- Financial test of self-insurance
- Corporate guarantee
- A surety bond in the required amount
- Letter of credit for the required amount
- A fully funded trust fund
- Another state-approved method, such as a risk retention group
- A combination of the above, with aggregate coverage in the required amount

(EPA 1998, p. IV-13,14).

The LUST Trust Fund

The Leaking Underground Storage Tank (LUST) Trust Fund was created by Congress in the 1986 amendments to RCRA Subtitle I and was reauthorized for 5 more years in 1990. The fund is financed by an excise tax on motor fuel sold in the U.S. After expiration in December 1995, the tax was reinstated by Congress in 1997 (*Environment Reporter,* August 6, 1999, p. 704). The fund provides money for:

- Overseeing corrective action taken by a responsible party — usually the owner or operator of the leaking UST
- Cleanups at UST sites where the owner or operator is unknown, unwilling, or unable to respond or that require emergency action

By March 1997, the fund had collected about $1.8 billion and had disbursed about $655 million to the EPA. The agency has passed through about $560 million to state programs for use in administration, oversight, and cleanup work. The remaining trust fund money has been used by the EPA for administrative activities; negotiating and overseeing cooperative agreements; implementing programs on Native American lands; and supporting EPA regional and state offices. States use trust fund money to oversee corrective action by a responsible party and to clean up sites where no responsible party can be found (EPA 1998a, pp. IV.17-18).

CLOSURE OF UNDERGROUND STORAGE TANK FACILITIES

In keeping with RCRA requirements for closure of hazardous waste sites, the UST regulations also require formal closure when use ends or is suspended.

Permanent Closure

Tanks that are not protected from corrosion and that are unused for more than 12 months or tanks to be permanently closed must conform to the required procedures for permanent closure. The requirements, in brief, are

- The regulatory agency must be notified at least 30 days prior to closure (§ 280.71).
- An assessment must be made to determine if leakage has occurred. The requirement may be satisfied if one of the external release detection methods (soil vapor or groundwater monitoring) is in operation at the time of closure and indicates that no release has occurred. If contamination is detected, corrective action must be taken in accord with Subpart F (§§ 280.71, 280.72).[6]
- The tank must be emptied and cleaned by removing all liquids, dangerous vapors, and accumulated sludge (§ 280.71).[7]

---

[6] Some states have taken exception to this federal regulation. State regulations should also be followed in a tank closure activity.

[7] An extremely hazardous activity (*see:* Chapter 15 this text; Bridge 1988).

- The tank must be removed from the ground or closed in-place. Closure in-place requires filling with an inert solid material such as sand (§ 280.71).

Exceptions to Permanent Closure

Requirements for permanent closure may not apply if:

- The tank meets requirements for a new or upgraded UST. It may remain "temporarily" closed indefinitely, provided it meets the requirements (below) for temporary closure (§ 280.71).
- The regulatory authority grants an extension beyond the 12-month limit on temporary closure of tanks that are unprotected from corrosion. In this case, a site assessment must be accomplished (§ 280.71).
- The stored contents are to be changed to an unregulated substance. The regulatory agency must be notified of the change; the cleaning and assessment procedures for permanent closure must be followed; and any release must be corrected per Subpart F (§§ 280.71, 280.72).

Temporary Closure

Tanks not in use for 3 to 12 months must meet requirements for temporary closure. These requirements, which are contained in the Subpart G regulations are, in brief:

- Operation and maintenance of corrosion protection equipment must be continued, and any detection of a release must be corrected per Subparts E and F.
- All vent lines must remain open and functioning.
- All other lines must be capped; pumps, manways, and other ancillary equipment must be secured (§ 280.70).

(*See also:* Bridge 1988.)

COMPLIANCE SUMMARY

Compliance Status As of September 30, 1999

The following status information is excerpted from the EPA Report to Congress referenced earlier:

… Since the inception of the program, when there were more than 2 million federally regulated USTs, more than 1.3 million substandard USTs have been closed. As a result of those closures, the substandard tanks are no longer sources of actual or potential releases which could harm human health and the environment. As of September 30, 1999, the federally regulated tank universe was about 760,000, of which states and EPA report approximately 85 percent were in compliance with the spill, overfill, and corrosion protection portion of the regulations (1998 requirements).

States have reported nearly 400,000 confirmed releases from USTs. Cleanups have been initiated for approximately 346,000 releases and almost 229,000 cleanups have been complete. More than 20,000 cleanups are completed annually. Even with this rate of success, many thousands of cleanups remain to be completed.

... Many states estimate that the operational compliance rate with the leak detection requirements is approximately 60 percent. (These requirements were phased in between 1989 and 1993.) Constant efforts, including increased inspections by states and EPA, will be necessary to improve this compliance rate.

... The population of active registered USTs is approximately 722,000.

... EPA estimates there are approximately 38,000 abandoned registered USTs.

... EPA estimates there are approximately 38,000 active unregistered USTs.

... EPA estimates there are approximately 152,000 abandoned unregistered ("orphaned") USTs.

(EPA 2000a, pp. 9–12.)

TOPICS FOR REVIEW OR DISCUSSION

1. Four factors result in releases from underground petroleum storage tanks. What are the factors? One of these is believed to be the most common cause of releases. Which one is that?
2. Describe/explain the galvanic corrosion process as it affects steel underground storage tanks.
3. How do sacrificial anodes protect underground storage tanks and piping? What other methods of corrosion protection/prevention are available for USTs.
4. Is the piping associated with underground storage tanks also subject to galvanic activity?
5. Why is testing of underground pressurized and suction piping associated with underground storage tanks considered to be so important?
6. The RCRA Subtitle I regulations (40 CFR 280) provide several leak detection options. What is meant by interstitial monitoring?
7. Discuss late and current findings regarding human health and environmental impacts of MTBE contamination of groundwater.
8. Under what circumstances is leak detection not required for UST piping? What is the rationale for that exception?
9. You've been notified by the attorney for your late uncle Harry that he left his old service station property to you. It has been padlocked since 1976. You decide to look the place over and find that there are at least two underground storage tanks that apparently have some petroleum product in them. The nearby, long-unused well contains water that has a strong odor of petroleum. What must be one of the first things that you do?

REFERENCES

Alleman, Bruce C. and Andrea Leeson. 1999. *Natural Attenuation of Chlorinated Solvents, Petroleum Hydrocarbons, and Other Organic Compounds.* Battelle Press, Columbus, OH.

Association of California Water Agencies. 2000. *ACWA's MTBE Page Summer 2000.*

Bauman, Bruce J. 1989. "Soils Contaminated by Motor Fuels: Research Activities and Perspectives of the American Petroleum Institute," Chapter 1, in *Petroleum Contaminated Soils: Remediation Techniques, Environmental Fate, and Risk Assessment,* Paul T. Kostecki and Edward J. Calabrese, Eds., Lewis Publishers, Inc., MI.

Bridge, Jennie. 1988. *Tank Closure without Tears: An Inspector's Safety Guide.* New England Interstate Water Pollution Control Commission, Boston, MA.

Cater, Stephen R., Bertrand W. Dussert, and Neal Megonnell. 2000. "Reducing the threat of MTBE-Contaminated Groundwater," *Pollution Engineering,* May 2000.

Cole, G. Mattney. 1992. *Underground Storage Tank Installation & Management.* Lewis Publishers, Chelsea, MI.

Environment Reporter, August 6, 1999, p. 704. Bureau of National Affairs, Washington, D.C.

Koenigsberg, Stephen S. and Robert B. Norris, Eds. 1999. *Accelerated Bioremediation Using Slow Release Compounds — Selected Battelle Conference Papers 1993–1999.* Regenesis Bioremediation Products, San Clemente, CA.

Koenigsberg, Stephen S. and Craig A. Sandefur. 1999. "The Use of Oxygen Release Compound for the Accelerated Bioremediation of Aerobically Degradable Contaminants: The Advent of Time Release Electron Acceptors," in *Remediation,* John Wiley & Sons, NY.

Leiter, Jeffrey L. Editor-in-Chief. 1989. *Underground Storage Tank Guide.* Thompson, Salisbury, MD.

Munter, Florence et al. 1995. "Hazardous Wastes," in *Accident Prevention Manual for Business and Industry — Environmental Management.* National Safety Council, Itasca, IL.

Regenesis Bioremediation Products. 1999. *Potential for the Bioremediation of Methyl Tertiary Butyl Ether (MTBE),* ORC Technical Bulletin 223, Regenesis, San Clemente, CA.

Robinson, Janet E., Paul Thompson, W. David Conn, and Leon Geyer. 1993. *Issues in Underground Storage Tank Management Tank Closure and Financial Assurance.* CRC Press, Boca Raton, FL.

Stocking, Andrew, Stephen Koenigsberg, and Michael Kavanaugh. 1999. "Remediation and Treatment of MTBE," in *Environmental Protection,* Stevens, Dallas, TX.

Swain, Walt. 2000. Methyl Tertiary-Butyl Ether (MTBE) Overview. U.S. Geological Survey, California.

U.S. Environmental Protection Agency. 1988a. Dollars and Sense. Office of Underground Storage Tanks, Washington, D.C., 530-UST-88-005.

U.S. Environmental Protection Agency. 1988b. Musts for USTs. Office of Underground Storage Tanks, Washington, D.C., 530-UST-88-008.

U.S. Environmental Protection Agency. 1990. RCRA Orientation Manual, 1990 Edition. Superintendent of Documents, U.S. Government Printing Office, Washington, D.C.

U.S. Environmental Protection Agency. 1994a. UST Program Facts Preventing Releases. Solid Waste and Emergency Response, Washington, D.C., EPA 510-F-94-004.

U.S. Environmental Protection Agency. 1994b. UST Program Facts Detecting Releases. Solid Waste and Emergency Response, Washington, D.C., EPA 510-F-94-005.

U.S. Environmental Protection Agency. 1994c. UST Program Facts Cleaning Up Releases. Solid Waste and Emergency Response, Washington, D.C., EPA 510-F-94-006.

U.S. Environmental Protection Agency. 1994d. UST Program Facts Overview of the UST Program. Solid Waste and Emergency Response, Washington, D.C., EPA 510-F-94-008.

U.S. Environmental Protection Agency. 1998. "Supplemental Information Regarding the August 10, 1998 Enforcement Strategy," Attachment to the December 9, 1998 Memorandum, Subject: EPA's Inspection and Compliance Assistance Priorities For Underground Storage Tank Systems Not Meeting The 1998 Deadline, From: Steven A. Herman and Timothy Fields, Jr., To: Regional Administrators, Regions I-X.

U.S. Environmental Protection Agency. 1998a. "Managing Underground Storage Tanks — RCRA Subtitle I," RCRA Orientation Manual Section IV, Office of Solid Waste, Washington, D.C., EPA 530-R-98-004.

U.S. Environmental Protection Agency. 1998b. Cathodic Protection. Office of Underground Storage Tanks, Office of Solid Waste and Emergency Response, Washington, D.C.

U.S. Environmental Protection Agency. 1998c. "Managing Underground Storage Tanks RCRA Subtitle I," RCRA Orientation Manual, Section IV. Office of Solid Waste and Emergency Response, Washington, D.C., EPA 530-R-98-004.

U.S. Environmental Protection Agency. 1998d. Technical Protocol for Evaluating Natural Attenuation of Chlorinated Solvents in Ground Water. Office of Research and Development, Washington, D.C., EPA 600-R-98-128.

U.S. Environmental Protection Agency. 1999. Monitored Natural Attenuation of Petroleum Hydrocarbons, U.S. EPA Remedial Technology Fact Sheet, Office of Research and Development, Washington, D.C., EPA 600-F-98-021.

U.S. Environmental Protection Agency. 2000. Natural Attenuation of MTBE in the Subsurface Under Methanogenic Conditions. Office of Research and Development, Washington, D.C., EPA 600-R-00-006.

U.S. Environmental Protection Agency. 2000a. Report to Congress on a Compliance Plan for the Underground Storage Tank Program. Solid Waste and Emergency Response, Washington, D.C., EPA 510-R-00-001.

Weaver, J. W., J. E. Haas, and C. B. Sosik. 1999. "Characteristics of Gasoline Releases in the Water Table Aquifer of Long Island," in *Proceedings of 1999 Petroleum, Petroleum Hydrocarbons and Organic Chemicals in Ground Water,* pp. 262 ff, American Petroleum Institute/National Ground Water Association, November 17–19, Houston, TX.

Wilcox, H. Kendall. 1990. "In-Tank Leak Detection Methodologies," in *Underground Storage Systems — Leak Detection and Monitoring,* Todd G. Schwendeman and H. Kendall Wilcox, Eds., Lewis Publishers, Chelsea, MI.

15 Hazardous Waste Worker Health and Safety

OBJECTIVES

At completion of this chapter, the student should

- Understand the types of hazards that may be encountered by workers on hazardous waste sites.
- Be familiar with actions and preventive measures that may or should be taken to minimize impacts of those hazards, during both routine and emergency conditions.
- Be familiar with regulatory requirements for protection of worker health and safety on hazardous waste sites.

INTRODUCTION

Item: *Labor Secretary Robert B. Reich proposes penalties of $1,597,000 against Rhone-Poulenc AG Co. of Institute, West Virginia for violations of the OSHA Chemical Process Safety Standard and the Hazardous Waste and Emergency Response Standard. One worker was killed and two others sustained lung and skin injuries as a result of a fire and explosion on August 18, 1993.* (From OSHA News Release, February 17, 1994.)

Item: *Cedric Jackson, a concrete finisher, decided to make a little extra money to support his wife and four children by helping Jerry Martin remove two 10,000 gallon tanks from Martin's Automotive Shop property. A Florida Department of Environmental Regulation official had earlier advised Martin to contact a pollution specialty contractor to remove the fuel from the tanks, dismantle the system and remove the tanks. Instead, Martin hired Jackson, who had never worked on underground storage tanks, at $5 per hour, to undertake the project. Jackson either slipped or was thrown between the tanks when one of the tanks rolled. It took fire and rescue teams more than five hours to secure the tanks and remove Jackson's body from beneath the concrete anchor in the muddy, fuel-contaminated tank hole.* (From Petroleum Equipment Institute, *Tulsaletter*, July 28, 1992.)

Item: *Workers using organic solvents and detergents to remove polychlorinated biphenyl contamination from a poorly ventilated factory basement experienced "grossly abnormal" neurologic symptoms. One worker developed headache, memory impairment, and acute confusion after three days of work with the solvents. His mental status — which was clinically normal before his employment — was judged abnormal by the same physician after the work with the solvents. Neuropsychiatric testing performed nine months and 20 months after job completion "demonstrated severe deficits in attention, memory, and concentration." Another worker who developed similar symptoms was tested 20 months later and showed deficits in attention, concentration, and memory. Complaints from both engineers and laborers indicated that work conditions were poor; ventilation was inadequate; respirator use was minimal; skin protection was ineffective; and cleaning agents were mixed together and used in higher-than-recommended concentrations.* (From *Occupational Safety and Health Reporter,* July 29, 1992.)

Item: *In a plea agreement filed September 30 in federal court, Lancaster Synthesis Inc. admitted to illegally transporting hazardous waste to a Cincinnati, Ohio storage facility, where it exploded and killed a man (U.S. v. Lancaster Synthesis, S.D. Ohio, No. CR 1-99-85, 9/30/99). … chemical company admitted that in 1994 it knowingly shipped hazardous waste containing sodium azide without the manifest required by the Resource Conservation and Recovery Act (RCRA). Lancaster Synthesis also admitted to willfully making a material false statement in the shipment's bill of lading by stating that the waste from a South Carolina facility it was closing, was non-hazardous and not regulated.* (From *Occupational Safety & Health Reporter,* October 6, 1999.)

Item: *OSHA proposed penalties greater than $2 million against Southern Scrap Metals, which employs 150 workers to process scrap and waste materials. The company was cited for 40 willful violations related to employee exposure to lead, 21 violations related to cadmium exposure, violations of various safety requirements, and four repeat safety violations. Many of the violations involve temporary Mexican workers who speak little English.*[1] *The alleged violations include overexposure of seven workers to lead, failure to monitor lead exposures, lack of a written compliance plan, work surfaces contaminated with lead, no change or shower facilities, no lunch room free from lead dust, no medical surveillance program, and no employee training. The OSHA Baton Rouge area director said the worst incident at the plant involved one worker who was exposed to 400 micrograms of lead per cubic meter of air, eight times OSHA's permissible limit.* (From *Occupational Safety and Health Reporter,* October 5, 1994.)

Workers face a formidable array of workplace hazards and potential hazards as they perform the many routine and nonroutine tasks associated with the practice of hazardous waste management. Whether collecting wastes from satellite collection points for transfer to a central collection point, remediating an abandoned chemical storage facility, or responding to a hazardous materials spill at a manufacturing facility, the hazardous waste worker is challenged by known and unknown hazards to an extent and extreme matched by few other workplace activities. In earlier times,

---

[1] For an exploration of ethnic populations in "high-hazard, low-wage" jobs, see Robinson (1991, Chapter 6).

the worker was characteristically ill-prepared, in terms of intellect, literacy, training, instruments, equipment, and supervision, to perform the required tasks with relative safety to him/herself, fellow workers, the public, and the environment. Commendable progress has been made toward improving workplace safety for the hazwaste worker, but the improvements are not consistent among employers and workplaces, as noted above, and much remains to be done to improve awareness and minimize the hazards on hazardous waste workplaces.

Owners and operators of hazardous waste facilities and managers and supervisors of hazardous waste workers are similarly on a rising curve, with respect to effective management, supervision, training, and equipping of workers and/or facilities. Some have taken the necessary steps to achieve the required compliance status. Others have demonstrated extraordinary leadership by going beyond mere compliance in terms of providing well-trained and experienced supervisors, adequate resources and equipment, and management emphasis. Unfortunately, some owners, operators, managers, and supervisors linger at the lower end of the curve. A frequent example of the latter is observed, by trainers and faculty, in the person of the employee having inadequate or no background or experience in worker safety and health who arrives at work one morning and is informed that he/she is the new health and safety (environment and safety, compliance, etc.) officer or specialist. Company and employee then initiate a hasty search for some quick training that will provide some legitimacy to the appointment. There is no satisfactory substitute for in-depth training in safety, industrial hygiene, hazardous waste/materials management, environmental compliance, and other disciplines related to the specific appointment.

HAZARDS ENCOUNTERED ON HAZARDOUS WASTE SITES

The designation of a site as a hazardous waste site leaves much unsaid insofar as worker health and safety is concerned. The hazards present include, but may also far exceed, those attributable to the specific hazardous waste which is cause for the site designation. A great variety of possible or potential hazards assert themselves, and it is difficult to construct an organized listing. The following ordering of on-site hazards is adapted and summarized from the Occupational Safety and Health Guidance Manual for Hazardous Waste Site Activities, prepared by the National Institute for Occupational Safety and Health (NIOSH), the Occupational Safety and Health Administration (OSHA), the U.S. Coast Guard (USCG), and the U.S. Environmental Protection Agency (EPA) (HHS 1985). This manual, frequently referred to as "the four agency manual," is an excellent resource and is here highly recommended for inclusion in the professional libraries of hazardous waste management practitioners.

Chemical Exposure

As discussed in Chapter 4, chemicals exert toxic effects on humans by gaining access to the tissues and cells. The three major routes of exposure are inhalation, dermal absorption, and ingestion. Entry may also occur in the form of a puncture wound or entry through mucous membranes of the eyes or nasal passages. Exposures may be chronic or acute, as discussed earlier, and may be temporary and reversible or may be permanent.

Inhalation is frequently the potential exposure route of greatest concern on hazwaste sites. The human respiratory system has the function of quickly facilitating the absorption of oxygen into the bloodstream, where it is efficiently distributed to the vital organs of the body. The toxic chemical, whether or not a threat to the lungs, may be absorbed and distributed in a similar manner. Particulates may coat the lung tissues, permanently limiting lung function.[2] Some toxic chemicals may not be detected by the human senses, i.e., they may be colorless, odorless, tasteless, or nonirritating, and their toxic effects may not produce immediate symptoms. Respiratory protection is therefore extremely important where the workplace atmosphere may contain hazardous substances.

Absorption by skin and mucous membrane is an important route of exposure. Chemicals may directly injure the skin or may pass through the skin and be transported to vulnerable organs. Skin absorption is enhanced by wounds, heat, and/or moisture. Contact with body orifices is an important route of entry. Airborne chemicals can dissolve in the moist surface of the eye, be absorbed by the near-surface capillaries, and be carried through the bloodstream. Workers must wear protective equipment, avoid using contact lenses in contaminated atmospheres, keep hands away from the face, and minimize skin contact with liquid and solid chemicals.

Ingestion is thought of as the least likely route of exposure at hazwaste sites, but workers should be aware of the possibility and the means. Personal habits such as chewing gum or tobacco, drinking, eating, or smoking cigarettes while on-site may provide a route of entry and should be prohibited. Particulate material may accumulate in the bronchial passages, be brought to the throat by the natural cleansing processes, and then be swallowed.

Injection of chemicals through puncture wounds may occur from stepping on or other contact with sharp objects. Protection from injection hazards can be improved by wearing safety footwear, avoiding physical hazards, following prescribed procedures when generating or handling infectious wastes or hazardous chemicals, and by taking common sense precautions.

Explosion and Fire

The potential causes of fires and explosions on hazardous waste sites are as listed in Chapter 4. They include

- Chemical reactions that produce explosion, fire, or heat, including those attributable to pyrophoric and water reactive substances
- Ignition of explosive or flammable chemicals
- Ignition of materials due to oxygen enrichment
- Agitation of shock-or friction-sensitive compounds
- Sudden release of material under pressure

---

[2] "Black lung," silicosis, asbestosis, etc.

Hazardous wastes may spontaneously ignite or explode. The more frequent causes include activities such as movement of drums, accidental mixing of incompatible chemicals, attainment of auto-ignition temperatures, or the introduction of an ignition source into an explosive or flammable environment. Such events not only pose the obvious hazards of intense heat, open flame, smoke inhalation, and destructive shock waves and flying objects, but may also release toxic and/or corrosive chemicals into the environment. Threats to on-site personnel, as well as the public, may be minimized by field monitoring for explosive atmospheres and flammable vapors; knowledge of ignitability potential of specific chemicals; identifying and verifying incompatible materials; keeping potential ignition sources away from flammable or explosive environments; using nonsparking, explosion-proof equipment; remotely handling unknown materials and suspect containers; and avoiding practices that might result in agitation or release of chemicals.

Oxygen Deficiency

The oxygen content of normal air is approximately 21%. Humans experience physiological effects when oxygen concentrations in the air are depressed to 16% at sea level. The effects include impaired attention, judgment, and coordination and increased breathing and heart rate. To provide for individual physiological responses and errors in measurement, the new Respirator Standard 29 CFR 1910.134 (d)(2)(b)(iii) states that all oxygen-deficient atmospheres (less than 19.5% oxygen) shall be considered to be IDLH (immediately dangerous to life and health). The application of the standard is discussed later in Respirator Selection Criteria (*see:* 29 CFR 1910.134, Table II).

Oxygen deficiency may result from the displacement of oxygen by another gas, by the consumption of oxygen by a chemical or biological reaction, or at higher altitudes as noted above. Confined spaces and low-lying areas are characteristically vulnerable to oxygen deficiency and should be monitored as entry operations begin and frequently thereafter. Workers in oxygen-deficient atmospheres must be trained in respirator use and wear atmosphere-supplying respirators. Air-purifying respirators should never be used in oxygen-deficient atmospheres and should only be used where the required conditions (discussed later) are met.

Ionizing Radiation

Health impacts and physiological effects of ionizing radiation on humans are summarized in Chapter 13. Use of protective clothing, coupled with scrupulous personal hygiene and decontamination, affords good protection against α and β radiation.

Chemical protective clothing affords no protection against γ radiation; however, use of respiratory and other protective equipment can provide some protection against entry of radiation-emitting materials from entering the body by inhalation, ingestion, injection, or skin absorption.

Sites having radiation greater than background levels should be entered only after consultation with a health physicist. At levels greater than 2 rem/hr, all site activities should cease until the site has been assessed by a health physicist.

Biologic Hazards

Medical and infectious wastes[3] as described in Chapter 12 are a significant hazard if encountered on-site and, like other wastes, are subject to wind and water dispersion. Other biologic hazards that may be present on hazardous waste sites include poisonous plants, insects, reptiles, animals, and indigenous pathogens (i.e., hanta virus). Protective clothing and respiratory equipment can help reduce the chances of exposure. Thorough washing of any exposed body parts and equipment will help protect against infection.

Bloodborne Pathogens

The hazards of percutaneous injury by contaminated needles and other sharps is a serious hazard to a variety of workers. According to a National Institutes of Occupational Safety and Health (NIOSH) publication, approximately 800,000 needlestick injuries (an average of 1 every 10 sec) occur annually *in hospitals* in the U.S. (NIOSH 1998). Emergency response workers may encounter infectious wastes, in general, and bloodborne pathogens, in particular, in response, rescue, and incident remediation situations. Exposure incidents can lead to infection from hepatitis B virus (HBV) or human immunodeficiency virus (HIV), which causes AIDS. Although few cases of AIDS are directly traceable to workplace exposure, about 8700 workers each year contract hepatitis B from occupational exposure and about 200 die from this bloodborne infection (OSHA 1998). Employers are required by the Bloodborne Pathogens Standard (29 CFR 1910.1030) to develop a written exposure control plan that identifies job classifications and tasks that involve exposure to blood and other infectious materials and to implement protective measures including hepatitis B vaccinations, protective clothing and equipment, engineering and work practice controls, housekeeping, and record keeping (OSHA 1998).

Safety Hazards

A wide variety of safety hazards are found on hazardous waste sites, including variations on the following:

- Holes or ditches
- Excavations and steep grades (cave-in hazards)
- Overhead and buried utilities
- Bins, silos, other containment structures (engulfment hazards)
- Confined spaces
- Underground storage tanks being lifted or positioned
- Precariously positioned objects, such as drums that may fall
- Sharp objects, such as nails, metal shards, and broken glass
- Slippery surfaces
- Steep grades

---

[3] *See also:* Bloodborne Pathogens, below.

- Uneven terrain
- Unstable surfaces, such as walls or floors that may fail

OSHA has promulgated health and safety standards for many of these workplace hazards (*see:* Appendix A to this chapter for a listing of the standards).

Safety hazards are also created as a result of the work in progress on the site. Movement of heavy equipment involves physical hazards as well as noise. Protective equipment can impair worker agility, hearing, and vision, in turn creating increased risk of accidents. Increased chemical exposure hazard is caused when protective equipment is damaged. Workers on-site must continually observe each other and the work area for potential safety hazards and immediately inform supervisors of any new or previously undiscovered hazards.

Electrical Hazards

Overhead power lines, downed electrical wires, and buried cables all pose a danger of shock or electrocution if workers contact or sever them during site operations. Electrical equipment used on-site may also be a hazard to workers. Strict adherence to the OSHA lockout/tagout[4] rules and procedures is a major preventive of electrical injuries and fatalities. Low-voltage equipment with ground-fault interrupters and water-tight, corrosion-resistant connecting cables should be used to minimize this hazard. Weather conditions should be monitored in order that work may be suspended during thunder storms, thereby eliminating the lightning hazard. Capacitors found on-site may retain a charge and should be grounded before handling. Underground storage tank removals frequently involve electrical cables and/or other electrical apparatus in the same trench or in close proximity to petroleum fuel or natural gas lines.

Heat Stress

Heat stress is a major hazard for workers wearing protective clothing. The protective clothing materials that serve to shield the body from chemical exposure also limit the dissipation of body heat and moisture. Depending upon the ambient conditions and the work being performed, heat stress can develop very rapidly — within a few minutes. It can pose danger to worker health as great as that of chemical exposure. Heat stress can initially cause rashes, cramps, discomfort, and drowsiness, resulting in impaired functional ability that threatens the safety of both the individual and co-workers. Continued heat stress can lead to heat stroke and death. Avoiding overprotection, careful training and frequent monitoring of personnel who wear protective clothing, shade and ventilation, judicious scheduling of work and rest periods, and frequent replacement of fluids can provide protection against this hazard. Employees and employers must be trained and alert to recognize symptoms of heat stress. The American Conference of Governmental Industrial Hygienists (ACGIH) handbook of *Threshold Limited Values* (TLVs) and *Biological Exposure Indices* (BEIs), pub-

---

[4] The Control of Hazardous Energy Standard (29 CFR 1910.147), more commonly known as the lockout/tagout standard.

lished annually, is an authoritative and detailed source for guidance regarding prevention of heat injury.[5] Prescribed Wet Bulb Globe Temperature (WBGT) monitoring procedures[6] are essential to maintenance of safe working conditions, where heat injury is a potential hazard.

Cold Exposure

Cold injury (frostbite and hypothermia) and impaired ability to work are dangers at low temperatures and when the wind chill factor is low. To guard against them, managers and supervisors should ensure that workers wear appropriate clothing, have warm shelter readily available, carefully schedule work and rest periods, and monitor workers' physical conditions.

Noise Hazard

On-site activity in proximity to heavy equipment and machinery may create a noise environment that is hazardous. Workplace noise is measured in decibels (dBA) on an "A-weighted" scale. The scale gives greater weight to the sound pressures in the more damaging frequencies (approximately 2000 Hz) and less weight to sound pressures outside this range (Martin 1994, p. 522). Effects of excessive noise may include

- Workers being startled, annoyed, or distracted
- Physical damage to the ear, pain, and temporary and/or permanent hearing loss
- Communication interference that may increase potential hazards due to the inability to warn of danger and the proper safety precautions to be taken

If employees are subjected to noise exceeding an 8-hr, time-weighted average sound level of 90 dBA, feasible administrative or engineering controls must be utilized. In addition, whenever employee noise exposures equal or exceed an 8-hr, time-weighted average sound level of 85 dBA, employers must administer a continuing, effective hearing conservation program as described in 29 CFR 1910.95.

Other Physical Hazards

A variety of other physical hazard encounters are possible on hazardous waste sites. Vibrations, misused or malfunctioning hand tools, falls from heights, highway accidents, MSDs[7] such as repetitive motion injury, excavation and engulfment hazards, and workplace violence are examples. Hazardous waste management activity requires intense focus on the primary objective. Employers and employees must be alert to the unexpected.

---

[5] The American Conference of Governmental Industrial Hygienists handbook of *Threshold Limit Values for Chemical Substances and Physical Agents and Biological Exposure Indices* (current edition).

[6] Performed by an industrial hygienist or person specifically trained in this discipline.

[7] Musculoskeletal disorders, discussed later herein.

Hazardous Waste Operations and Emergency Response

Background

The U.S. Occupational Safety and Health Administration[8] was created in December 1970 by enactment of the Occupational Safety and Health Act (PL 91-596) and began operations in April 1971 (Miller, 1985, Chapter 8). OSHA (the agency), under authorities of the original Act and subsequent amendments, undertook the promulgation of workplace health and safety standards as specified by Section 6 (g) based upon the needs of specific "industries, trades, crafts, occupations, businesses, workplaces, or work environments." In the years to follow, OSHA issued a variety of proposed standards, and some were made final. Standards dealing incidentally with activities of hazardous waste workers were promulgated, e.g., exposure standards for specific chemicals, standards governing handling of compressed gases, etc.[9] In 1986, as Congress deliberated the Superfund Amendments and Reauthorization Act (SARA), Section 126 was added to Title I, requiring the Secretary of Labor to promulgate a hazardous waste worker health and safety standard. Interim final standards were issued on December 19, 1986. The final Hazardous Waste Operations and Emergency Response standards were published on March 6, 1989 (54 FR 9317) and were codified at 29 CFR 1910.120.

The Hazardous Waste Operations and Emergency Response standard, frequently referred to as the HAZWOPER, became effective on March 6, 1990. It is intended to protect hazardous waste workers who are private employees, federal employees, and state and local government employees in states having delegated OSHA programs. The similar EPA standard (40 CFR 311) covers state and local government employees engaged in hazardous waste operations and emergency response in states that do not have an OSHA-approved state plan (Levine et al. 1994, p. 3). The scope of the HAZWOPER encompasses three clearly defined groups of workers engaged in:

- Clean-up sites, whether being cleaned up as a Superfund site, a RCRA Corrective Action site, or a voluntary clean-up site, are subject to paragraphs (a) through (o) of the standard.
- Treatment, Storage, and Disposal facilities (TSD) (RCRA permitted or interim status facilities) are subject to paragraph (p) of the standard.
- Emergency response operations for releases of, or substantial threats of releases of, hazardous substances without regard to the location of the hazard are subject to paragraph (q) of the standard.

Generators who store hazardous wastes for less than 90 days and small quantity generators having emergency response teams that respond to releases of (or sub-

---

[8] OSHA was created by amendment to an existing statute, during the same month that the EPA was created by President Nixon's Reorganization Order No. 3 of 1970 (an executive order). OSHA was buried in the Department of Labor bureaucracy; the EPA was made an independent agency in the Executive Department (the Administrator reports to the President). OSHA was organized primarily as an enforcement organization, with most of the staff as inspectors; the EPA was to be staffed with a mix of administrative, program management, research, and enforcement personnel.

[9] *See:* Appendix A to this chapter.

stantial threats of releases of) hazardous substances are required to comply only with paragraph (p)(8) of the standard. This requirement for an emergency response plan does not apply to generators and small quantity generators who do not have emergency response teams if they provide an emergency action plan complying with 29 CFR 1910.38(a). The Department of Labor has issued letters,[10] interpretations, and policy statements to the effect that employees who conduct leaking underground storage tank remediation are required to comply with 29 CFR 1910.120, including the training requirements.

In the following summary, the salient features of the standard are covered within the framework of the three groupings noted earlier. Space does not permit detailed explanation or discussion. The intent here is, as in previous chapters, to provide an introduction and oversight to and of the practice of hazardous waste management. The beginning practitioner should carefully read, at a minimum, the HAZWOPER, the applicable standards referred to therein, and the four-agency manual.

THE HAZWOPER SUMMARIZED

Standards Applicable to Clean-up Sites

(a) Scope, Application, and Definitions. The standard applies to mandatory clean-up operations involving hazardous substances at *uncontrolled* hazardous waste sites such as NPL sites; RCRA Corrective Action sites; voluntary clean-up operations at sites that are uncontrolled; emergency response operations involving hazardous substances without regard to location. See paragraph (a)(2) for specific definitions.

(b) Safety and Health Program. Employers are required to develop and implement a written safety and health program, which must incorporate the following:

- An organizational structure
- A comprehensive workplan
- A site-specific safety and health plan, including an emergency response plan
- The safety and health training program
- The medical surveillance program
- The employer's standard operating procedures for safety and health
- Coordination of general safety and health program and site-specific activities

Contractors and subcontractors must be informed regarding all hazards on-site. The written health and safety plan must be made available to contractors and regulatory personnel having authority over the site.

---

[10] Department of Labor memorandum of August 31, 1990, to OSHA Regional Administrators states in part: Activity under subtitle I of RCRA could fall under the following scope sections of 29 CFR 1910.120: (1) clean-up operations, 1910.120 (a)(1)(i) and (a)(1)(iii); (2) corrective actions, 1910.120 (a)(1)(ii); (3) emergency response operations, 1910.120 (a)(1)(v). Leak detection, leak prevention, tank cleaning, and closure activity are covered by 29 CFR 1910.120 if any of the following apply: (1) a government body is requiring the tank to be removed because of the potential threat to the environment or the public; (2) the activities are necessary to complete a corrective action; (3) a governmental body has recognized the site to be uncontrolled hazardous waste; (4) there is a need for emergency response procedures.

(c) Site Characterization and Analysis. Sites where clean-up operations are planned must be evaluated to identify specific hazards and to determine safety and health control procedures needed to protect employees from the identified hazards. The process proceeds with a preliminary evaluation, in which a qualified person determines initial levels of personal protection necessary for entry and/or beginning operations. Thereafter a more detailed evaluation of the site's specific hazards is conducted using, to the extent practicable, nonintrusive methods and technologies, e.g., ground penetrating radar, historical aerial imagery, etc. The next step involves development of detailed physical, chemical, biological, and toxicological data on the site. Monitoring of radiation and air quality is accomplished with direct reading instruments. Risk identification associated with the identified substances is then determined and communicated to all employees involved in the project.

(d) Site Control. The site is closely controlled with respect to work zones, the use of a "buddy system" in the exclusion zone, on-site communications, standard operating procedures, and identification of the nearest medical assistance. Continuous or periodic air quality monitoring is performed to detect changes that may have occurred since initial entry. A site map is used to communicate current and new information regarding the site as shifts change or as new contractors arrive.

(e) Training. All employees working on-site must receive training about:

- Names of personnel responsible for site safety and health
- Safety, health, and other hazards on-site
- Use of personal protective equipment
- Work practices by which the employee can minimize risks from hazards
- Safe use of engineering controls and equipment, including instrumentation, on-site
- Medical surveillance
- Contents of the site safety and health plan

General site workers (such as equipment operators, general laborers, and supervisors) engaged in hazardous substance removal or other activities that expose or potentially expose them to hazardous substances and health hazards must receive 40 hr[11] of instruction off-site and an additional 3 days[11] of actual experience under the direct supervision of a trained, experienced supervisor. Workers on-site only occasionally or regularly on-site in areas that are characterized as having minimal exposure hazards must receive 24 hr[11] of training off-site and 1 day[11] of actual experience under a trained, experienced supervisor. On-site managers and supervisors must receive the 40 or 24 hr (as above) of off-site training and 3 days or 1 day (as above) of supervised field experience, plus 8 additional hours[12] of specialized training pertaining to their duties. All employees must receive 8 hr of refresher training annually.

Trainers must meet the qualifications of 29 CFR 1910.120(e)(5) requiring specific training in the subjects taught or appropriate academic credentials. On January 28,

---

[11] Employees who may be required to perform emergency response tasks at hazardous waste cleanup sites must receive training in appropriate response to emergencies that may arise on the site.

[12] Hourly classroom and field experience requirements are stated as minimum requirements.

1990, the EPA proposed a new accreditation standard under a new 29 CFR 1910.121 (55 FR 2790). The rule was never finalized, but an expanded, nonmandatory Training Curriculum Guideline was published as Appendix E to § 910.120 on August 22, 1994 (59 FR 43270). Employers or others seeking the required training should ascertain that prospective training sources are in substantial accord with the guidelines.

(f) Medical Surveillance. Employers of employees engaged in hazardous waste operations who:

- Are or may be exposed above permissible exposure limits, without regard to the use of respirators, for more than 30 days per year
- Wear a respirator for 30 days or more per year
- Are injured due to overexposure from an emergency incident involving hazardous substances or health hazards
- Are members of HAZMAT teams

should institute a medical surveillance program including a pre-assignment examination, annual or more frequent medical examinations and consultations, and medical examinations and consultations at the time of termination or transfer of the employee to an assignment which would not be subject to these requirements. These examinations and consultations must also be provided, at no cost to the employee, as soon as possible upon notification by an employee of detection of signs or symptoms of overexposure to hazardous substances or health hazards or that the employee has been injured or exposed above permissible limits. The frequency of examinations may be increased or decreased as determined by the examining physician, but may not exceed 2 years.

Employees of *excepted* [§ 1910.120(p)] generator facilities, who have no emergency response assignments and who may be injured, may develop health impairments or symptoms of exposure to hazardous substances, or are exposed to concentrations above permissible or published limits, while not using appropriate personal protective equipment, must be provided the required examinations and consultations. The content of the examinations is to be determined by the physician,[13] but must include a medical and work history. The employer must furnish a copy of the physician's written opinion regarding the examination, but the opinion may not reveal specific findings or diagnoses unrelated to occupational exposures.

(g) Engineering Controls, Work Practices, and Personal Protective Equipment for Employee Protection. The title phrases of this subparagraph are the three elements of a hierarchy of preferable approaches to hazardous waste worker protection. The preferred solution to an exposure or injury hazard is to reduce or "engineer" the problem out of existence by preventing, containing, isolating, or removing the hazard. Examples include enhanced ventilation, remotely operated devices for operating material handling equipment, use of pressurized cabs or control booths on equipment, elimination of sources of excess noise, or smoothing the paths of forklifts

---

[13] It is important that the employer and the examining physician refer to the specific chemical standards (§§ 1910.1001 thru .1052), which are applicable to the possible employee exposure, e.g., § 1910.1025. Lead (exposure) requires a blood lead test, etc.

carrying hazardous chemicals. Work practices (also referred to as administrative controls) are also considered preferable to the use of protective clothing and equipment. Examples include removing all nonessential personnel from a worksite while drums are being opened, scheduling operations to take advantage of cooler temperatures to reduce heat stress hazards, wetting down dusty operations, or locating employees upwind of airborne hazards. Only if the hazard cannot be eliminated by engineering controls and/or work practices should protective clothing and equipment (PPE) be the protective option [§ 1910.134(a); *see also:* Wallace 1994, p. 208)].

Selection of levels of PPE must be based upon an evaluation of the performance characteristics of the PPE *relative to the identified and potential hazards on-site.*[14] In Level A, totally encapsulating chemical protective suits and self-contained or supplied air respirators are required where skin absorption of a hazardous substance may result in a substantial possibility of death, immediate serious illness or injury, or impairment of the ability to escape. Level B is worn in situations where the highest level of respiratory protection is required, but a lesser level of skin protection is adequate. Level B protection consists of self-contained or supplied air respirator and chemical resistant protective clothing (not fully encapsulating), inner and outer gloves, chemical resistant safety boots (boot covers are optional), hard hat or face shield. In Level C, PPE consisting of an air purifying respirator (APR) and clothing similar to that of Level B may be worn when concentration(s) and types of airborne substance(s) are known and all criteria for use of APRs are met (*see:* § 1910.134, as amended, for detailed requirements). *In all cases, the chemical resistance characteristics of the protective clothing, as provided by the manufacturer, must be compatible with the known chemical hazards and with the solvent(s) to be used in decontamination.*

The level of PPE decisions must balance protection, worker productivity, worker comfort, and cost. Neither overcautiousness, overconfidence, nor indifference have a place in the decision. For example, the degree of worker protection achieved by a supervisor's decision to require wearing of Level A vs. Level B in many scenarios is primarily in the degree of skin protection achieved. The supervisor must balance the reality of the splash or vapor hazard against the extreme stresses and limitations imposed on the worker by a Level A outfit. In all cases, however, the supervisor should be guided by the PPE selection criteria of Chapter 8 of the four-agency manual, Subtitle I of 29 CFR 1910, and the NIOSH Pocket Guide to Chemical Hazards.

The appendices to 29 CFR 1910.120, the applicable standards of Subpart I, and the four-agency manual must be read and understood before the use of PPE.[15]

---

[14] The importance of this linkage between site characterization and level of PPE selection cannot be overemphasized.

[15] For a listing of other workplace standards that may apply to a particular site or set of conditions, *see:* Appendix A to this chapter.

FIGURE 15.1 Workers in Level A fully encapsulating protective clothing. (Courtesy of URS Corporation, 100 California Street, San Francisco, CA 94111.)

Employees must not be assigned to on-site tasks requiring PPE before receiving the required training of § 1910.120 (e) (*see also:* Schwope and O'Leary 1994, Chapter 9; Goldman 1994, Chapter 10).

(h) Monitoring. Initial and periodic air quality monitoring are performed where there may be a question of employee exposure to hazardous concentrations of hazardous substances, in order to assure proper selection of engineering controls, work practices, and PPE. Upon initial entry, air monitoring is conducted to identify any "Immediately Dangerous to Life or Health" (IDLH) condition, exposure over permissible or published exposure levels, exposure over a radioactive material's dose limits, or other danger such as the presence of flammable atmospheres or oxygen deficient environments. Periodic monitoring is conducted when there is the possibility (or actuality) of chemical concentrations in excess of a ceiling, an IDLH condition or flammable atmosphere, or an indication that exposures may rise over permissible limits. Individual high-risk employees are monitored during the actual cleanup operations, e.g., when soil, surface water or containers are moved or disturbed.

(i) Informational Programs. Employees, contractors, and subcontractors must be informed of the nature, level, and degree of exposure likely in their participation in hazardous waste operations.

(j) Handling Drums and Containers. The subparagraph (j) standards pertaining to drums are lengthy and do not lend themselves to summarization. In general, drums and other containers used during clean-up operations must meet appropriate DOT, OSHA, and EPA regulations for the wastes to be contained. Drums must be inspected and their integrity assured before being moved. Leaking or damaged drums must be overpacked or have the contents transferred prior to being moved. Drums with old labels and unlabeled drums should be considered to contain hazardous substances and be handled accordingly until the contents are positively identified and labeled. Containers suspected of containing radioactive materials must not be handled by workers until evaluated by an expert. Site operations must be organized to minimize movement

FIGURE 15.2 Workers in Level B protective clothing with self-contained breathing apparatus (SCBA). (Courtesy of URS Corporation, 100 California Street, San Francisco, CA.)

FIGURE 15.3 Workers in Level B protective clothing with supplied air respirators (SAR). (Courtesy of URS Corporation, 100 California Street, San Francisco, CA.)

FIGURE 15.4 Workers in Level C protective clothing. (Courtesy of URS Corporation, 100 California Street, San Francisco, CA.)

of drums. Exhumation of buried drums must be done with caution in order to prevent rupture (see Figure 11.7), and provision must be made for containing spills. Drums that are bulged must be opened remotely or with the operator shielded and must not be moved until the cause of the bulging has been determined. Drums that show signs of crystalline material must be treated as shock-sensitive until identification of the contents can be made and should be opened remotely or with operator shielded.

(k) Decontamination. A decontamination procedure must be developed, communicated to employees, and implemented before any employees or equipment enter the exclusion zone (or areas where potential for exposure exists). The decontamination area or corridor must be located to provide a transition from contaminated to noncontaminated areas, without exposing noncontaminated employees or equipment. Employees leaving a contaminated area must be decontaminated; contaminated clothing and equipment must be properly decontaminated or disposed of. Decontamination procedures must be monitored by the site safety officer to determine their effectiveness and to modify or correct them as necessary. Shower and change rooms must be provided where the decontamination procedure indicates need for regular showers. The shower must be used immediately where nonimpermeable clothing becomes wetted with hazardous substances or impermeable clothing becomes compromised. As indicated in (g) above, solvent(s) used in decontamination must be chemically compatible with protective clothing worn and with the contaminant(s) encountered. Solvents used in decontamination must be managed as hazardous waste until it can be shown that they are nonhazardous.

(l) Emergency Response by Employees at Uncontrolled Hazardous Waste Sites. Clean-up site employers must develop and implement an emergency response plan,

which is a separate section of the Site Safety and Health Plan. The plan must be in writing and available for inspection and copying by employees, employee representatives, and regulatory agencies having relevent purview. Employers who will evacuate employees from the workplace when an emergency occurs and who do not permit any of their employees to assist in handling the emergency are exempt from this requirement if they provide an emergency action plan that complies with 29 CFR 38(a). Minimum requirements for an emergency response plan include

- Pre-emergency planning
- Personnel roles, lines of authority, and communication
- Emergency recognition and prevention
- Safe distances and places of refuge
- Site security and control
- Evacuation routes and procedures
- Decontamination procedures not covered by the site safety and health plan
- Emergency medical treatment and first aid
- Emergency alerting and response procedures
- Critique of response and follow-up
- PPE and emergency equipment

In addition to the listed minimum requirements, emergency response plans must include site topography, layout, and prevailing weather conditions and procedures for reporting incidents to appropriate local, state, and federal agencies. The plan must be compatible and integrated with disaster, fire, and/or emergency response plans of local state and federal agencies and can be integrated with RCRA contingency, spill prevention control and countermeasures, and process safety management plans. The plan must be rehearsed regularly and reviewed and amended as needed. An employee alarm system, as prescribed by 29 CFR 1910.165, must be installed and operated to inform employees of an emergency situation.

(m) Illumination. This standard identifies minimum criteria ranging from 3 foot-candles in excavation and waste areas to 30 foot-candles for first aid stations. The standard is summarized in Table H-120.1 of subparagraph (m).

(n) Sanitation at Temporary Workplaces. The sanitation requirements cover potable water, containers, drinking cups, nonpotable water systems, toilets, food handling, temporary sleeping quarters, washing facilities, and showers and change rooms.

(o) New Technology Programs. This subparagraph requires employers to stay abreast of new technologies and equipment developed for protection of employees on clean-up sites, to provide procedures for the introduction of new technologies, and to implement them.

Standards Applicable to Treatment, Storage, and Disposal Sites

(p) Certain Operations Conducted under the Resource Conservation and Recovery Act of 1976 (RCRA). The employer conducting operations at RCRA permitted or interim status treatment, storage, and disposal (TSD) facilities is required to provide and implement many of the same, or similar, standards as required of

clean-up site employers. The TSD facility is presumed to be controlled as differentiated from the clean-up site that may be, or potentially is, uncontrolled. The site is presumed to be characterized, i.e., the hazards are known, and the employee is theoretically less likely to be exposed. The more apparent difference between the clean-up site and TSD facility requirements are those pertaining to training. The TSD facility employee must have 24 hr of initial training, plus an 8-hr annual refresher. The trainer providing the initial training must have satisfactorily completed a training course for teaching the required subjects or have equivalent academic credentials and instructional experience.

Standards Applicable to Emergency Response Teams

The separate and distinct "first responder" standards of subparagraph (q) are frequently the basis for confusion or misunderstanding. The first few sentences of the subparagraph attempt to make the distinction between the general site worker, who may have emergency response duties during an in-house or on-site incident, and the first responder, who responds to incidents or releases regardless of location. Martin (1994, p. 551) define emergency response as: "A response effort by employees from outside the immediate release area or by other designated responders (e.g., mutual-aid groups or local fire departments) to a situation that results, or is likely to result, in an uncontrolled release of a hazardous substance." These responders are required to have the training of 29 CFR 1910.120(q), but most state and local governments require first responders to have training far in excess of that specified in subparagraph (q).

(q) Emergency Response to Hazardous Substance Releases. OSHA defines five levels of response training, each of which is specific to assigned duties of the employee:

1. First Responder, Awareness Level — These are individuals likely to witness or discover a hazardous substance release and initiate the emergency response. They must demonstrate competency in such areas as recognizing the presence of hazardous materials in an emergency and have the ability to identify the hazardous material (if possible); the risks involved; and the outcomes associated with an emergency involving hazardous materials. Although no minimum hours of training are specified for the first responder "awareness," the individual must understand that role in the employers emergency response plan and must be able to recognize the need for additional resources and make the appropriate notification(s).

2. First Responder, Operations Level — These are individuals who respond for the purpose of containing the release from a safe distance, keeping it from spreading, and preventing exposures, i.e., a defensive posture. In addition to demonstrating the competencies of the "awareness level," the operations level responder must receive at least 8 hours of training or demonstrate competency in basic hazard and risk assessment techniques; selection and use of appropriate PPE; hazmat terminology; basic control, containment, and confinement operations; decontamination procedures; and standard operating procedures.

3. Hazardous Materials Technician — These are individuals who respond aggressively to releases or potential releases for the purpose of stopping the release. Hazardous materials technicians must receive at least 24 hours of training equal to the first responder operations level and must know how to implement the employer's emergency response plan; know the classification, identification, and verification of hazardous materials by using field instruments and equipment; know how to select and use specialized chemical protective equipment; understand hazard and risk assessment techniques; be able to perform advanced control, containment, and confinement operations; understand and implement decontamination and termination procedures; and understand basic chemistry and toxicology.

4. Hazardous Materials Specialist — These are individuals with duties parallel to those of the technician, but requiring more advanced knowledge. They may also be required to respond aggressively. The specialist must have at least 24 hours of training equal to the technician level; know the local and state emergency response plan and know how to implement the local plan; understand classification and identification of hazardous materials using advanced survey instruments and equipment; understand and use specialized chemical protective equipment; understand chemical, radiological, and toxicological terminology and behavior as well as in-depth hazard and risk assessment techniques; be able to develop a site safety and control plan; and be able to determine and implement decontamination procedures.

5. On-Scene Incident Commander — This individual must have at least 24 hours of training equal to the operations level and, in addition, be able to implement the employer's incident command system, emergency response plan, and the local emergency response plan; know of the state emergency response plan and of the Federal Regional Response Team; and know and understand the hazards and risks associated with employees working in chemical protective clothing as well as the importance of decontamination procedures.

Employers must certify the training and/or competence of each individual assigned to one of the above levels. Moreover, the employer must also refer to specific OSHA standards pertaining to hazardous materials and operations listed in Appendix A to this chapter.

OTHER IMPORTANT TOPICS AND COMPLIANCE ISSUES

In this section we attempt to alert managers and supervisors to several issues which are not clearly defined, applicable, and/or explained by OSHA. They are issues which require a degree of monitoring by responsible individuals. In a more general context, managers and supervisors should be alert to a trend of the ever-widening scope and detail of the OSHA regulatory structure. The trend is clearly toward more emphasis on hazard assessment, employee training, employee participation in health and safety planning, amelioration of stress (physical and emotional),[16] and ever-more

---

[16] *See:* Stress at Work, DHHS (NIOSH) Publication No. 99-101.

prescriptive[17] treatment of workplace operations. The responsible employer, manager, or supervisor must constantly be alert to hidden, poorly defined, or undefined hazards, not only for the well-being of his/her employees, but to avoid the devastating impact of citation under the "general duty clauses" of the federal[18] and state worker safety and health laws and regulations.

Respirator Selection Criteria

Respiratory protection is of primary importance because inhalation is one of the major routes of exposure to chemical toxicants. As before, space does not permit a detailed presentation on respirator selection. However, widespread misuse of respirator equipment by hazardous waste workers is cause for concerns regarding emphasis or adequacy of training, or both. All concerned with respirator selection should be thoroughly familiarized with the most recent updates[19] of 29 CFR 1910.134, the OSHA respirator standard; the source standard, ANSI[20] Z88.2; and the 29 CFR 1910.1000 exposure limits for air contaminants.

Respirators that supply air to the user are called atmosphere-supplying respirators and consist of two types:

- A self-contained breathing apparatus (SCBA) that supplies air from a source carried by the user (see Figure 15.2).
- A supplied-air respirator (SAR) (or air line respirator) that supplies air from a source located some distance away, through an air line, to the user (see Figure 15.3).

Atmosphere-supplying respirators are also further classified as positive or negative pressure respirators. Positive pressure respirators are either pressure-demand or continuous-flow types. The pressure-demand system supplies air on demand (inhalation by the wearer) while maintaining a slight positive pressure inside the facepiece. The pressure-demand system is the most commonly used atmosphere-supplying system, favored for economy of air supply and sufficient positive pressure to deter leakage of ambient air into facepiece. A negative pressure atmosphere-supplying respirator may be allowed in oxygen-deficient atmospheres under special conditions (*see:* § 1910.134, Table I).

Air-purifying respirators (APRs) do not provide air from a separate source. They provide ambient air which has been "purified" by a filtering element, e.g., a cartridge or canister (see Figure 15.4). Negative pressure APRs depend upon the negative pressure created inside the respirator when the user inhales; however, a powered air purifying respirator (PAPR) may maintain positive pressure in the facepiece.

---

[17] For example, the newly promulgated "Work-Related Musculoskeletal Disorders" (WMSDs) (or ergonomics) Standard, discussed later herein.
[18] Occupational Safety and Health Act, Public Law 91-596 (December 29, 1970) and PL 101-552 (November 5, 1990), § 5(a)1.
[19] As this is written, January 8, 1998, 1992, and July 1, 1999, respectively.
[20] The American National Standards Institute, 11 West 42nd Street, New York, NY 10036.

APR cartridges have finite life spans based upon the saturation rates of the absorbent materials and must be replaced before breakthrough occurs. The service life of a cartridge also depends upon respiratory rate, contaminant concentration, cartridge efficiency, and humidity. Some cartridges are equipped with end-of-service-life (ESLI) indicators. If cartridges for the contaminant of concern are not so equipped, the employer/supervisor must establish a cartridge change schedule based upon "objective information or data that will ensure that canisters and cartridges are changed before the end of their service life." The employer/supervisor, in this case, must meet requirements for including a cartridge change schedule, but basic decision-making factors may be absent or unclear particularly on old or abandoned sites. The new Respiratory Protection, Final Rule deals with this situation in § 1910.134(d)(1)(iii), ... "Where the employer cannot identify or reasonably estimate the employee exposure, the employer shall consider the atmosphere to be IDLH." Section 1910.134(d)(2) requires that employees use either a full facepiece, pressure-demand SCBA, or SAR (the latter with an auxiliary self-contained air supply) in IDLH atmospheres. Although the new standard is clear on this point, workers are regularly seen wearing APRs while engaged in hazardous waste activity, where it is clear that one or more of the standards are not met. The manager/supervisor should carefully consider the ramifications of anything less than full adherence to §§ 1910.134 and 1910.1001 through 1910.1052 (*see also:* ANSI Z88.2, 1992; four agency manual 1985, pp. 8–7; Schwope and O'Leary 1994, pp. 223ff; Jones 1994, Chapter 4; Maslansky and Maslansky 1997, Chapter 5).

Applicable Air Contaminant Standards

Soon after enactment of the Occupational Safety and Health Act in 1970, OSHA promulgated Permissible Exposure Limits (PELS) for many substances, per Section 6(a) of the Act. The standards can be traced back to 1968 Threshold Limit Values (TLVs) of the American Conference of Governmental Industrial Hygenists (ACGIH) and to the American Standards Association, the predecessor of ANSI. By 1989, significant data had accumulated, to the effect that the existing 400 substances regulated were inadequate, but OSHA lacked resources to rigorously develop substance-by-substance rulemaking and elected to engage in "generic" rulemaking to achieve the desired improvements. The 1989 rulemaking covered a total of 600 substances including PELs for 164 new substances, adoption of more protective PELs for 212 substances, no changes for 160 substances, and lesser adjustments (Introduction to OSHA Publication 3112, 1989).

A July 1992 federal appeals court decision vacated the 1989 rulemaking and forced OSHA to roll back exposure limits for many hazardous chemicals to less protective 1971 levels and eliminate exposure limits for dozens of other substances that had been unregulated prior to 1989 (*Environment Reporter,* June 30, 1993, p. 108). The enforceable PELs are those now listed in Tables Z-1, Z-2, and Z-3 of 29 CFR 1910.1000. The vacated 1989 PELs are listed in Appendix G of the NIOSH Pocket Guide to Chemical Hazards (HHS 1997). OSHA can be expected to promulgate new limits on new and presently regulated substances, in the earlier substance-by-substance mode, or on a few substances in any given action. The manager/supervisor is thus faced with at least two additional quandaries: (1) the necessity to maintain close observation of seemingly insignificant promulgations pertaining to single or a few substances,[21] and (2) the question of adequacy of the earlier (1968/1971) PELs, which have yet to be updated, to protect his/her employees.

Work-Related Musculoskeletal Disorders

Based upon an estimated 600,000 cases of workers affected each year, OSHA has been engaged with the work-related musculoskeletal disorders (WMSDs, i.e., ergonomics) issue since 1979. The agency issued guidelines, provided education, and issued citations under the general duty clause. OSHA published an advance notice of proposed rulemaking (ANPRM) in 1990, but withdrew it in the face of industry and congressional opposition. Beginning in 1995, Congress has passed appropriations riders that have delayed development of the standard. In 1998, Congress barred work on a final rule, but allowed the agency to work on a proposal or guidelines, and approved an $890,000 study of the issue to be conducted by the National Academy of Sciences (NAS) (*Environment Reporter,* November 24, 1999, p. 664). Nevertheless, OSHA published the proposed ergonomics standard on November 23, 1999, in the *Federal Register* (64 FR 65767-66078) and has encountered harsh reactions from Congress in numerous hearings. The new standard was published in final form on November 14, 2000 (65 FR 68261).

Meanwhile the ACGIH may quietly adopt a TLV for use of the hand, wrist, and forearm in jobs performed for 4 or more hours per day (*Occupational Safety and Health*, August 17, 2000, p. 757). Managers and supervisors should be alert to the fact that OSHA can cite violations of professional and consensus standards such as this under the General Duty Clause of the Occupational Safety and Health Act.

Bloodborne Pathogens Standard

The standard covers ... all employees who could be "reasonably anticipated," as the result of performing their job, to face contact with blood and other potentially infectious materials. OSHA has not attempted to list all occupations where exposures could occur. Infectious materials include practically all body fluids, unfixed tissue, or organ other than intact skin from a human, human immunodeficiency virus (HIV)-containing cell or tissue cultures, organ cultures and HIV or hepatitis B (HBV)- containing culture medium or other solutions as well as blood, organs, or other tissues from experimental animals infected with HIV or HBV (29 CFR 1910.1030). Managers of employees which are covered must, without question, receive the training prescribed by the standard.

The applicability phrase places an obvious quandary on the agenda for the hazardous waste manager involved in cleanup or remediation activity where the described exposure is *unlikely* or *unanticipated*. At a bare minimum, the manager of such activities must train his employees to recognize the universal biohazard symbol (Figure 12.1) and the current domestic and international labeling practice. The careful and conscientious manager may well reason that his employees could encounter infectious materials in unanticipated situations or in the event of cleanup work made necessary by unlawful activity. The exposure plan, training, and other

---

[21] OSHA has indicated intention to promulgate PELs for four chemicals — carbon disulfide, glutaraldehyde, hydrazine, and trimetallic anhydride (OSHA 2000).

compliance requirements are another set of expensive and time-consuming activities that must be periodically repeated in order to deal with employee turnover and the need for drills and updates. The alternative is to scrupulously avoid dealing with any situation, including life-saving activity, that might bring the employees into contact with infectious waste (and hope for the best).

Chemical Hazard Communication

Employer recognition of need as well as OSHA requirements have brought hazard communication to the forefront of workplace safety management. The Hazard Communication Standard, or HAZCOM (29 CFR 1910.1200), establishes uniform requirements to ensure that the hazards of all chemicals imported into, produced, or used in workplaces are evaluated and that this hazard information is transmitted to affected employers and to employees that are at risk of exposure. The requirements of the standard focus on use of the Material Safety Data Sheet (MSDS) and labels as the primary means of communicating chemical hazard information. The standard requires training of employees regarding their rights under the standard, health effects of chemicals being used, and how to interpret and use labels and the MSDS in terms of available controls and use of PPE. Competent and conscientious owners, operators, managers, and supervisors of facilities using toxic chemicals have generally met or exceeded requirements of the HAZCOM.[22] Some go beyond the requirement by using the MSDS to screen out highly hazardous chemicals and find less toxic substitutes.

For hazardous waste workers, hazard communication takes on a different meaning. The HAZCOM [29 CFR 1910.1200(b)(6)(i),(ii)] excludes RCRA hazardous waste and CERCLA hazardous substances from applicability of the standard. Nevertheless, hazard communication requirements are threaded throughout the HAZWOPER and worker health and safety courses regularly include discussion of the MSDS, the types of information contained, responsibilities of the chemical manufacturer/importer, the distributor, and the employer. MSDSs may often be located and obtained for identified hazardous waste constituents. Moreover, waste chemicals such as used solvents may be chemically similar to the virgin product whereby the original MSDSs would suffice for HAZCOM purposes. However, clean-up sites frequently involve chemical mixtures, decay products, reaction products, etc. Exposure threats on abandoned sites or sites for which historical use data are unavailable are frequently unknown until preliminary field work has been accomplished. Thus, MSDSs for hazardous wastes found on clean-up or abandoned sites may be nonexistent, and the manager/supervisor is then faced with the necessity to perform the required risk assessments using very limited chemical data. This circumstance may lead to a false sense of security on the part of the worker and unwarranted relaxation of the PPE regimen.

Workplace Violence

The 1998 Department of Justice (DOJ) National Crime Victimization Survey (NCVS) revealed that assaults and threats of violence against Americans at work

---

[22] Yet the HAZCOM is the most frequently violated (based upon numbers of citations) of all § 1910 standards.

numbered almost 2 million. The most common type of workplace violent crime was simple assault, with an average of 1.5 million per year. There were 396,000 aggravated assaults, 51,000 rapes and sexual assaults, 84,000 robberies, and 1000 homicides in 1998 (OSHA 1999). The report then ranks job categories from police officer with 306 per 1000 officers, through college teachers with 3 per 1000. Hazardous waste workers are not listed, nor is there an industrial category that might shed some light on the waste management category. Employers, managers, and supervisors nevertheless need to be vigilant for warning signs and conditions and take appropriate action to head off violent incidents.

Risk factors that may increase a worker's risk for workplace assault as identified by NIOSH are

- Contact with the public
- Exchange of money
- Delivery of passengers, goods, or services
- Having a mobile workplace, such as a taxicab or police cruiser
- Working with unstable or volatile persons in health care, social services, or criminal justice settings
- Working alone or in small numbers
- Working late at night or during early morning hours
- Working in high crime areas
- Guarding valuable property or possessions
- Working in community-based settings

Employers, managers, and supervisors may be able to adjust working conditions to minimize some of the listed risk factors; however, more detailed guidance is available from professional consultants, some state agencies, and OSHA. OSHA and NIOSH have developed prevention programs, engineering and administrative controls, and post-incident response and evaluation procedures. These measures can be accessed in a variety of formats and detail at: <http://www.osha.gov/oshinfo.html> and <http://www.cdc.gov/niosh.html>. In addition, the OSHA Home Page Index is useful in locating publications on this and other OSHA topics. The Home Page can be accessed at <http://www.osha.gov>.

OSHA has not promulgated a specific standard for prevention of workplace violence. In the manner of other workplace hazards that are difficult or impossible to regulate, as discussed in previous sections, employers can be cited for failure to provide a safe workplace by invocation of the General Duty Clause of the OSH Act (OSHA 1999).

The OSHA Unified Agenda

Those wishing to maintain knowledge of OSHA standards development, revision, and promulgation; program elements; policy developments; and other agenda items can access the useful and orderly OSHA Unified Agenda, Table of Contents, at <http://www.osha-slc.gov.html>. The OSHA Home Page at <www.osha.gov> is similarly helpful.

APPENDIX A
OSHA Workplace Standards That May Apply to Hazardous Waste Sites

PART 1910 — OCCUPATIONAL SAFETY AND HEALTH STANDARDS
Subpart A — General

Subpart B — Adoption and Extension of Established Federal Standards

Subpart C [Removed and Reserved]

Subpart D — Walking — Working Surfaces

Subpart E — Means of Egress

Appendix to Subpart E — Means of Egress
Subpart F — Powered Platforms, Manlifts, and Vehicle-Mounted Work Platforms

APPENDIX A   *(Continued)*
OSHA Workplace Standards That May Apply to Hazardous Waste Sites

Subpart G — Occupational Health and Environmental Control

§ 1910.94	Ventilation.
§ 1910.95	Occupational noise exposure.
§ 1910.96	[Redesignated as 1910.1096]
§ 1910.97	Nonionizing radiation.
§ 1910.98	Effective dates.

Subpart H — Hazardous Materials

§ 1910.101	Compressed gases (general requirements).
§ 1910.102	Acetylene.
§ 1910.103	Hydrogen.
§ 1910.104	Oxygen.
§ 1910.105	Nitrous oxide.
§ 1910.106	Flammable and combustible liquids.
§ 1910.107	Spray finishing using flammable and combustible materials.
§ 1910.108	Dip tanks containing flammable or combustible liquids.
§ 1910.109	Explosives and blasting agents.
§ 1910.110	Storage and handling of liquified petroleum gases.
§ 1910.111	Storage and handling of anhydrous ammonia.
§ 1910.112	[Reserved]
§ 1910.113	[Reserved]
§ 1910.119	Process safety management of highly hazardous chemicals.
§ 1910.120	Hazardous waste operations and emergency response.
§ 1910.121	[Reserved]
§ 1910.122	Table of contents.
§ 1910.123	Dipping and coating operations: coverage and definitions.
§ 1910.124	General requirements for dipping and coating operations.
§ 1910.125	§ Additional requirements for dipping and coating operations that use flammable or combustible liquids.
§ 1910.126	Additional requirements for special dipping and coating applications.

Subpart I — Personal Protective Equipment

§ 1910.132	General requirements.
§ 1910.133	Eye and face protection.
§ 1910.134	Respiratory protection.
§ 1910.135	Head protection.
§ 1910.136	Foot protection.
§ 1910.137	Electrical protective devices.
§ 1910.138	Hand Protection.
§ 1910.139	Respiratory protection for M. tuberculosis

Subpart J — General Environmental Controls

§ 1910.141	Sanitation.
§ 1910.142	Temporary labor camps.
§ 1910.143	Nonwater carriage disposal systems. [Reserved]
§ 1910.144	Safety color code for marking physical hazards.
§ 1910.145	Specifications for accident prevention signs and tags.

APPENDIX A *(Continued)*
OSHA Workplace Standards That May Apply to Hazardous Waste Sites

§ 1910.146	Permit-required confined spaces.
§ 1910.147	The control of hazardous energy (lockout/tagout).

Subpart K — Medical and First Aid
§ 1910.151	Medical services and first aid.
§ 1910.152	[Reserved]

Subpart L — Fire Protection
§ 1910.155	Scope, application and definitions applicable to this subpart.
§ 1910.156	Fire brigades.

Portable Fire Suppression Equipment
§ 1910.157	Portable fire extinguishers.
§ 1910.158	Standpipe and hose systems.

Fixed Fire Suppression Equipment
§ 1910.159	Automatic sprinkler systems.
§ 1910.160	Fixed extinguishing systems, general.
§ 1910.161	Fixed extinguishing systems, dry chemical.
§ 1910.162	Fixed extinguishing systems, gaseous agent.
§ 1910.163	Fixed extinguishing systems, water spray, and foam.

Other Fire Protective Systems
§ 1910.164	Fire detection systems.
§ 1910.165	Employee alarm systems.

Appendices to Subpart L
Appendix A — Fire Protection
Appendix B — National Concensus Standards
Appendix C — Fire Protection References for Further Information
Appendix D — Availability of Publications Incorporated by Reference in Section 1910.156, Fire Brigades
Appendix E — Test Methods for Protective Clothing

Subpart M — Compressed Gas and Compressed Air Equipment
§ 1910.166	[Reserved]
§ 1910.167	[Reserved]
§ 1910.168	[Reserved]
§ 1910.169	Air receivers.

Subpart N — Materials Handling and Storage
§ 1910.176	Handling material — general.
§ 1910.177	Servicing multi-piece and single-piece rim wheels.
§ 1910.178	Powered industrial trucks.
§ 1910.179	Overhead and gantry cranes.
§ 1910.180	Crawler locomotive and truck cranes.
§ 1910.181	Derricks.
§ 1910.183	Helicopters.
§ 1910.184	Slings.

APPENDIX A *(Continued)*
OSHA Workplace Standards That May Apply to Hazardous Waste Sites

**Appendix A to 1910.178 — Stability of Powered Industrial Trucks
(non-mandatory Appendix to Paragraph (l) of this section).**

Subpart O — Machinery and Machine Guarding

§ 1910.211	Definitions.
§ 1910.212	General requirements for all machines.
§ 1910.213	Woodworking machinery requirements.
§ 1910.214	Cooperage machinery.
§ 1910.215	Abrasive wheel machinery.
§ 1910.216	Mills and calenders in the rubber and plastics industries.
§ 1910.217	Mechanical power presses.
§ 1910.218	Forging machines.
§ 1910.219	Mechanical power-transmission apparatus.

Subpart P — Hand and Portable Powered Tools and Other Hand-Held Equipment

§ 1910.241	Definitions.
§ 1910.242	Hand and portable powered tools and equipment, general.
§ 1910.243	Guarding of portable powered tools.
§ 1910.244	Other portable tools and equipment.

Subpart Q — Welding, Cutting, and Brazing

§ 1910.251	Definitions.
§ 1910.252	General requirements.
§ 1910.253	Oxygen-fuel gas welding and cutting.
§ 1910.254	Arc welding and cutting.
§ 1910.255	Resistance welding.

Subpart R — Special Industries

§ 1910.261	Pulp, paper, and paperboard mills.
§ 1910.262	Textiles.
§ 1910.263	Bakery equipment.
§ 1910.264	Laundry machinery and operations.
§ 1910.265	Sawmills.
§ 1910.266	Logging operations.
§ 1910.267	[Reserved]
§ 1910.268	Telecommunications.
§ 1910.269	Electric power generation, transmission, and distribution.
§ 1910.272	Grain handling facilities.

Subpart S — Electrical

General

§ 1910.301	Introduction.

Design Safety Standards for Electrical Systems

§ 1910.302	Electric utilization systems.
§ 1910.303	General requirements.
§ 1910.304	Wiring design and protection.
§ 1910.305	Wiring methods, components, and equipment for general use.

APPENDIX A   *(Continued)*
OSHA Workplace Standards That May Apply to Hazardous Waste Sites

§ 1910.306   Specific purpose equipment and installations.
§ 1910.307   Hazardous (classified) locations.
§ 1910.308   Special systems.
§ 1910.309–1910.330   [Reserved]

Safety-Related Work Practices
§ 1910.331   Scope.
§ 1910.332   Training.
§ 1910.333   Selection and use of work practices.
§ 1910.334   Use of equipment.
§ 1910.335   Safeguards for personnel protection.
§ 1910.336–1910.360   [Reserved]

Safety-Related Maintenance Requirements
§ 1910.361–1910.380   [Reserved]

Safety Requirements for Special Equipment
§ 1910.381–1910.398   [Reserved]

Definitions
§ 1910.399   Definitions applicable to this subpart.

Appendices to Subpart S
Appendix A — Reference Documents
Appendix B — Explanatory Data [Reserved]
Appendix C — Tables, Notes, and Charts [Reserved]

Subpart T — Commercial Diving Operations
General
§ 1910.401   Scope and application.
§ 1910.402   Definitions.

Personnel Requirements
§ 1910.410   Qualifications of dive team.

General Operations Procedures
§ 1910.420   Safe practices manual.
§ 1910.421   Pre-dive procedures.
§ 1910.422   Procedures during dive.
§ 1910.423   Post-dive procedures.

Specific Operations Procedures
§ 1910.424   SCUBA diving.
§ 1910.425   Surface-supplied air diving.
§ 1910.426   Mixed-gas diving.
§ 1910.427   Liveboating.

Equipment Procedures and Requirements
§ 1910.430   Equipment.

APPENDIX A   *(Continued)*
OSHA Workplace Standards That May Apply to Hazardous Waste Sites

Record Keeping
§ 1910.440   Record keeping requirements.
§ 1910.441   Effective date.

Appendices to Subpart T
Appendix A — Examples of Conditions That May Restrict or Limit Exposure to Hyperbaric Conditions
Appendix B — Guidelines for Scientific Diving

Subparts U — Y [Reserved]
§ 1910.442–1910.999   [Reserved]

Subpart Z — Toxic and Hazardous Substances
§ 1910.1000   Air contaminants.
§ 1910.1001   Asbestos.
§ 1910.1002   Coal tar pitch volatiles; interpretation of term.
§ 1910.1003   13 Carcinogens (4-Nitrobiphenyl, etc.).
§ 1910.1004   alpha-Naphthylamine.
§ 1910.1005   [Reserved]
§ 1910.1006   Methyl chloromethyl ether.
§ 1910.1007   3,3'-Dichlorobenzidine (and its salts).
§ 1910.1008   bis-Chloromethyl ether.
§ 1910.1009   beta-Naphthylamine.
§ 1910.1010   Benzidine.
§ 1910.1011   4-Aminodiphenyl.
§ 1910.1012   Ethyleneimine.
§ 1910.1013   beta-Propiolactone.
§ 1910.1014   2-Acetylaminofluorene.
§ 1910.1015   4-Dimethylaminoazobenzene.
§ 1910.1016   N-Nitrosodimethylamine.
§ 1910.1017   Vinyl chloride.
§ 1910.1018   Inorganic arsenic.
§ 1910.1020   Access to employee exposure and medical records.
§ 1910.1025   Lead.
§ 1910.1027   Cadmium.
§ 1910.1028   Benzene.
§ 1910.1029   Coke oven emissions.
§ 1910.1030   Bloodborne pathogens.
§ 1910.1043   Cotton dust.
§ 1910.1044   1,2-Dibromo-3-chloropropane.
§ 1910.1045   Acrylonitrile.
§ 1910.1047   Ethylene oxide.
§ 1910.1048   Formaldehyde.
§ 1910.1050   Methylenedianiline.
§ 1910.1051   1,3-Butadiene.
§ 1910.1052   Methylene chloride.
§ 1910.1096   Ionizing radiation.
§ 1910.1200   Hazard communication.

[57 FR 42389, September 14, 1992; 58 FR 4549, January 14, 1993; 58 FR 35308, June 30, 1993; 59 FR 4437, January 31, 1994; 59 FR 36695, July 19, 1994; 59 FR 51672, October 12, 1994; 61 FR 5507, February 13, 1996; 61 FR 9227, March 7, 1996; 61 FR 31427, June 20, 1996; 62 FR 1493, January 10, 1997; 62 FR 42666, August 8, 1997; 62 FR 48175, September 15, 1997; 62 FR 66275, December 18, 1997; 63 FR 1152, January 8, 1998; 63 FR 13338, March 19, 1998; 63 FR 17093, April 8, 1998; 63 FR 20098, April 23, 1998; 63 FR 33450, June 18, 1998; 63 FR 66270, December 1, 1998; 64 FR 13908, March 23, 1999; 65 FR 46818, July 31, 2000.]

REFERENCES

American Conference of Governmental Industrial Hygienists. 1993. *Threshold Limit Values for Chemical Substances and Physical Agents and Biological Exposure Indices,* Technical Affairs Office, Cincinnati, OH.

American National Standards Institute. 1992. *American National Standard for Respiratory Protection ANSI Z882.2-1992,* NY.

Bollinger, Nancy J. and Robert H. Schutz. 1987. NIOSH Guide to Industrial Respiratory Protection, Division of Safety Research, U.S. National Institute for Occupational Safety and Health, Cincinnati, OH.

Environment Reporter. June 30, 1993. Bureau of National Affairs, Washington, D.C.

Environment Reporter. November 24, 1999, p. 664. Bureau of National Affairs, Washington, D.C.

Goldman, Ralph F. 1994. "Heat Stress in Industrial Protective Encapsulating Garments," Chapter 10, in *Protecting Personnel at Hazardous Waste Sites,* William F. Martin and Steven P. Levine, Eds., Butterworth-Heinemann, Stoneham, MA.

International Commission on Radiological Protection (ICRP). 1977. *Recommendations of the International Commission on Radiological Protection,* ICRP Publication 26. Pergamon Press, Oxford, England.

Jones, Frank E. 1994. *Toxic Organic Vapors in the Workplace.* CRC Press, Boca Raton, FL.

Levine, Steven P., Rodney D. Turpin, and Michael Gochfeld. 1994. "Protecting Personnel at Hazardous Waste Sites: Current Issues, " Chapter 1, in *Protecting Personnel at Hazardous Waste Sites,* William F. Martin and Steven P. Levine, Eds., Butterworth-Heinemann, Stoneham, MA.

Martin, William F. 1994. "Site Health and Safety Plans," Chapter 16, in *Protecting Personnel at Hazardous Waste Sites,* William F. Martin and Steven P. Levine, Eds., Butterworth-Heinemann, Stoneham, MA.

Maslansky, Steven P. and Carol J. Maslansky. 1997. *Health and Safety at Hazardous Waste Sites,* Van Nostrand Reinhold, NY.

Miller, Marshall Lee. 1985. "Occupational Safety and Health Act," Chapter 8, in *Environmental Law Handbook*, Government Institutes, Inc., Rockville, MD.

Occupational Safety and Health Reporter. July 29, 1992. Bureau of National Affairs, Washington, D.C.

Occupational Safety and Health Reporter. October 5, 1994. Bureau of National Affairs, Washington, D.C.

Occupational Safety and Health Reporter. November 24, 1999. Bureau of National Affairs, Washington, D.C.

Occupational Safety and Health Reporter. August 17, 2000. Bureau of National Affairs, Washington, D.C.

Petroleum Equipment Institute. 1992. *Tulsaletter,* July 28, Tulsa, OK.

Robinson, James C. 1991. *Toil and Toxics.* University of California Press, Los Angeles.

Schwope, Arthur D. and Christopher C. O'Leary. 1994. "Personal Protective Equipment," Chapter 9, in *Protecting Personnel at Hazardous Waste Sites*, William F. Martin and Steven P. Levine, Eds., Butterworth-Heinemann, Stoneham, MA.

U.S. Department of Health and Human Services. 1985. Occupational Safety and Health Guidance Manual for Hazardous Waste Site Activities. Superintendent of Documents, U.S. Government Printing Office, Washington, D.C. (cited in text as the "four-agency manual").

U.S. Department of Health and Human Services. 1997. NIOSH Pocket Guide to Chemical Hazards. U.S. National Institute for Occupational Safety and Health, U.S. Government Printing Office, Washington, D.C.

U.S. Department of Health and Human Services. 1998. Selecting, Evaluating, and Using Sharps Disposal Containers. U.S. National Institute for Occupational Safety and Health (NIOSH) Publication No. 97-111, Washington, D.C.

U.S. Department of Labor. 1989. Air Contaminants — Permissible Exposure Limits (Title 29 Code of Federal Regulations Part 1910.1000). Occupational Safety and Health Administration, Washington, D.C., OSHA 3112.

U.S. Department of Labor. 1994. Occupational Safety and Health Administration (OSHA) News Release, February 17, 1994, Washington, D.C.

U.S. Department of Labor. 1998. Occupational Exposure to Bloodborne Pathogens — Precautions for Emergency Responders, Occupational Safety and Health Administration (OSHA), Washington, D.C., OSHA 3106.

U.S. Department of Labor. 1999. Workplace Violence, Occupational Safety and Health Administration, Washington, D.C.

U.S. Department of Labor. 2000. OSHA Unified Agenda Proposed Rule Stage, 1943. Permissible Exposure Limits (PELS) for Air Contaminants. Occupational Safety and Health Administration, Washington, D.C.

Wallace, Lynn P. 1994. "Site Layout and Engineered Controls," Chapter 8, in *Protecting Personnel at Hazardous Waste Sites,* William F. Martin and Steven P. Levine, Eds., Butterworth-Heinemann, Stoneham, MA.

Glossary

We adopt three conventions with this glossary:

1. Definitions using the word "means" are those taken directly from the Resource Conservation and Recovery Act (RCRA), the implementing regulations, and other pertinent statutes and regulations. Definitions void of the word "means" are those commonly used and accepted in the hazardous waste management field.
2. Acronyms are mixed throughout, rather than being compiled separately, because many require definition as well as explanation.
3. Words in the masculine gender also include the feminine and neuter gender; words in the singular include the plural; and words in the plural include the singular.

Abandoned means, for purposes of defining a material as a solid waste under RCRA Subtitle C, a material that is disposed of, burned, or incinerated.

Absorption — the process whereby one or more gaseous contaminants are dissolved into a relatively nonvolatile liquid and may be characterized as chemical or physical. Chemical absorption occurs when there is a reaction between the absorbed gas and the liquid solvent. Physical absorption occurs when the absorbed gas merely dissolves in the liquid solvent.

Accumulated speculatively means storage of a material in lieu of expeditious recycling. Materials are usually accumulated speculatively if the waste being stored has no viable market or if a facility cannot demonstrate that at least 75% of the material has been recycled in a calendar year.

Actinide — an element with an atomic number from 89 to 103 inclusive. All are radioactive.

Activated sludge — sludge resulting from the mixing of primary effluent with bacteria-laden sludge and then agitated and aerated to promote biological treatment. This speeds breakdown of organic matter in primary effluent undergoing secondary waste treatment.

Active life means the period from the initial receipt of hazardous waste at a facility until the Regional Administrator receives certification of final closure.

Active portion means that portion of a facility where treatment, storage, or disposal operations are being or have been conducted after the effective date of 40 CFR 261, and which is not a closed portion (*see also:* **closed portion** and **inactive portion**).

Acute effect — an adverse effect on the receptor organism, with symptoms of severity coming quickly to a crisis.

Acutely Hazardous Waste (AHW) — wastes listed in 40 CFR 261.31 and which are followed by the symbol (H) and all of the "P" wastes listed in 40 CFR 261.33(e).

Administrator means the Administrator of the Environmental Protection Agency or his designee.

Adsorption — a physical phenomenon that refers to the ability of certain solids to attract and collect organic substances from the surrounding medium. Granular activated carbon is widely used to adsorb organic components from liquid and gaseous waste streams.

Aerobic decomposition — the natural decay and breakdown of organic matter by bacteria which utilize oxygen in respiration.

ALARA — an acronym for the phrase "as low as reasonably achievable," meaning that exposures to a source or sources of radioactivity should be kept as low as reasonably achievable, with economic and social factors being taken into account.

Alpha particle — a positively charged particle composed of two neutrons and two protons released by some atoms undergoing radioactive decay. The particle is identical to the nucleus of a helium atom.

Anaerobic decomposition — the natural decay and breakdown of organic matter by bacteria that do not require oxygen for respiration.

Aquifer means a geologic formation, group of formations, or part of a formation capable of yielding a significant amount of groundwater to wells or springs.

Artesian — the occurrence of groundwater under greater than atmospheric pressure.

Artificial recharge — the addition of water to the groundwater reservoir by activities of man.

Asphyxiant — a chemical that can deny oxygen to cells of the host organism, thereby slowing or halting metabolism.

Audit (environmental) — a systematic, documented, periodic, and objective review by regulated entities of facility operations and practices relate to meeting environmental requirements. The audit may verify compliance with environmental requirements; evaluate the effectiveness of environmental management systems already in place; or assess risks from regulated and unregulated materials and practices.

Automatic Tank Gauging (ATG) — a release detection method for **USTs** that uses a probe in the tank, connected to a monitor, to provide information on product level and temperature.

Base flow — the portion of the flow, in surface streams, that has been discharged to the stream as influent groundwater.

Basel Convention — the international treaty that establishes standards for global trade of hazardous waste, municipal waste, and municipal incinerator ash. The U.S. is not a party to the convention. U.S businesses can only export waste to those countries with which the U.S. government has negotiated a separate waste trade agreement.

Beta particle — negatively charged particles (electrons) emitted by radioactive decay processes. Beta particles travel at 30 to 99% of the speed of light, but have low mass and relatively low penetrating power.

Bentsen Amendment — RCRA Section 3001(b)(2)(A-C), by then-Senator Lloyd Bentsen, enacted on October 21, 1980, which postponed **RCRA** regulation of drilling fluids, produced waters, and other wastes associated with the exploration, development, or production of crude oil or natural gas, pending a study by the **Administrator**. The exclusions remain in effect at this writing. The implementing regulations are found at 40 CFR 261.4(b)(5).

Bevill Amendment — RCRA Section 3001(b)(3)(A-C), by Representative Tom Bevill of Alabama, enacted on October 21, 1980, which postponed regulation of fly ash waste, bottom ash waste, slag waste, flue gas emission control waste, solid waste from the extraction, beneficiation, and processing of ores and minerals, including phosphate rock and overburden from the mining of uranium ore, and cement kiln dust, pending a study by the **Administrator**. The exclusions remain in effect at this writing. The implementing regulations are found at 30 CFR 261.4(b)(2-4).

Bioaccumulation — the retention and concentration of a substance by an organism.

Biodegradation — decomposition of a substance into more elementary compounds by the action of microorganisms such as bacteria.

Biological treatment — a treatment technology that uses bacteria to consume waste constituents in municipal or industrial wastewaters.

Bioreclamation — a technique for treating contaminated areas by enhancing the natural microbial degradation of organic contaminants, thereby reducing the toxicity of the target compounds. The term was originally meant to imply the use of microbial degradation technology to "reclaim" the contaminated area, but is now apparently used interchangeably with **bioremediation**.

Bioremediation — a treatment process in which organic wastes in soils or other media may be seeded with soil microorganisms to alter or destroy the waste; or specific nutrients may be added to an organic waste to enhance naturally occurring (or extant) microorganisms and stimulate the activity.

Boiler — an inclosed device using controlled flame combustion, which is designed to recover and export thermal energy in the form of steam, heated fluids, or heated gases and which meets the specifications of 40 CFR 260.10. Boilers and industrial furnaces (**BIFs**) are regulated by the EPA when used to combust hazardous wastes for destruction of the waste, or when burning "hazardous waste fuel" (*see:* **industrial furnace**).

Brownfields means abandoned, idled, or under-used industrial and commercial facilities where expansion or redevelopment is complicated by real or perceived contamination.

BTEX — an acronym for benzene, toluene, ethylbenzene, and xylene, which are the major volatile aromatic compounds in fuel hydrocarbons.

CAA — Clean Air Act of 1970, 42 USC 7401 et seq. The Act has been amended many times (*see:* **CAAA**).

CAAA — Clean Air Act Amendments of 1990, 42 USC 7407(d).

Canister — a closed or sealed container for nuclear fuel or other radioactive material, which isolates and contains the contents. A canister is usually placed in a **cask** to provide shielding (*see:* Figure 13.7).

Cask — a massive container (usually of concrete) used in transport or storage of spent nuclear fuel rods or other radioactive material. It provides chemical, nuclear, and radiological protection and dissipates heat generated by radioactive decay (*see:* Figure 13.7).

Carbon Regeneration Unit (CRU) means any enclosed thermal treatment device used to regenerate spent activated carbon.

Carcinogen —an agent that has the potential to induce the abnormal, excessive, and uncoordinated proliferation of certain cell types or the abnormal division of cells, i.e., a material that causes cancer cells to develop and proliferate.

Capacity Assurance Plan (CAP) — a written statement ensuring that a state has a hazardous waste treatment and disposal capacity. This capacity must be for facilities that are in compliance with RCRA Subtitle C requirements and must be adequate to manage hazardous wastes projected to be generated within the state over 20 years.

Cathodic protection — a form of corrosion protection for **USTs** that uses sacrificial anodes or a direct current source to protect steel by halting the naturally occurring electrochemical process that causes corrosion.

CNS — the central nervous system; the portion of the nervous system consisting of the spinal cord and the brain.

CERCLA — the Comprehensive Environmental Response Compensation and Liability Act of 1980, 42 U.S.C. 9601 et seq. (*see also:* **Superfund** and **SARA Title III**).

Certification means a statement of professional opinion based upon knowledge and belief.

CEP — *see:* **Completed Environmental Pathway**.

CFCs — chlorofluorocarbon compounds, such as Freon-12 (CCl_2F_2), used chiefly as refrigerants. Their use as propellants for aerosols was prohibited in 1979 because of their depleting effect on stratospheric ozone.

CFR — *see:* **Code of Federal Regulations**.

Chronic effect — an adverse effect upon a receptor organism, with symptoms that develop slowly over a long period of time or recur frequently.

Clean closure means removal, or treatment, of all contaminated soils, liquids, equipment, or structures from a **TSDF**, until testing shows no contamination greater than specified in the approved closure plan [*see:* 40 CFR 270.1(c)(5)].

Closed portion means that portion of a facility closed by an owner or operator in accordance with the approved facility closure plan and all applicable closure requirements (*see also:* **active portion** and **inactive portion**).

Code of Federal Regulations (CFR) — the cumulation of executive agency regulations published in the *Federal Register* combined with regulations issued previously and remaining in effect. It is published annually by the Office of the Federal Register, National Archives and Records Service, Washington, D.C. The CFR is divided into 50 titles, each associated with a broad subject area (e.g., Title 40 contains most of the environmental regulations; Title 49, transportation, etc.). Citations may be by title and chapter, but most frequently are by title and part (e.g., 40 CFR 261.1). The citation is definitive in either form. Individual volumes of the Code are revised each year, but are issued on a staggered quarterly basis. Thus, availability of a complete and current regulation may follow publication in the *Federal Register* by many months. Commercial publications and newsletters, which provide timely prints of newly revised or promulgated regulations, are available.

Completed Exposure Pathway (CEP) — an indicator that competent research has established a statistically consistent cause and effect linkage between an environmental hazard and a human health impact.

Conditionally Exempt Small Quantity Generator (CESQG) — a generator who generates no more than 100 kg of hazardous waste or no more than 1 kg of acutely hazardous waste in any calendar month [*see:* 40 CFR 261.5(a) and (e)].

Confined aquifer means an aquifer bounded above and below by impermeable beds or by beds of distinctly lower permeability than that of the aquifer itself; an aquifer containing confined groundwater.

Container means any portable device in which a material is stored, transported, treated, disposed of, or otherwise handled.

Contained-In Policy — EPA policy that determines the health threats posed by contaminated environmental media and debris and whether such materials must be managed as **RCRA** hazardous waste.

Containment building means a hazardous waste management unit that is used to store or treat hazardous waste under the provisions of subpart D.D. of 40 CFR 264 or 265.

Contamination — the degradation of natural water quality as a result of man's activities, to the extent that its usefulness is impaired.

Contingency plan means a document setting out an organized, planned, and coordinated course of action to be followed in case of a fire, explosion, or release of hazardous waste or hazardous waste constituents that could threaten human health or the environment.

Corrective Action Management Unit (CAMU) means an area within a facility that is used only for managing rededication wastes for implementing corrective action or cleanup at the facility.

Corrosion expert means a person who, by reason of his knowledge of the physical sciences and the principles of engineering and mathematics, acquired by a professional education and related practical experience, is qualified to engage in the practice of corrosion control on buried or submerged metal piping systems and metal tanks. Such a person must be certified by the National Association of Corrosion Engineers (NACE) or be a registered engineer who has certification or licensing that includes education and experience in corrosion control on buried or submerged metal piping systems and metal tanks.

Curie (Ci) means that quantity of radioactive material producing 37 million nuclear transformations per second. (The amount of radioactivity contained in 1 g of radium.)

CWA — Clean Water Act of 1977, 33 USC 1251 et seq. The Clean Water Act is an amendment to the Federal Water Pollution Control Act of 1972.

Deactivation — as used with the **LDRs**, means any of the recommended treatment technologies listed in 40 CFR 268, Appendix IV, or other technologies not listed, which are capable removal of the characteristics of ignitability, corrosivity, and reactivity from the wastes listed (*see:* 40 CFR 268.42 and Appendix VI).

Debris — a broad category of large manufactured and naturally occuring objects that are commonly discarded (e.g., construction materials, decommissioned industrial equipment, discarded manufactured objects, tree trunks, boulders).

Decontamination — *infectious waste:* the use of physical or chemical processes to remove, inactivate, or destroy living organisms to some lower level(not necessarily zero). *Chemical inactivation:* physical removal and/or dissolution of chemical contaminants until they can be rinsed away. *Radioactive contaminants:* the removal of radioactive contaminants with the objective of reducing the residual radioactivity level in or on materials, persons, or equipment.

DDT — dichlorodiphenyltrichloroethane, which is a highly persistent insecticide that causes liver damage; is a suspect carcinogen and mutagen; and moves through the food chain to threaten higher forms of wildlife. DDT has been banned for use in the U.S.

Designated facility — a hazardous waste treatment, storage, or disposal facility that has received an EPA permit (or a facility with interim status) and has been designated as such on the manifest pursuant to 40 CFR 262.20.

Destruction and Removal Efficiency (DRE) — the formula by which hazardous waste (and other) incinerators performance is measured and regulated. The DRE (in percent) for each principal organic hazard constituent (POHC) is determined from the following equation:

$$DRE = \frac{(W_{in} - W_{out})}{W_{in}} \times 100$$

Where W_{in} = the mass feed rate of one POHC in the waste stream feeding the incinerator and W_{out} = the mass feed rate of the same POHC present in exhaust emissions prior to release to the atmosphere (*see:* 40 CFR 264, Subpart O).

Dike means an embankment or ridge of either natural or man-made materials used to prevent the movement of liquids, sludges, solids, or other materials.

Discarded material means any material which is abandoned, recycled, or considered inherently waste-like (*see:* 40 CFR 261.2).

Discharge or hazardous waste discharge means the accidental or intentional spilling, leaking, pumping, pouring, emitting, emptying, or dumping of hazardous waste into or on any land or water.

Disposal means the discharge, deposit, injection, dumping, spilling, leaking, or placing of any solid waste or hazardous waste into or on any land or water so that such solid waste or hazardous waste or any constituent thereof may enter the environment or be emitted into the air or discharged into any waters, including any groundwaters.

Disposal facility means a facility or part of a facility at which hazardous waste is intentionally placed into or on any land or water and at which waste will remain after closure.

Dioxin — a term used to identify a group of polychlorinated compounds which vary greatly in the degree of toxicity. Those with four to six chlorinated atoms are the most active and have the greatest potential toxicity. 2,3,7,8-Tetrachlorodibenzo-*p*-dioxin (2,3,7,8-TCDD) has the greatest acute toxicity. The compound occurs as a side product contaminant in the pesticide 2,4,5-T (2,4,5-trichlorophenoxyacetic acid) and as an incomplete combustion product of the burning of chlorine-bleached materials.

Dose — the quantity of a chemical, or of radiation, absorbed by an organism. *Quantitative:* the mass of toxicant per mass of receptor (mg toxicant/kg animal). In radiology, the quantity of energy or radiation absorbed.

Electron acceptor — a compound capable of accepting electrons during oxidation-reduction reactions. Microorganisms obtain energy by transferring electrons from electron donors such as organic compounds (or sometimes reduced inorganic compounds such as sulfide) to an electron acceptor. Electron acceptors are compounds that are relatively oxidized and include oxygen, nitrate iron (III), manganese (IV), sulfate, carbon dioxide, or in some cases the chlorinated aliphatic hydrocarbons such as perchloroethene PCE, TCE, DCE, and vinyl chloride.

Electron donor — a compound capable of supplying (giving up) electrons during oxidation–reduction reactions. Microorganisms obtain energy by transferring electrons from electron donors such as organic compounds (or sometimes reduced inorganic compounds such as sulfide) to an electron acceptor. Electron donors are compounds that are relatively reduced and include fuel hydrocarbons and native organic carbon.

Environmental Audit (EA)* — an *investigative* process to determine if the operations of an existing facility meet documented expectations, usually reg-

---

* There are broad variations in the use and meaning of these terms, e.g., some include testing of environmental media in an ESA, implying greater rigor than an audit. Parties to implementation of these tools should reach agreement on objectives and procedures before substantial work begins.

ulations, standards, or policies, by conducting interviews, reviewing records, and making first-hand observations or measurements.

Environmental Site Assessment (ESA)* — an *estimation or judgment* process that seeks to verify expectations, regarding a parcel of real property, by conducting interviews, reviewing records, and making first-hand observations. An ESA is generally less rigorous than an **EA**.

EPA hazardous waste number means the number assigned by the EPA to each hazardous waste listed in Subpart D of 40 CFR 261 and to each characteristic identified in Subpart C of 40 CFR 261.

EPA identification number means the number assigned by the EPA to each generator, transporter, and treatment, storage, or disposal facility.

EPA Region means the states and territories found in any one of the ten standard federal regions of the U.S.

EPCRA — the Emergency Planning and Community Right-to-Know Act is Title III of the Superfund Amendments and Reauthorization Act (SARA) of 1986. EPCRA requires emergency planning by local committees; notification of local and state authorities in the event of accidental releases of hazardous substances; community right-to-know reporting; and toxic chemical release reporting.

Existing hazardous waste management (HWM) facility or **existing facility** means a facility that was in operation or for which construction commenced on or before November 19, 1980 (*see:* 40 CFR 260.10 for more detail).

Existing portion means that land surface area of an existing waste management unit, included in the original Part A permit application, on which wastes have been placed prior to the issuance of a permit.

Existing tank system or **existing component** means a tank system or component that is used for the storage or treatment of hazardous waste and that is in operation or for which installation has commenced on July 14, 1986 (*see:* 40 CFR 260.10 for more detail).

Exposure pathway — the course a chemical takes from the source to the exposed individual. The exposure pathway analysis links the sources, locations, and types of environmental releases with population locations and activity patterns to determine the significant pathways of human exposure.

Facility means all contiguous land, and structures, or other appurtenances, and improvements on the land, used for treating, storing, or disposing of hazardous waste. A facility may consist of several treatment, storage, or disposal operational units (e.g., one or more landfills, surface impoundments, or combinations of them).

Feasibility Study (FS) means a study undertaken by the lead agency to develop and evaluate options for remedial action. The FS emphasizes data analysis and is generally performed concurrently and in an interactive fashion with

---

* There are broad variations in the use and meaning of these terms, e.g., some include testing of environmental media in an ESA, implying greater rigor than an audit. Parties to implementation of these tools should reach agreement on objectives and procedures before substantial work begins.

the **Remedial Investigation (RI)**, using data gathered during the **RI**. The term also refers to a report that describes the results of the study.

Federal Register (FR) — the medium by which the federal government agencies make regulations and other legal documents of the executive branch available to the public; found in most public libraries. The FR, published daily by the Office of the Federal Register, National Archives and Records Service, Washington, D.C., includes both proposed and final regulations. Citations are by volume and page number (e.g., 43 FR 27736). Some writers prefer to include the date; however, the citation as shown is definitive. Once a regulation has completed the proposal and public comment phases, it is published in final form. If the published regulatory matter is a change to an existing regulation, the FR will publish only the change(s). To view the changed regulation in full, the reader must consult the **Code of Federal Regulations**. This is a cause of frustration, since the **CFR** publications generally follow the FR publications by a year or more. There are a variety of commercial newsletters and trade publications which provide timely prints of regulations in full form.

Final closure means the closure of all hazardous waste management units at the facility in accordance with all applicable closure requirements so that hazardous waste management activities under 40 CFR 264 and 265 are no longer conducted at the facility.

FIFRA — Federal Insecticide, Fungicide, and Rodenticide Act of 1972, 7 USC 136 et seq.

Food chain crops means tobacco, crops grown for human consumption, and crops grown as feed for animals whose products are consumed by humans.

Free liquids means liquids that readily separate from the solid portion of a waste under ambient temperature and pressure.

Freeboard means the vertical distance between the top of a tank or surface impoundment dike and the surface of the waste contained therein.

Gamma radiation — electromagnetic energy waves, without mass, such as X-rays, which have great penetrating power and can penetrate and damage critical organs of the human body.

Generator means any person, by site, whose act or process produces hazardous waste identified or listed in 40 CFR 261 or whose act first causes a hazardous waste to become subject to regulation.

Gradient (groundwater or water table) — the slope of the piezometric surface. A positive slope is referred to as "upgradient"; a negative slope is called "downgradient." (There is lack of consistency in hyphenation.)

Groundwater — water below the land surface in a zone of saturation. In some early literature, a distinction was made between the terms *groundwater* (or *ground-water,* the hyphenated form) and the combined term, *groundwater.* The former was used as the noun and the latter as an adjective. Writers and users of the terms now use both forms without distinction.

Half-life — the time required for the amount (concentration, strength, quantity, emission rate, etc.) of a substance to decrease, due to natural decay processes, to half of the original value.

Hazard Ranking System (HRS) means the method used by the EPA to evaluate the relative potential of hazardous substance releases to cause health or safety problems, or ecological or environmental damage. (The HRS is found at 40 CFR 300, Appendix A.)

Hazardous Air Pollutants (HAPs) — any of 189 chemicals listed pursuant to Title III of the **Clean Air Act Amendments (CAAA)** of 1990. HAPs are suspected cancer-causing agents or other chemicals that are health or environmental hazards. The EPA must set technology-based standards for HAPs.

Hazmat — short or trade term for hazardous material(s).

Hazmat employee means a person who is employed by a hazmat employer and who in the course of employment directly affects hazardous materials transportation safety. This term includes an owner-operator of a motor vehicle that transports hazardous materials in commerce. This term invludes an individual, including a self-employed individual, employed by a hazmat employer who, during the course of employment, (1) loads, unloads, or handles hazardous materials; (2) tests, reconditions, repairs, modifies, marks, or otherwise represents containers, drums, or packagings as qualified for use in the transportation of hazardous materials; (3) prepares hazardous materials for transportation; (4) is responsible for safety of transporting hazardous materials; or (5) operates a vehicle used to transport hazardous materials*.

Hazmat employer means a person who uses one or more of his/her employees in connection with transporting hazardous materials in commerce; causing hazardous materials to be transported in commerce; or representing, marking, certifying, selling, offering, reconditioning, testing, repairing, or modifying containers, drums, or packagings as qualified for use in the transportation of hazardous materials. This term includes an owner-operator of a motor vehicle who transports hazardous materials in commerce. This term also includes any department, agency, or instrumentality of the United States, a state, a political subdivision of a state, or an Indian tribe engaged in an activity described in the first sentence of this definition*.

Hazardous materials — the DOT defines hazardous *materials* as: "a substance or material which has been determined by the Secretary of Transportation to be capable of posing an unreasonable risk to health, safety and property when transported in commerce, and which has been so designated. The term includes hazardous substances, hazardous wastes, marine pollutants, and elevated temperature materials as defined in this section, materials designated as hazardous under the provisions of 49 CFR 172.101 of this subchapter, and materials that meet the defining criteria for hazard classes and divisions in 49 CFR 173 of this subchapter (49 CFR 173)."

Hazardous substance — includes **RCRA** hazardous wastes, as well as substances regulated under the Clean Air Act, Clean Water Act, and Toxic Substances Control Act. The DOT defines hazardous substances as any

---

* 49 CFR 171.8; 57 FR 20952, May 15, 1992.

material, including mixtures, that is listed in Appendix A of the Hazardous Materials Table (49 CFR 172.101), is in a quantity, in one package, that equals or exceeds the **reportable quantity (RQ)** listed in Appendix A, and is in concentrations that meet or exceed those listed on a hazardous substance **RQ** table.

Hazardous waste — the Solid Waste Disposal Act, as amended by the **Resource Conservation and Recovery Act of 1976,** [42 USC 6903(5)] defines "hazardous waste" as: ... "a solid waste, or combination of solid wastes, which because of its quantity, concentration, or physical, chemical, or infectious characteristics may —

(A) cause, or significantly contribute to an increase in mortality or an increase in serious irreversible, or incapacitating reversible, illness: or
(B) pose a substantial present or potential hazard to human health or the environment when improperly treated, stored, transported, or disposed of, or otherwise managed." The implementing regulations (40 CFR 261) define hazardous wastes as those solid wastes that:
  * Exhibit one or more of the four characteristics described in Subpart C (ignitability, corrosivity, reactivity, or toxicity)
  * Are listed in Subpart D
  * Are mixtures of solid and hazardous wastes
  * Are derived from hazardous wastes

Hazwaste — short or trade term for hazardous waste(s).

Hazardous waste constituent (HWC) — a constituent of a waste that causes the waste to be listed in Subpart D of 40 CFR 261 or a constituent listed in Table 1 of 40 CFR 261.24.

Hazardous waste management unit (HWMU) means a contiguous area of land on or in which hazardous waste is placed, or the largest area in which there is significant likelihood of mixing hazardous waste constituents in the same area. Examples of hazardous waste management units include a surface impoundment, a waste pile, a land treatment area, a landfill cell, an incinerator, a tank and its associated piping and underlying containment system, and a container storage area. A container alone does not constitute a unit; the unit includes containers and the land or pad upon which they are placed.

Henry's Law Constant (H) — Henry's law states that *at constant temperature the solubility of a gas in a liquid is proportional to the partial pressure of the gas in contact with the liquid.* The constant H equals the vapor pressure of a substance divided by the aqueous solubility. The constant is the air-water partition coefficient and the rate a substance will evaporate from water. Higher H indicates a greater tendency for the gas to volatilize from the water.

High-Level Waste (HLW) — means the highly radioactive material resulting from the reprocessing of spent nuclear fuel, including liquid waste produced directly in reprocessing and any solid material derived from such other highly radioactive material that the Nuclear Regulatory Commission, by rule, requires permanent isolation of the material.

HMTA — Hazardous Materials Transportation Act of 1975, 49 USC 1801 et seq. HMTA has been amended frequently, including **HMTUSA**.

HMTUSA — Hazardous Materials Transportation and Uniform Safety Act of 1990, 49 USC 1802 et seq.

Holiday — a separation of cladding or coating on the outer surface of an underground storage tank.

HSWA — the Hazardous and Solid Waste Amendments of 1984, which are major amendments to the Solid Waste Disposal Act (*see also:* the **Resource Conservation and Recovery Act of 1976** or **RCRA**.)

Inactive portion means the portion of a facility that is not operated after the effective date of 40 CFR 261 (*see also:* **active portion** and **closed portion**.)

Incident — a broad catch-all term used by government agencies and practitioners to denote an unscheduled event in which conditions hazardous or potentially hazardous to human health or the environment are present. Examples include collision, derailment, chemical spill or release, accident, cave-in, explosion, fire, building collapse, storm, flood, etc.

Incinerator means any enclosed device using controlled flame combustion that neither meets the criteria for classification as a boiler nor is listed as an industrial furnace.

Incompatible waste means a hazardous waste that is unsuitable for:

1. Placement in a particular device or facility because it may cause corrosion or decay of containment materials (e.g., container inner liners or tank walls)
2. Commingling with another waste or material under uncontrolled conditions because the commingling might produce heat or pressure, fire or explosion, violent reaction, toxic dusts, mists, fumes, or gases, or flammable fumes or gases

(*see:* Appendix V of 40 CFR 265 for examples).

Industrial furnace — enclosed devices that are integral components of manufacturing processes and that use controlled flame devices to accomplish recovery of materials or energy, such as cement kilns, lime kilns, aggregate kilns, coke ovens, and blast furnaces. Boilers and industrial furnaces (**BIFs**) are regulated by the EPA when used for destruction of hazardous wastes or for combustion of "hazardous waste fuel" (*see:* 40 CFR 260.10 for more detail).

Inherently waste-like — for purposes of defining a material as a **RCRA** solid waste, materials that are ordinarily managed as hazardous wastes, that pose a substantial hazard when recycled, or that are always considered a solid waste because of their intrinsic threat to human health and the environment, e.g., dioxin-containing wastes.

Injection well means a well into which fluids are injected (*see also:* **underground injection** and **well**).

Inner liner means a continuous layer of material placed inside a tank or container which protects the construction materials of the tank or container from the contained waste or reagents used to treat the waste.

Interim Status — hazardous waste treatment, storage, and disposal facilities that were in existence on November 19, 1980 and met certain conditions were allowed to continue operating until their permit was issued or denied. Such facilities are said to have interim status.

IRIS — Integrated Risk Information System; an electronic database prepared and maintained by the EPA, containing both cancer and non-cancer chronic health hazard information on specific chemicals. IRIS is available to the public online through the National Library of Medicine's Toxicology Data Network (TOXLINE) and can be accessed by calling (301) 496-6531. The service is also available through the National Technical Information Service (NTIS) on diskette, furnished quarterly. The NTIS can be reached at (703) 487-4650.

Landfill means a disposal facility or part of a facility where hazardous waste is placed in or on land. It is not a pile, a land treatment facility, a surface impoundment, an underground injection well, a salt dome formation, a salt bed formation, an underground mine, or a cave (40 CFR 260.10); a disposal facility in which waste is deposited on land or in trenches and is compacted and covered in layers or daily cells (author).

Landfill cell means a discrete volume of a hazardous waste landfill that uses a liner to provide isolation of wastes from adjacent cells or wastes. Examples of landfill cells are trenches and pits.

Land treatment facility means a facility or part of a facility at which hazardous waste is applied onto or incorporated into the soil surface; such facilities are disposal facilities if the waste will remain after closure.

Leachate means any liquid, including any suspended components in the liquid, that has percolated through or drained from hazardous waste.

Leak-detection system means a system capable of detecting the failure of either the primary or secondary containment structure or the presence of a release of hazardous waste or accumulated liquid in the secondary containment structure (*see:* 40 CFR 260.10 for more detail).

Liner means a continuous layer of natural or man-made materials, beneath or on the sides of a surface impoundment, landfill, or landfill cell, which restricts the downward or lateral escape of hazardous waste, hazardous waste constituents, or leachate.

LOAEL — the Lowest Observed Adverse Effect Level; the lowest dose in an experiment which produces an observable adverse effect.

Local Emergency Planning Committee (LEPC) — LEPCs are appointed by the State Emergency Planning Commission to carry out emergency planning at the local district (or county) level as required by the **Emergency Planning and Community Right-to-Know Act (EPCRA).**

Lower Explosive Limit (LEL) — the concentration of a flammable compound, in air, below which a flame will not propagate if an ignition source is present. LEL is expressed as the percent of the flammable vapor, by volume, in air.

Low-Level Waste (LLW) — radioactive waste generated by uranium enrichment processes, reactor operations, isotope production, and medical and research activities. LLW often contains small amounts of radioactivity dispersed in large amounts of material.

Management or hazardous waste management means the systematic control of the collection, source separation, storage, transportation, processing, treatment, recovery, and disposal of hazardous waste.

Manifest means the shipping document EPA Form 8700-22 and, if necessary, EPA Form 8700-22A, originated and signed by the generator in accordance with the instructions included in the Appendix to 40 CFR 262.

Maquiladora — (twin) industrial facilities that are established in Mexico and in the U.S. or another country, usually to take advantage of significantly less stringent regulatory burdens and labor costs in Mexico. Work that is labor intensive, involves use of toxic chemicals, or produces hazardous waste is performed in the Mexican facility. Subassemblies or intermediates are shipped to the U.S. side for final assembly, inspection, distribution, etc.

Methyl Tertiary Butyl Ether (MTBE) — an oxygenate used as a constituent in gasoline since 1979, initially to increase octane levels, but more recently to meet fuel oxygen requirements mandated by the Clean Air Act.

Mixed waste (MW)* — **MW** contains both **RCRA** *hazardous* waste and radioactive waste (as defined by the Atomic Energy Act and amendments thereto). It is jointly regulated by the Nuclear Regulatory Commission or NRC's Agreement States and the EPA or EPA's RCRA Authorized States. The Federal Facilities Compliance Act (FFCA), § 4001(41), statutory definition of MW is "The term 'mixed waste' means waste that contains both hazardous waste and source, special nuclear, or byproduct material subject to the Atomic Energy Act of 1954."

Monitored Natural Attenuation (MNA) means reliance on natural attenuation processes (within the context of a carefully controlled and monitored clean-up approach) to achieve site-specific remedial objectives within a time frame that is reasonable compared to other methods. The "natural attenuation processes" that are at work in such a rededication approach include a variety of physical, chemical, or biological processes that, under favorable conditions, act without human intervention to reduce the mass, toxicity, mobility, volume, or concentration of contaminants in soil and groundwater. These *in situ* processes include biodegradation, dispersion, dilution, sorption, volatilization, chemical or biological stabilization, and transformation or destruction of contaminants (OSWER Directive 9200.4-17, 1997; *see also:* **Natural Attenuation**).

Monitoring well — a well used to measure groundwater levels or to obtain water samples for water quality analysis.

MPRSA — Marine Protection, Research, and Sanctuaries Act of 1972, 33 USC 1401 et seq. The MPRSA has been amended several times, most importantly by the Ocean Dumping Ban Act.

---

* The acronyms MLLW, MHLW, and MTRU refer to Mixed Low-Level, Mixed High-Level, and Mixed Transuranic Waste, respectively.

Mutagen — an agent that causes a permanent genetic change in a cell other than that which occurs during normal genetic recombination (i.e., causes mutations).

National Contingency Plan (NCP) — the "blueprint" for remedial actions taken under Superfund (*see:* 40 CFR 300).

National Priorities List (NPL) means the list, compiled by the EPA pursuant to CERCLA Section 105 of uncontrolled hazardous substance releases in the U.S. that are identified and prioritized for long-term remedial evaluation and response. (The list is frequently updated and now identifies more than 1200 sites.)

Natural Attenuation — allowing or relying upon natural processes to retain, degrade, or destroy contaminants that have been released into the subsurface. Natural attenuation in groundwater systems results from the integration of several subsurface mechanisms that are classified as either destructive or nondestructive. Destructive attenuation mechanisms are mainly biodegradation in one form or another (also variously known as *intrinsic bioremediation, natural bioremediation, passive bioremediation, or bioattenuation*), although abiotic destruction does take place. Nondestructive natural attenuation processes include dispersion, dilution from recharge, sorption, and volatilization.

NEPA — National Environmental Policy Act of 1969, 42 USC 4321 et seq.

NESHAPS — National Emission Standards for Hazardous Air Pollutants, mandated by the Clean Air Act and promulgated by the EPA. By early 1991, the EPA had issued standards for seven NESHAPS: benzine, arsenic (two forms), asbestos, vinyl chloride, mercury, and beryllium. The 1990 **Clean Air Act Amendments (CAAA)** requirements for technology-based standards for 189 **Hazardous Air Pollutants (HAPs)** is an attempt by Congress to accelerate the standards-setting process for hazardous air pollutants.

Notify — RCRA Section 3010(a) requires that any person who manages a hazardous waste (i.e., generators, transporters, owners or operators of treatment, storage, or disposal facilities) must notify the EPA of that activity on EPA Form 8700-12.

NPDES Permit System — the National Pollutant Discharge Elimination System, established by Section 402 of the Clean Water Act, requires that any person responsible for the discharge of a pollutant or pollutants into any waters of the U.S. from any point source must apply for and obtain a permit.

Nuclear Waste — wastes that are highly radioactive, including **high-level waste (HLW)** from national defense activities, **spent nuclear fuel (SNF)** from commercial reactors, and **transuranic waste (TRU)** from weapons production facilities.

ODBA — Ocean Dumping Ban Act of 1988, 33 USC 1414(b), (c).

On-site means the same or geographically contiguous property that may be divided by public or private right-of-way, provided the entrance and exit between the properties is at a crossroads intersection and access is by crossing as opposed to going along the right-of-way. Noncontiguous properties owned by the same person but connected by a right-of-way that he controls and to which the public does not have access is also considered on-site property.

Open burning means the combustion of any material without the following characteristics:

1. Control of combustion air to maintain adequate temperature for efficient combustion
2. Containment of the combustion-reaction in an enclosed device to provide sufficient residence time and mixing for complete combustion
3. Control of emission of the gaseous combustion products

(*see also:* **incineration** and **thermal treatment**).

Operator means the person responsible for the overall operation of a facility.

Organic — being, containing, or relating to compounds that contain carbon in combination with one or more elements, whether derived from living organisms or not.

OSHA — the acronym identifies both a statute and an agency. The Occupational Safety and Health Act of 1970, 29 USC 651, authorized the Occupational Health and Safety Administration to be established in the Department of Labor.

Owner means the person who owns a facility or part of a facility.

Oxidation — a chemical reaction in which there is an increase in valence resulting from a loss of electrons.

PPA — Pollution Prevention Act of 1990, 42 USC 13101, et seq.

Partial closure means the closure of a hazardous waste management unit in accordance with the applicable closure requirements of 40 CFR 264 and 265 at a facility that contains other active hazardous waste management units.

PCA — tetrachloroethane, $C_2H_2C_{14}$.

PCE — perchloroethylene; also tetrachloroethylene, C_2C_{14}.

Percolate — the water moving by gravity or hydrostatic pressure through the interstices of unsaturated rock or soil.

Person means an individual, trust, firm, joint stock company, federal agency, corporation (including a government corporation), partnership, association, state, municipality, commission, political subdivision of a state, or any interstate body.

Personnel or **facility personnel** means all persons who work at or oversee the operations of a hazardous waste facility and whose actions or failure to act may result in noncompliance with the requirement of 40 CFR 264 or 265.

Phase I Environmental Site Assessment — a structured **ESA** typically including a records review, interviews, and a site reconnaissance. The American Society for Testing Materials (ASTM) publishes the *Standard Practice for Environmental Site Assessments: Phase I, Environmental Site Assessment Process,* designation E 1527-93, which has become a widely used format for a Phase I ESA.

Phase II Environmental Site Assessment — a more detailed **ESA** which may include air, soil, and surface and/or groundwater sampling on and near the property, as needed to characterize the site. The Federal National Mortgage Association (FNMA) publishes a guide for the conduct of a Phase II ESA.

Phytoextraction — the use of metal-accumulating plants that translocate and concentrate metals from the soil in roots and above ground shoots or leaves. Phytoextraction offers significant cost advantages over alternative schemes of soil excavation and treatment or disposal. An important issue is whether the metals can be economically recovered from the plant tissue or whether disposal of the waste is required.

Phytoremediation — the use of vegetation for *in situ* treatment of contaminated soils, sediments, and water. It is best applied at sites with shallow contamination of organic, nutrient, or metal pollutants that are amenable to one of five applications: phyotransformation, rhizosphere bioremediation, phytostabilization, phytoextraction, or rhizofiltration. It is an emerging technology that should be considered for remediation of contaminated sites because of its cost effectiveness, aesthetic advantages, and long-term applicability. Phytoremediation **is** well-suited for use at very large sites where other methods of remediation are not cost effective or practicable; at sites with low concentration of contaminants where only "polishing treatement" is required over long periods of time; and in conjunction with other technologies where vegetation is used as a final cap and closure of the site.

Phytotransformation — the uptake of organic and nutrient contaminants from soil and groundwater and the subsequent transformation by plants. Phytotransformation depends on the direct uptake of contaminants from soil water and the accumulation of metabolites in plant tissue. It is important that the metabolites that accumulate in vegetation be nontoxic or at least significantly less toxic that the parent compound.

Phytostabilization — the holding of contaminated soils and sediments in place by vegetation and the immobilization of toxic contaminants in soils. Establishment of rooted vegetation prevents windblown dust, an important pathway for human exposure at hazardous waste sites. Hydraulic control is possible, in some cases, due to the large volume of water that is transpired through plants, thereby preventing migration of leachate toward groundwater or receiving waters. Phytostabilization is especially applicable for metal contaminants at waste sites where the best alternative is often to hold contaminants in place.

Pile means any noncontainerized accumulation of solid, nonflowing hazardous waste that issued for treatment or storage.

Piezometric surface — the surface defined by the levels to which groundwater will rise in tightly cased wells that tap an artesian aquifer.

Point source means any discernible, confined, and discrete conveyance, including, but not limited to any pipe, ditch, channel, tunnel, conduit, well, discrete fissure, container, rolling stock, concentrated animal feeding operation, or vessel or other floating craft, from which pollutants may be discharged. This term does not include the return flows from irrigated agriculture.

Potentially Responsible Party (PRP) — any individual or company including owners, operators, transporters, or generators potentially responsible for, or contributing to, the contamination problems at a **Superfund** site.

Publicly Owned Treatment Works (POTW) — any device or system used in the treatment (including recycling and reclamation) of municipal sewage or industrial wastes of a liquid nature, which is owned by a state or municipality. This definition includes sewers, pipes, or other conveyances only if they convey wastewater to a POTW providing treatment.

Pretreatment — treatment given domestic sewage which contains industrial waste components that would, if not removed or treated, interfere with (or pass through untreated) the treatment processes in a sewage treatment plant.

Pyrolysis — the chemical decomposition or change brought about by heating in the absence of oxygen.

Radiation means any or all of the following: alpha, beta, gamma, or X-rays; neutrons; and high-energy electrons, protons, or other atomic particles; but not sound or radio waves, nor visible, infrared, or ultraviolet light.

Radiological Waste — generally, waste material which has a radioactive component. *See:* nuclear waste, **high-level waste (HLW), spent nuclear fuel (SNF), transuranic waste (TRU), low-level waste (LLW), uranium mining**, and **mill tailings**.

Radwaste — short or trade term for radiological waste.

RCRA — the acronym refers both to a law and a program. The Resource Conservation and Recovery Act of 1976, as amended (42 U.S.C. Sec. 6901 et seq.), and the program and regulations, which implement the Act. The Act originated as the Solid Waste Disposal Act of 1965, 42 USC 3251 et seq.

RCRA/Superfund Hotline — a telephonic information system, manned by an EPA contractor, which can be accessed by the public. The hotline provides a wide range of information pertaining to the RCRA and Superfund programs and can provide some EPA publications upon request. The number is 800-424-9346.

Recharge — the addition of water to the groundwater system by natural or artificial processes.

Reduction — a chemical reaction in which there is a decrease in valence as a result of gaining electrons.

Regional Administrator — the administrator of an EPA (or other Federal government agency) geographical region.

Remedial design (RD) means the technical analysis and procedures that follow the selection of remedy for a site and result in a detailed set of plans and specifications for implementation of the remedial action.

Remedial investigation (RI) is a process undertaken by the lead agency to determine the nature and extent of the problem presented by (a) release.

The RI emphasizes data collection and site characterization and is generally performed concurrently and in an interactive fashion with the feasibility study. The RI includes sampling and monitoring, as necessary, and includes the gathering of sufficient information to determine the necessity for remedial action and to support the evaluation of remedial alternatives.

Representative sample means a sample of a universe or whole (e.g., waste pile, lagoon, groundwater), which can be expected to exhibit the average properties of the universe or whole.

Roentgen (r) — the amount of gamma or X-radiation that will produce in 1 cubic centimeter of dry air, at 0°C and 760-mm pressure, 1 electrostatic unit (esu) of electricity. The roentgen is a unit of the total quantity of ionization produced by gamma or X rays. Dosage rates are expressed in terms of roentgens per unit time.

Rem (roentgen-equivalent-man) — the amount of radiation that will produce an energy dissipation in the human body that is equivalent to 1 roentgen of radiation of X rays.

Reportable quantity (RQ) — the quantity of a hazardous substance (listed in 40 CFR 302, Table 302.4) or extremely hazardous substance (listed in 40 CFR 355, Appendix A or B) that requires reporting under **CERCLA**. If a substance is released in a quantity that exceeds its RQ, the release must be reported to the National Response Center, as well as to the **State Emergency Response Commission (SERC),** and to the community emergency coordinator for areas likely to be affected by the release. The practitioner should carefully study **CERCLA** Sections 101 thru 104, regarding the RQs and their application.

Rhizofiltration — the use of plant roots to sorb, concentrate, and precipitate metal contaminants from surface or groundwater. Roots of plants are capable of sorbing large quantities of lead and chromium from soil water or from water that is passed through the root zone of densely growing vegetation. Rhizofiltration is being tested at two sites contaminated with radionuclides.

Rhizosphere — the portion of soil intimately associated with the roots of growing plants.

Rhizosphere bioremediation — **phytoremediation** of the rhizosphere increases soil organic carbon, bacteria, and mycorrhizal fungi, all factors that encourage degradation of organic chemicals in soil. Research has shown that the numbers of beneficial bacteria increased in the root zone of hybrid poplar trees relative to an unplanted reference site. Denitrifiers, *Pseudonomad* spp., **BTEX** degrading organisms, and general heterotrophs were enhanced. Also, plants may release exudates to the soil environment that help stimulate the degradation of organic chemicals by inducing enzyme systems of existing bacterial populations, stimulating growth of new species that are able to degrade the wastes, and/or increasing soluble substrate concentration for all microorganisms.

Risk Based Corrective Action (RBCA) — an integration of exposure and risk assessment practices with traditional components of the corrective action process to ensure that appropriate and cost-effective remedies are selected and that limited resources are properly allocated.

Run-off means any rainwater, leachate, or other liquid that drains over land from any part of a facility.

Run-on means any rainwater, leachate, or other liquid that drains over land onto any part of a facility.

SARA Title III — Title III of the **Superfund Amendments and Reauthorization Act of 1986 (SARA)** embodies the Emergency Planning and Community Right-to-Know Act.

Saturated zone or **zone of saturation** means that part of the earth's crust in which all voids are filled with water.

SDWA — Safe Drinking Water Act of 1974, 42 USC 300f et seq.

Sewage — domestic and industrial wastes and wastewaters discharged into sewers.

Sewerage — the system of sewage collection, treatment, and disposal.

Site inspection (SI) means an on-site investigation to determine whether there is a release or potential release and the nature of the associated threats. The purpose is to augment the data collected in the preliminary assessment and to generate, if necessary, sampling and other field data to determine if further action or investigation is appropriate.

Sludge means any solid, semi-solid, or liquid waste generated from a municipal, commercial, or industrial wastewater treatment plant, water supply treatment plant, or air pollution control facility exclusive of the treated effluent from a wastewater treatment plant.

Small Quantity Generator (SQG) — a generator who generates more than 100 kg, but less than 1000 kg of hazardous waste, and no more than 1 kg of acutely hazardous waste, in a calendar month.

Solid waste — the Solid Waste Disposal Act [42 USC 6903(27)] defines solid waste as … "any garbage, refuse, sludge from waste water treatment plant, water supply treatment plant, or air pollution control facility and other discarded material, including solid, liquid, semisolid, or contained gaseous material resulting from industrial, commercial, mining, and agricultural operations, and from community activities, but does not include solid or dissolved material in domestic sewage, or solid or dissolved materials in irrigation return flows or industrial discharges which are point sources subject to permits under section 1342 of title 33, or source, special nuclear, or byproduct material as defined by the Atomic Energy Act of 1954, as amended." The **RCRA** regulations define solid waste as … any discarded material that is not excluded by § 261.4(a) or that is not excluded by variance under §§ 260.30 and 260.31; *see:* **Discarded**; *see:* 40 CFR 261.2.

SNF — **spent nuclear fuel** means fuel that has been withdrawn from a nuclear reactor following irradiation. Within the regulatory framework, SNF is a subset of **HLW**.

State means any of the several states, the District of Columbia, the Commonwealth of Puerto Rico, the Virgin Islands, Guam, American Samoa, and the Commonwealth of the Northern Mariana Islands.

State Emergency Response Commission (SERC) — SERCs are appointed by the governor of each state, per the **Emergency Planning and Community Right-to-Know Act (EPCRA)**. The SERCs designate local emergency planning districts, appoint **Local Emergency Planning Committees (LEPC),** coordinate and supervise the activities of the **LEPCs**, review the local emergency response plans, and notify the EPA of all facilities that are subject to the emergency planning requirements.

Storage means the holding of hazardous waste for a temporary period, at the end of which the hazardous waste is treated, disposed of, or stored elsewhere.

Sump means any pit or reservoir that meets the definition of **tank** and those troughs/trenches connected to it that serve to collect hazardous waste for transport to hazardous waste storage, treatment, or disposal facilities.

Superfund — a "nickname" given the program which implements the Comprehensive Environmental Response, Compensation, and Liability Act (*see:* **CERCLA**).

Surface impoundment or **impoundment** means a facility or part of a facility which is a natural topographic depression, man-made excavation, or diked area formed primarily of earthen materials (although it may be lined with man-made materials), which is designed to hold an accumulation of liquid wastes or wastes containing free liquids, and which is not an injection well.

SW 846 (EPA/SW-846) — *Test Methods for Evaluating Solid Waste: Physical/Chemical Methods; Third Edition; Volumes IA, IB, IC, and II.* The document is the EPA guidance document for test procedures that "... may be used to evaluate properties of solid waste which determine whether waste is hazardous within the definition of Section 3001, RCRA." The document is available from the U.S. Government Printing Office, order number 955-001-00000-1. The NTIS number is PB88-239223. It can also be obtained commercially on CD. The document is updated periodically and a recent update is designated *Final Update Package I,* (EPA/SW-846.3-1). The update can be obtained from the U.S. Government Printing Office, order number 955-001-00000-1.

SWDA — Solid Waste Disposal Act of 1965, 42 USC 3251 (*see:* **RCRA**).

Tank means a stationary device, designed to contain an accumulation of hazardous waste. It is constructed primarily of non-earthen materials (e.g., wood, concrete, steel, plastic), which provide support.

Tank system means a hazardous waste storage or treatment tank and its associated ancillary equipment and containment system.

TCA — trichloroethane, $C_2H_3Cl_3$.

TCE — trichloroethylene, C_2HCl_3.

TCLP — the Toxicity Characteristic Leaching Procedure, a laboratory procedure designed to produce an extract simulating the leachate that may be produced in a land disposal situation. The extract is then analyzed to determine if it includes any of the toxic contaminants listed in Table 2.1. If the concentrations of any to the Table 2.1 constituents exceed the levels listed in the table, the waste is classified as hazardous (*see:* 40 CFR 261.24).

Teratogen — a physical or chemical agent that is capable of causing nonhereditary congenital malformations (birth defects) in offspring.

Thermal treatment means the treatment of hazardous waste in a device that uses elevated temperatures as the primary means to change the chemical, physical, or biological character or composition of the hazardous waste. Examples of thermal treatment processes are incineration, molten salt, pyrolysis, calcination, wet air oxidation, and microwave discharge (*see also:* **incinerator** and **open burning**).

Totally enclosed treatment facility means a facility for the treatment of hazardous waste that is directly connected to an industrial production process and is constructed and operated in a manner that prevents the release of any hazardous waste or any constituent thereof into the environment during treatment.

Toxicity — the ability of a material to produce injury or disease upon exposure, ingestion, inhalation, assimilation by living organism.

Threshold — the lowest dose of a chemical at which a specified measurable effect is observed and below which it is not observed.

Transfer facility means any transportation-related facility including loading docks, parking areas, storage areas, and other similar areas where shipments of hazardous waste are held during the normal course of transportation.

Transporter — a transporter, subject to 40 CFR 263, is any person engaged in the off-site transportation of manifested hazardous waste by air, rail, highway, or water.

Transport vehicle means a motor vehicle or rail car used for the transportation of cargo by any mode. Each cargo-carrying body (trailer, railroad freight car, etc.) is separate transport vehicle.

Transportation means the movement of hazardous waste by air, rail, highway, or water.

Transuranic waste (TRU) means waste material containing radionuclides with an atomic number greater than 92, which are excluded from shallow burial by the federal government. Transuranic elements are those having atomic numbers greater than 92. **TRU** waste is defined, in the U.S., as **radwaste** that is not classified as **HLW** but contains an activity of more than 100 nanocuries per gram from α-emitting transuranic isotopes having **half-lives** greater than 20 years (typically clothing, equipment, tools, and scrap contaminated with plutonium from laboratory and facility operations).

Toxics Release Inventory National Report (TRI) — the compilation of chemical releases to air, water, and land from selected manufacturing facilities, as authorized and required by the **Emergency Planning and Community Right-to-Know Act (EPRCRA)**. The **TRI** is published in odd-numbered years by the Environmental Protection Agency.

Treatment means any method, technique, or process including neutralization designed to change the physical, chemical, or biological character or composition of any hazardous waste so as to neutralize such waste, or so as to recover energy or material resources from the waste, or so as to render such waste nonhazardous, or less hazardous; safer to transport, store, or dispose of; or amenable for recovery, amenable for storage, or reduced in volume.

TSCA — Toxic Substances Control Act of 1976, 15 USC 2601 et seq.

Underground injection means the subsurface emplacement of fluids through a bored, drilled, or driven well, or through a dug well, where the depth of the dug well is greater than the largest surface dimension (*see also: injection well*).

Underground tank means a device meeting the definition of **tank**, whose entire surface area is totally below the surface of and covered by the ground.

Unsaturated zone or **zone of aeration** means the zone between the land surface and the water table.

United States means the 50 states, the District of Columbia, the Commonwealth of Puerto Rico, the U.S. Virgin Islands, Guam, American Samoa, and the Commonwealth of the Northern Mariana Islands.

Uppermost aquifer means the geologic formation nearest the natural ground surface that is an aquifer, as well as lower aquifers that are hydraulically interconnected with this aquifer within the facility's property boundary.

Uranium mill tailings — the sandy residues from the extraction and refining of uranium ore. The tailings contain low concentrations of thorium and radium and radioactive radon gas from the decay of the radium. The emissions of radon are the major health hazard associated with **uranium mining** and mill tailings.

Uranium mine tailings — waste rock and granular material resulting from uranium ore mining operations. The characteristics are similar to those of **uranium mill tailings**.

USGS —U.S. Geological Survey.

Volatile organic compound (VOC) — materials such as gasoline, paint solvents, and other organic compounds, which evaporate and enter the air in a vapor state, as well as fragments of molecules resulting from incomplete oxidation of fuels and wastes. In general, higher vapor pressures relate to higher volatility.

Wastewater treatment unit — device that:

1. Is part of a wastewater treatment facility that is subject to regulation under either Section 402 or 307(b) of the Clean Water Act; *and*
2. Receives and treats or stores an influent wastewater that is a hazardous waste as defined in 40 CFR 261.3 or that generates and accumulates a wastewater treatment sludge that is a hazardous waste; *or*
3. Treats or stores a wastewater treatment sludge that is a hazardous waste and meets the definition of tank or tank system in 40 CFR 260.10.

Well means any shaft or pit dug or bored into the earth, generally of a cylindrical form, and often walled with bricks or tubing to prevent the earth from caving in. (Some early literature and regulatory issues used a definition attributed to A. E. Meinzer. "A well is a hole in the ground having depth greater than its diameter.")

Underground Storage Tank (UST) — any one or combination of **tanks** (including underground pipes connected thereto) that is used to contain an accumulation of regulated substances, the volume of which (including the volume of the underground pipes connected thereto) is 10% or more beneath the surface of the ground.

Upgrade means the addition or retrofit of some systems such as cathodic protection, lining, or spill and overfill controls to improve the ability of an **underground storage tank** system to prevent the release of a product.

Yellow boy — the metallic precipitate formed when acidic, metal-bearing mine drainage, leachate from ore piles, or process wastes are introduced to alkaline receiving waters.

Index